Urban Biodiversity

Urban Biodiversity

The Natural History of the New Jersey Meadowlands

Erik Kiviat and Kristi MacDonald

LEXINGTON BOOKS
Lanham • Boulder • New York • London

Published by Lexington Books
An imprint of The Rowman & Littlefield Publishing Group, Inc.
4501 Forbes Boulevard, Suite 200, Lanham, Maryland 20706
www.rowman.com

86-90 Paul Street, London EC2A 4NE

British Library Cataloguing in Publication Information Available

Library of Congress Cataloging-in-Publication Data

Names: Kiviat, Erik, author. | MacDonald-Beyers, Kristi, author.
   Title: Urban biodiversity : the natural history of the New Jersey
      Meadowlands / Erik Kiviat, Kristi MacDonald.
   Other titles: Natural history of the New Jersey Meadowlands
   Description: Lanham, Maryland : Lexington Books, [2022] | Includes
      bibliographical references and index.
   Identifiers: LCCN 2022013908 (print) | LCCN 2022013909 (ebook) | ISBN
      9781498599917 (cloth) | ISBN 9781498599924 (ebook)
   Subjects: LCSH: Natural history--New Jersey--Hackensack Meadowlands. |
      Biodiversity--New Jersey--Hackensack Meadowlands. | Hackensack
      Meadowlands (N.J.)
   Classification: LCC QH105.N5 K57 2022 (print) | LCC QH105.N5 (ebook) |
      DDC 508.749/21--dc23/eng/20220413
   LC record available at https://lccn.loc.gov/2022013908
   LC ebook record available at https://lccn.loc.gov/2022013909

∞™ The paper used in this publication meets the minimum requirements of American
National Standard for Information Sciences—Permanence of Paper for Printed Library
Materials, ANSI/NISO Z39.48-1992.

For Elaine, who shares the wetlands of the world with me—Erik

For my parents, Robert and Mary MacDonald; my husband, John; and my son Gavin—Kristi

Hudsonia's research in the Meadowlands, which is included in this book, has been underwritten by:

The Geraldine R. Dodge Foundation

Furthermore: a program of the J.M. Kaplan Fund

Metropolitan Conservation Alliance (founder, Dr. Michael W. Klemens)

Geoffrey C. Hughes Foundation

The DHR Foundation

Emma Barnsley Foundation

Meadowlands Environmental Research Institute

Will Nixon

New Jersey Division of Fish & Wildlife's Conserve Wildlife Matching Grants

Conserve Wildlife Foundation New Jersey

Rutgers Environmental Law Clinic

Bergen County Audubon Society

Contents

List of Figures

List of Tables

Introduction

With more than 55% of the world's people living in cities and suburbs, humans are now officially an urban species (World Bank 2019). The United States leads this pattern with 82% of the population urban. Yet only 3.7% of the area of the conterminous 48 states are in urban land use with 400,000 ha added annually (Merrill and Leatherby 2018). These metropolitan areas include urban, suburban, and industrial environments that we combine in our definition of urban. Urban-fringe and rural lands continue to be developed for residence, commerce, industry, and transportation, adding to the growth of metropolitan regions. The New Jersey Meadowlands are located in the New York City metropolitan region, the most densely populated region in the United States, in the heart of the Washington, DC, to Boston Megalopolis. The Meadowlands, with their history of urban residential and industrial development in connection with major ports in New York Harbor, interspersed with a mosaic of the Hackensack River and its tributaries, expansive estuarine wetlands, and upland greenspaces, provide a case study of how urban regions may support, or not support, ecosystem services and their requisite biodiversity, which is applicable to other large urban regions in North America and perhaps throughout the world.

Many U.S. cities are associated with seacoasts, rivers, or the shores of large lakes because rich natural resources favored settlement and these locations provide ports for trade and commerce. Half of the U.S. population lives within 80 km of a seacoast. Coastal areas, tidal river mouths, and their associated wetlands are also hotspots of biological diversity. Thus, large, intensely developed metropolitan regions tend to be situated against a backdrop of immense biological wealth. Therein lie the challenge and the opportunity to maintain viable and diverse biological communities where the majority of the human population lives, works, and plays. The methodologies of surveying for, assessing, and managing biodiversity in urban areas lag behind the need for immediate, sound decision-making. Although we often do not know which species occur or what their ecological affinities and tolerances are, in many cases we will have to manage individual species as well as their habitats if we want to increase the chances of a rich panoply of biodiversity surviving for future human generations to use and enjoy.

The *Greater Meadowlands* constitute an approximately 150 km^2 urban complex largely composed of extant and former wetlands 4 km west of extreme urban development in Manhattan. The Meadowlands are part of the Piedmont Lowland, the region of New Jersey with the "greatest density of population, highways, industry, and pollution" (Montgomery and Fairbrothers 1992, 42). We use the Meadowlands as a case study in urban biodiversity and its management because it is a familiar region with a long history of bio-logical observation and human use and abuse. The Meadowlands are near the extreme end of the spectrum of wetland filling, waste dumping, contamina-tion, fragmentation, hydrological alteration, salinization, and construction; thus, the region clearly illustrates the effects of human activities on the biota.

There is an enormous literature on the Meadowlands, the great majority of it gray literature (e.g., agency and consulting reports, planning documents, and theses), popular literature, and Web pages. However, most organisms using habitats in the Meadowlands receive little or no scientific, conservation, or political attention. These underappreciated groups include most inverte-brates, the amphibians, mosses, lichens, fungi, and many of the seed plants. Large trees, for example (Figure 0.1), are important to many other organ-isms yet their species, sizes, and numbers in the Meadowlands are unknown.

Figure 0.1 Tidal *Phragmites* reedbed and large trees bordering an industrial site on Bellman's Creek. Photo by Erik Kiviat.

Hudson County, including the substantial portion in the Meadowlands, has been largely overlooked by professional field biologists until very recently. Many of the habitats of the Meadowlands, including the odd array of upland types, are essentially unstudied. Biological resources need identification and description before they can be protected, used, or interpreted. We aim to identify both the typical and the unusual species and ecological features of the Meadowlands for further survey and management, compare this diversity to that of surrounding rural areas and other cities, and discern lessons the Meadowlands offer for the study and management of biodiversity in other urban areas, and in urbanizing rural areas. This analysis, of course, is limited by the ecological and taxonomic breadth of the extant information and our own surveys.

In 1999, a development project was proposed to fill 80 ha of the most extensive remaining wetland in the Meadowlands, and we were asked to join a group of experts reviewing the environmental assessment documents for this project. Subsequently, we compiled a technical report (Kiviat and MacDonald 2002) on the biology of the region to aid environmental professionals and scientists faced with a large, fragmentary, and mostly informal body of information. Portions of this report were adapted for a booklet (Kiviat and MacDonald 2006) for the general public. We conducted a variety of field surveys in the Meadowlands and nearby areas and became increasingly fascinated by questions about which animals and plants were present, how they arrived and survived in the Meadowlands, how their habitats were being developed or conserved, and how the habitat management being performed did or did not benefit various species and ecosystem services.

In addition to value for natural history and pure science, the story of Meadowlands biodiversity bears study for two reasons. It helps us understand how to manage the marvelous biological resources that exist in metropolitan areas in spite of, or even because of, urbanization. It also alerts us to mistakes that we can avoid in environmental planning and nature management in areas that are still rural or wild yet are undergoing urban development. We present a lot of information about the Meadowlands biota, detail that is necessary where almost all of the environment will be developed or directly managed in some way. Despite federal, state, and local laws, regulations, policies, and public opinion protecting certain habitats and species, small and large portions of the Meadowlands and other urban areas can still be converted to developed land uses almost overnight. For example, 20 ha of wetlands and wetland fill were legally developed into a private truck parking lot in Secaucus in about 2007.

We focus on biodiversity and its management from a species and habitat viewpoint. There is virtually no information on ecological processes, such as nutrient cycling, in the Meadowlands, whereas we know a moderate amount about the biota. Our analyses are based primarily on the habitat and species levels, although biodiversity is also important below these levels (e.g., the levels of genotypes and subspecies) and above them (communities, ecosystems, landscapes, and regions). The species is best known and easiest for most people to understand, and the species of an area are the basic elements of biodiversity. In order to manage that diversity, much information is needed on distribution, abundance, ecological interactions, and history. Ultimately, conservation must be done at the species level or below if we are to avoid loss of much diversity, because one taxonomic group of species does not necessarily predict another (e.g., conserving trees does not conserve all macrofungi or birds), and one habitat or biological community does not predict the presence of all other species.

We address common and rare species because they both play an important role in maintaining the function and diversity within ecosystems and their patterns tell a story of which traits favor persistence in an urban setting and which do not. Nonetheless, we emphasize the importance of uncommon and rare species. What is so important about rare animals, plants, and other organisms? Species that have become rare due to urbanization provide insights into how management or restoration can help certain species recover. Rare species have economic value because people are willing to pay more or exert more effort for things that there are less of. Birders, for example, love to discover a rare bird that they have not seen before; botanists and mycologists also enjoy finding rarities. Rare species as well as common ones contain unique genes and have practical uses for pharmaceuticals, industrial compounds, crop improvements, research, aesthetics, and other purposes because each species is different. Lichens and mosses contain novel organic chemicals that have potential for the treatment of cancer and other diseases, and even weeds like mugwort and sweet Annie have important medicinal uses. For many species, there is too little information to know if they are really rare or just poorly surveyed (Raphael and Molina 2007), and many species of, for example, invertebrates, have not been documented in the Meadowlands due to lack of study. To date, wasps, ants, grasshoppers, moths, flies, springtails, mites, fungi, protists, bacteria, and many others have not been surveyed at all. Moreover, because so little research has been conducted on most of these taxa, we do not even know which species are common or rare statewide in New Jersey, or around the northeastern states in general.

Rare species that thrive in the Meadowlands, like the Atlantic Coast leopard frog or least bittern, may represent source populations that can contribute dispersers to repopulate restoration sites or other areas inside or outside the Meadowlands. Patterns of species populations, distributions, and trends in these variables may tell us something about the effects of climate change, air and water quality improvement, other changes in habitat quality, management practices, larger scale population changes, and other factors on urban biotas.

Genetic adaptations to urban conditions or contaminants may produce local forms of organisms that teach special lessons about human impacts on the environment, and some of these locally-adapted species may even be useful in remediating pollution (such as bacteria that evolve to break down organic compounds, or plants that resist and accumulate metals thus can potentially be used for removing metals from soil or water).

Our goal is to understand, through this case study of the Meadowlands, how some organisms thrive in urban areas and others do not. We are also interested in the ways that land use change and climate change (including sea level rise) affect biodiversity. And we aim to learn what can be done to help rare and common native organisms and their habitats remain viable in the Meadowlands and other cities without unnecessarily large expenditures of funds and time. The first task is to describe the occurrence of habitats and species in the region, and that is the matter of this book. In addition to describing patterns of diversity of different groups of organisms, in a future book we will address urban ecology, biogeography, pollution, overabundant species, climate change, and sea level rise in the Meadowlands, and further analyze the success and sustainability of habitat and species management methods.

In general, many principles of urban ecology apply to the Meadowlands flora and fauna; organisms living in urban regions are exposed to novel human-dominated environments and multiple stressors imposed by people and their activities on the landscape. Most large cities, in part because of their association with major water bodies, possess a variety of habitats. However, some wetlands, forests, and other habitats are lost, shrunk, or damaged during urbanization, and hydrology is often greatly altered. Remnant habitats are islands surrounded by a built matrix. Long narrow bands of habitat, such as riparian buffers, are small habitats because of the high ratio of edge to interior. Many small forests and shrublands are on abandoned industrial sites. Cities have artificial barriers that impede movement of animals, as well as potential movement corridors for organisms. For populations of some species, habitats in urban areas act as ecological sinks or traps. Certain organisms evolve adaptations to urbanization whereas some others are preadapted. The type, extent and frequency of ecological disturbance are different in cities. Predation and competition pressures may be higher or lower depending on the organisms involved. In some habitats, nonnative plants dominate vegetation. Species diversity and population density often peak at the midpoint of a gradient between least and most developed (built or manicured) areas. The fauna is dominated by species with generalized diets and broad ranges of environmental tolerances. Organisms are subject to more or less sunlight and wind, higher temperatures, artificial night lighting, and anthropogenic noise. Soils may be low in organic matter, lacking in natural structure, high or low in plant nutrients, elevated in calcium, contaminated with synthetic chemicals and refuse, and compacted. The fauna is continually subject to direct human disturbance. These multiple, urban stressors, combined with the global stressors of climate change and other factors, shape urban biodiversity in a multitude of ways we are just beginning to understand.

The first chapter of this book (Environmental Setting) introduces the environment and urban context of the Meadowlands. There follow chapters on particular types of habitats and on major groups of organisms. Finally, in the conclusions, we relate our findings about the Meadowlands to other urban areas. At the end of the book are species checklists for seed plants and birds, and the references cited.

Certain ecological terms are ambiguous or may imply inappropriate concepts to some readers. Moreover, nature is complex and sloppy, and it is important not to assume too much organization and predictability. This is how we use the following terms. *Biodiversity* is the variety in nature, from genes to species to ecosystems to regions. Biodiversity is more than species diversity, and we emphasize species of conservation concern as components of biodiversity. A *taxon* (plural *taxa*) refers to organism(s) belonging to a particular taxonomic level of biological and evolutionary organization, such as a variety, species, genus, family, order, or class. The *biota* (or the *flora* and *fauna*) is the list of species in an area; the *vegetation* is the total plant cover of an area. A *community* or *assemblage* is the group of species of plants, animals, fungi, and other organisms occupying a habitat. Although communities are structured by predation, competition, and facilitation among species, fungal interconnections, and other processes, chance, and influences from outside the habitat, are also important and communities are highly variable in space and time. *Vegetation development* refers to temporal change that may be caused by external factors (such as a wind storm, logging, nutrient input, or arrival of new species) or by processes within the ecosystem (including soil development and competition). We avoid the term *succession* because it has multiple meanings and may imply unproven causes (see Drury and Nisbet 1973, Pickett et al. 2009). Photographs are chosen to illustrate Meadowlands habitats; images of most of the species we mention can readily be found on the Web.

Species of conservation concern encompass species that are rare, declining, vulnerable, ecologically specialized, or less tolerant of human influences. These are, of course, partly value judgments concerning wild species and their importance for conservation. We use the state list of *Species of Greatest Conservation Need* (SGCN; NJDEP 2018a, Appendices B and C, for animals only), and the New Jersey Natural Heritage Program state rankings (NJDEP 2019), both with many taxa excluded for lack of state review, as a general guide to the conservation status of species.

An *ecosystem* is the combined biological community and the *abiotic* (non-living) environment with which living species interact. Ecosystem is more a concept for analysis and discussion than a reality of nature, particularly because boundaries in the wild are usually imprecise. A *habitat*, strictly speaking, is the place where an organism (or community) occurs; however, we also use habitat to indicate an area that differs from other areas, such as a tidal marsh or a forest, independent of the species there.

Waste ground is a habitat that has been highly altered by human activities, usually lacks topsoil, and is often sparsely vegetated or bare of seed plants (Figure 0.2). Waste grounds are not necessarily places where wastes have been disposed of. We refer to *semi-natural* habitats, or semi-natural soil and vegetation, to denote the less-managed areas. *Greenspaces* are those areas dominated by vegetation, whether managed or not, including forests, cemeteries, and marshes ("unbuilt areas in an urban region, i.e., areas without continuous closely spaced buildings" [Forman 2008, 7]).

We use standardized common names of organisms where those exist or have been recommended. We mention the currently accepted scientific name the first time in the book or a chapter that a common name is mentioned, as well as in the tables listing seed plants, ferns, mosses and liverworts, lichens, mammals, birds, reptiles and amphibians, fishes, butterflies, bees, and dragonflies. Scientific and common names of many organisms are in a confusing

Figure 0.2 Sparsely vegetated wetland fill and freshwater pond, Meadowlark Marsh. Waste ground habitat in foreground. Photo by Erik Kiviat.

state of flux, and we give a few alternate names that many naturalists may be familiar with—other such *synonyms* may be found on scientific and natural history Web sites. We have adopted scientific names from standard sources as best we could; ambiguities are indicated.

Throughout this book, we supplement the knowledge of taxa in the Meadowlands with details about their habitat use, diet, and other information that have been gleaned about them in other places, in the scientific and natural history literature. Although this is a necessary step for understanding urban biodiversity, caution is needed in generalizing from one place or time to another. The habitat, abundance, behavior, and other aspects of the ecology of a species can be similar or different between a published study and the current situation in the Meadowlands or in another urban-industrial area at another time where the soil structure, water chemistry, predators, herbivores, human activities, or other features are likely to be different. The patterns we describe or hypothesize here need to be tested elsewhere for their applicability and generality.

Biodiversity and the various ecosystem services it provides are often discounted in urban areas. Insufficient consideration in planning, preeminence of other values (e.g., aesthetics, history, recreation, politics, profits), and the perception that there is nothing important to conserve often drive land use and management decisions. We hope the findings of our case study in one of the most human-altered regions on the U.S. East Coast will help planners, landscape designers, horticulturists, engineers, educators, public officials, naturalists, and ecologists to think and look more carefully at urban biological resources and their effective conservation.

Chapter 1

The Environmental Setting of the Meadowlands[1]

Before we consider the habitats and organisms of the Meadowlands or another urban region, we need to understand the context of their environment. The biota (plants, animals, fungi, and other organisms) exists in close relationship to the *abiotic* (non-living) environment, including bedrock, soils, surface waters, groundwater, air, and climate. The abiotic environment shapes many aspects of biodiversity. Here we describe aspects of the physical and chemical environment of the Meadowlands that we believe are most important to the distribution, abundance, and health of organisms.

The Meadowlands are in the western edge of an extensive estuarine delta region approximately centered on New York Harbor (Figure 1.1). Large portions of our study region have been called the Hackensack Meadows, Hackensack Meadowlands, Hackensack Marsh, Newark Meadows, Newark Marshes, Newark-Elizabeth Meadows, Bergen Meadows, Meadowlands District, New Jersey Meadowlands, and other names. Various ecological boundaries have been shown (e.g., Tedrow 1986, 209). We use a broad geographic definition of the New Jersey Meadowlands (*Greater Meadowlands*): the entire region that is now or formerly estuarine or nontidal wetland, from the Oradell Dam south along the Hackensack River to the southern edge of the former Newark-Elizabeth Meadows (now Newark Liberty International Airport and Port Newark–Port Elizabeth) and adjacent Newark Bay, including biologically relevant uplands within or close to the estuarine and wetland areas. These uplands include Laurel Hill (Figure 1.6), Little Snake Hill, Granton Quarry (Schuberth 1968), Ridgefield Nature Center, Borg's Woods, Matthew Feldman Nature Preserve in Teaneck, Lincoln Park in Jersey City, the escarpments and mine faces (see Puffer 1984) overlooking the wetlands in North Arlington and Lyndhurst and continuing north less distinctly through Carlstadt, Wood-Ridge, and Hasbrouck Heights (Figures 1.2 and 1.3), and the bordering cemeteries, parks, and landfills. The entirety

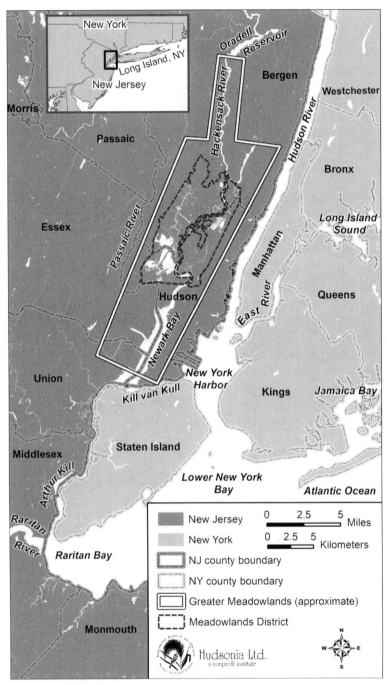

Figure 1.1 The New York City estuarine delta. Map by Hudsonia staff.

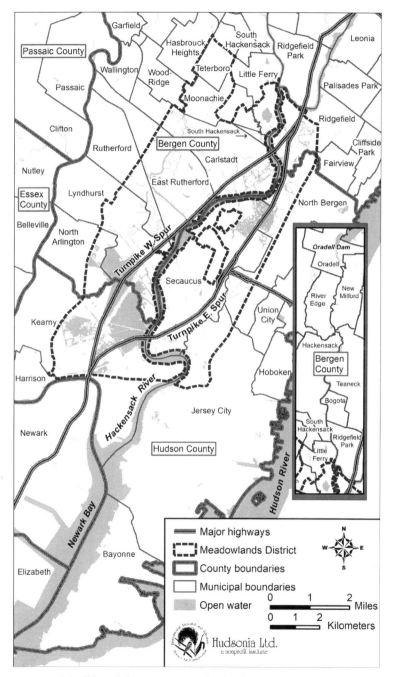

Figure 1.2 Municipalities of the Greater Meadowlands. Most municipalities encompass both elevated uplands and estuary-dominated lowlands. Map by Hudsonia staff.

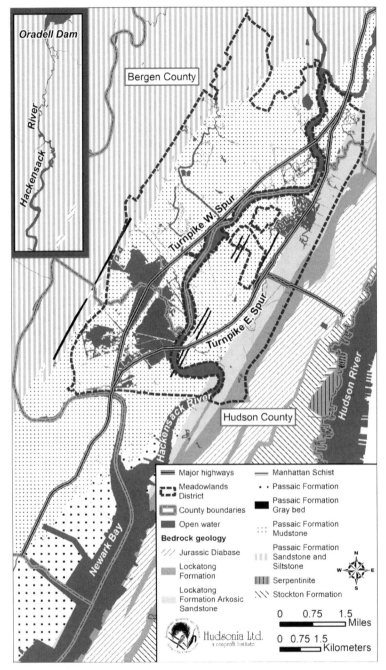

Figure 1.3 Bedrock geology of the Greater Meadowlands. Data from New Jersey Geological Survey Digital Geodata Series. Map by Hudsonia staff.

of the estuarine-influenced Overpeck Creek, Berry's Creek, Bellman's Creek, Cromakill Creek, Losen Slote, and other Hackensack River estuary tributaries are also included, as well as nontidal lower reaches of a few upland tributaries such as Wolf Creek and Cole's Brook, and the estuarine lower end of the Passaic River. Using this broad definition we discuss the gradients and interactions between the Core Meadowlands (the broad estuarine and formerly estuarine wetlands and wetland fill that are mostly in the official Meadowlands District) and the adjoining areas. The Core Meadowlands constitute the area for which the largest amount of information is available; the Meadowlands District itself covers 78.7 km², about half of the Greater Meadowlands. There are about 266 km of river, streams, and channels in the Meadowlands District (NJDEP 2015). NJDEP data for broad categories of land use and land cover within the Meadowlands District indicate that about 45% is urban, 21% is wetland (tidal, nontidal, freshwater, estuarine, and forested wetlands), 20% is open water (impoundments, ponds, streams, canals), 8% is barren land (altered and transitional lands of variable vegetation), and 6% is forest (upland forest, shrubland, old field); most of the urban land use is non-residential (95%; including commercial-industrial, schools, parks, cemeteries, utilities and transportation corridors). These numbers are not representative of the Greater Meadowlands; for example, we expect residential urban land use and forest cover to be higher in some areas outside the Core Meadowlands.

Humans and many other animals characteristically move between estuary and bordering areas, and these interactions are germane to the biodiversity of the Meadowlands. Our Greater Meadowlands region is mostly in Bergen and Hudson counties and includes small areas of Essex County and Union County (part of the former Newark Meadows; see county and municipal boundaries in Figure 1.2).

The estuarine delta includes the nominal mouth of the Hudson River at the southern tip of Manhattan, as well as extant or former concentrations of wetlands associated with the western end of Long Island Sound and the East River, Jamaica Bay, Raritan Bay and the Raritan River, and both the Staten Island and New Jersey sides of the Arthur Kill (Cox et al. 2002). Within this delta, the Meadowlands are a subunit aligned north-northeast to south-southwest (Figure 1.1). The Hackensack River and the Passaic River flow together into the upper (northern) end of Newark Bay. Staten Island nearly plugs the lower (southern) end of Newark Bay, making the Meadowlands an estuary that is sheltered from the ocean and into which the tides flow through the tortuous channels of the Kill van Kull and Arthur Kill. The Oradell Dam, at the northern or inland end of the Hackensack River estuary, is only 10 km south of the New Jersey–New York state line.

Elevations of wetlands in the Meadowlands range from sea level to about 3 m above sea level (with the exception of a pond at 36 m on top of the 1-E Landfill in North Arlington). Bedrock outcrops at Laurel Hill rise to about 58 m; several landfills reach 30–40 m above sea level. Depth to bedrock ranges from 8 to 81 m below the sea level surfaces of the wetlands (Widmer 1964).

BEDROCK GEOLOGY

Bedrock is the solid, continuous rock underlying the landscape and exposed in some places as ledges or slabs. Different types of bedrock vary in physical and chemical properties and thus affect habitats for organisms. Several bed-rock units occur in or at the edges of the Greater Meadowlands (Figure 1.3) and within these units the rocks vary in *mineral* (chemical compound) and elemental composition. Both the natural high ground and human-made fill areas are quite variable upland environments for the biota. This may explain why, for example, Mattox's clam shrimp (*Cyzicus gynecius*) was found only in intermittent pools on two sections of dirt roadbed that seem to have been built from a particular type of rock material.

The Hackensack River Valley, in which the Meadowlands lie, is separated from the Passaic River Valley on the west by a low sandstone ridge, and from the Hudson River on the east by a high ridge of igneous rock—the Palisades diabase, sometimes called *traprock* (Facciolla 1981, Day et al. 1999). The Hackensack Meadowlands are within the Northern Triassic Lowlands (Newark Basin), a subdivision of the Piedmont physiographic province in northeastern New Jersey. The entire Newark Basin extends from southeastern New York through northern and central New Jersey into eastern Pennsylvania (Puffer and Husch 1996). The Newark Basin is underlain by Late Triassic to Early Jurassic rocks about 220 million years old. Most of this bedrock is sedimentary in origin and has been described as red shale, siltstone, and sand-stone (Schuberth 1968). Within this category is the Passaic Formation (Drake et al. 2002), and four Passaic Formation units underlie the Meadowlands. The Passaic formation mudstone, silty to sandy mudstone and siltstone, comprise the bedrock unit underlying most of the Core Meadowlands area (roughly the Meadowlands District and extending northward along Overpeck Creek and southward through the western portion of the Newark Meadows). Between the Passaic formation mudstone and Newark Bay (i.e., the eastern portion of the Newark Meadows) lie variously colored siltstones and shales. West of the Core Meadowlands and extending northward on both sides of the upper Hackensack River estuary are reddish sandstone and siltstone. Very narrow, north-south aligned slivers of shale, siltstone, mudstone (locally calcium-rich), and silty mudstone occur in a few places in Secaucus, Kearny, and North Arlington (Figures 1.2 and 1.3). This unit probably includes portions of the old quarry faces forming the western border of the Core Meadowlands (western bluffs) in North Arlington and Lyndhurst.

From the north end of Newark Bay northward along the eastern edge of the Core Meadowlands is a 1 km wide strip of the Lockatong Formation. The bedrock is gray to brown arkosic sandstone. Farther east at the very edge of the Core Meadowlands is a narrower strip of gray to black sedimentary bedrock that includes dolomitic or analcimic silty argillite, mudstone, silty or calcareous argillaceous pyritic sandstone and siltstone, and silty limestone. These are rocks formed from clayey, muddy, silty, or sandy sediments that may be rich in magnesium, calcium, sodium, or iron.

The Triassic and Jurassic geological periods are familiar as two of the ages of the dinosaurs. The sedimentary rocks of the eastern and western edges of the Meadowlands have yielded important dinosaur fossils. At Granton Quarry, the fossils were recovered from mudstone and arkose beds (Colbert and Olsen 2001). (See locality maps, Figures 1.4 and 1.5)

Laurel Hill (Snake Hill) and Little Snake Hill in Secaucus, a sliver of land west of the Core Meadowlands, other portions of Granton Quarry in North Bergen, and two small areas in Teaneck, as well as the approximately 2 km wide Palisades ridge to the east, are underlain by diabase (see Figures 1.4 and 1.5, localities). The Palisades Sill and Granton Quarry dip beneath the eastern edge of the Meadowlands. The Palisades diabase is an igneous rock intruded below ground between layers of sedimentary rocks; it contains variable proportions of the silicate minerals plagioclase, pyroxene, and olivine (Wolfe 1977). These minerals vary in their content of calcium, magnesium,

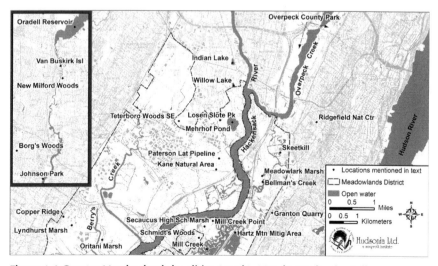

Figure 1.4 Greater Meadowlands localities (north). Map by Hudsonia staff.

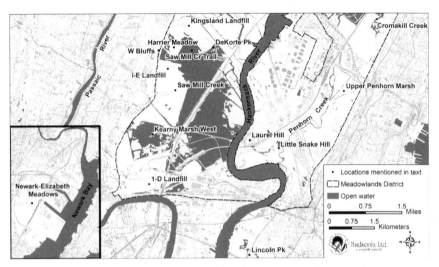

Figure 1.5 Greater Meadowlands localities (south). Map by Hudsonia staff.

iron, sodium, and aluminum (Klein and Hurlbut 1985). The texture and chemistry of the bedrock influences plants, lichens, and other organisms that colonize rock fragments or outcrops. Elemental composition also affects the suitability of soils for support of certain plants and animals—for example, some organisms have an affinity for calcium-rich (*calcareous*) soils. At least at Laurel Hill and Granton Quarry (van Houten 1969, Butler et al. 1975), *hornfels* occurs at the zone of contact between the sedimentary rocks and the igneous diabase where the sedimentary rocks were thermally metamorphosed by contact with the *magma* (hot molten rock) that eventually cooled to form the diabase. The mineral and elemental composition of hornfels is variable. Laurel Hill has a high diversity of minerals, including magnetite, hematite, calcite, analcime, galena, malachite, titanite, and zircon (Facciolla 1981), and petersite, a mineral first described from Laurel Hill (Peacor and Dunn 1982); geochemical diversity is great (Puffer and Benimoff 1997).

The Granton diabase sill, at the eastern edge of the Meadowlands in North Bergen, is a small igneous intrusion similar to the Palisades sill but separated from it belowground by 107 m of Stockton sandstones and shales and Lockatong shales and argillites (Wolfe 1977) (or siltstone and arkose; Butler et al. 1975). The diabase has columnar jointing (Butler et al. 1975) much like the cliffs of the Palisades that overlook the Hudson River. In sum, the Meadowlands are underlain by a variety of rocks with different mineral compositions and potentially variable chemical contributions to the soil and groundwater, providing diverse habitats for organisms.

Because the bedrock layers *dip* (slope down) westward, the land at the eastern border of the Meadowlands tends to slope gently upward to the east

Figure 1.6 Chestnut oak stand, Laurel Hill (see Chapter 4 concerning this area). Photo by Erik Kiviat.

(a *dip slope*), whereas at the western border of the Meadowlands there is in many places a more abrupt, broken-off escarpment (a *scarp slope*). Mining evidently steepened the natural escarpment in North Arlington, Lyndhurst, and northward, where the cliffs are as high as 30+ m.

EARTHQUAKES

Earthquakes occur when bedrock masses shift in relation to each other along large cracks in the earth called *faults*. The Ramapo Fault borders the western edge of the Newark Basin in northern New Jersey and is 20–28 km west of the Meadowlands (Drake et al. 2002). Little *seismic* (earthquake) activity has been associated with the Newark Basin in recent history, but it is generally located within a seismic hotspot that includes central and northern New Jersey, southern New York, and much of Connecticut, within eastern North America (Soto-Cordero et al. 2018). Moreover, Cameron's line, a boundary between tectonic plates, crosses the Greater Meadowlands. The epicenter of one earthquake during the period 1962–1977, shown by Aggarwal and Sykes (1978), was in, or close to, the Meadowlands. From 1737 to 1927, several earthquakes caused damage in New Jersey (von Hake 2009, Dombroski 2020). The strongest occurred in 1884 between Sandy Hook and Brooklyn, about 20 km southeast of Newark Bay, and damaged brick buildings and chimneys in New Jersey and New York City (NYCEM 2019). The regional (including Manhattan and Trenton) probability of a quake of similar magnitude during a 50-year period is estimated to be 22%; for stronger quakes, the probabilities are less (Sykes et al. 2008). Seismic effects at that level could cause a temporary loss of cohesion and bearing strength of some of the wetland fill and garbage landfill substrates beneath buildings and streets in the lowland areas of Secaucus, Carlstadt, Lyndhurst, and other parts of the Meadowlands. Soft soils over bedrock, or artificial fill, can exacerbate damage caused by an earthquake (NYCEM 2019). Some of the garbage landfills have been capped, but an earthquake or subsidence beneath can cause a landfill cap to crack or erode (Ezyske and Deng 2012). An earthquake could also cause the natural and artificial talus slopes at Laurel Hill and Little Snake Hill to shift, and the bedrock mine faces at Laurel Hill and the western scarps to shed fragments. Movement of landfill material, mine rubble, or natural talus could have dramatic effects on the wetlands and waterways bordering those features as well as on some of the local infrastructure.

SURFICIAL GEOLOGY AND SOILS

Surficial materials (*regolith*) constitute the unconsolidated sediments and soils that overlie the bedrock in most places. The Meadowlands have several types of regolith that differ according to their texture (particle size) and mode of formation (Figure 1.7; see Stone et al. 2002). Surficial materials include different size particles, from large to small: *boulders, cobbles, gravel, sand, silt*, and *clay*. Clay is composed of microscopic particles of silicate minerals. Clay particles remain suspended for a long time in quiet waters, and clayey soils have a distinctive appearance and odor. In the U.S. Department of Agriculture classification, clay particles are < 0.002 mm in diameter. Silt particles are larger but still microscopic, 0.002–0.05 mm. Sand particles, 0.05–2 mm, are still larger, the individual particles are visible to the naked eye, and they feel gritty between the fingers. Gravel is larger still, 2–75 mm; these are the familiar pebbles. Natural cobbles and boulders are rare (e.g., in talus) in the Meadowlands, but many boulders have been dumped, placed for landscaping or for vehicle barriers, or have been left in quarry rubble.

Surficial materials originated from the weathering or erosion of bedrock, and surficial geology in and around the Meadowlands was substantially shaped by the Pleistocene glaciations, subsequently by water erosion, and then by modern human activities. Glacial *till* is an unsorted mixture of various particle sizes, typically deposited beneath advancing glaciers or at the southern edge of the glacier as the ice front stagnated or retreated. Some tills

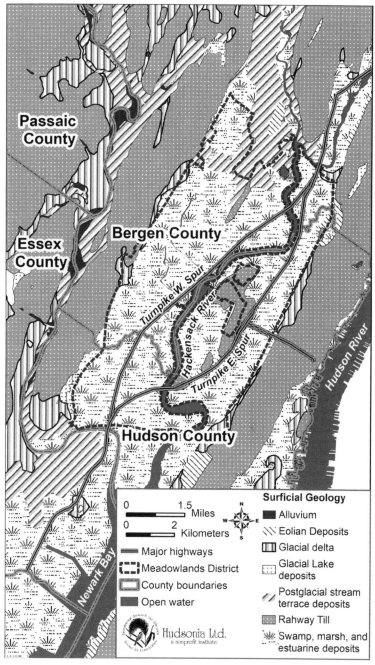

Figure 1.7 Surficial geology of the Greater Meadowlands. In many places the wetland deposits have been covered by fill. Map by Hudsonia staff.

are predominantly clayey, sandy, or gravelly. The deposits adjoining the Core Meadowlands on the east and west sides are mostly till. Unlike till, glacial *outwash* was sorted by flowing water in meltwater streams beneath or south of the glacier. Outwash is generally sandy or gravelly in texture, sometimes silty, and may be deposited in a variety of features such as terraces paralleling the stream. *Eolian* (wind) deposits are silty or sandy and occur in small areas of the Greater Meadowlands. Wetland and estuarine deposits predominate in the Core Meadowlands; on Figure 1.7 these mapped areas include anthropogenic wetland fill.

During the greatest extent of the *Wisconsinan* (the last) glaciation, ice sheets extended from the Far North to what is now the Perth Amboy terminal moraine just southwest of the Meadowlands (10–15 km west and south of Elizabeth). When the glaciers were melting and retreating northward at the end of the Pleistocene, the terminal moraine—extending from Staten Island in the east and through Perth Amboy—impounded glacial meltwater, creating proglacial Lake Hackensack and proglacial Lake Bayonne (Heusser 1963, Stone et al. 2002); we refer to this complex as Lake Hackensack for convenience. Lake Hackensack covered the low-lying area that extended from the terminal moraine north to Tappan, New York (Reeds 1933), including the Core Meadowlands and most of the rest of the Greater Meadowlands.

Deep clay and silt deposits formed in Lake Hackensack. *Glaciolacustrine* (glacial lake) clay and silt were transported into lake waters by streams and deposited in the lake bed as *varved* clay and silt. The varves are thin layers resulting from annual cycles of deposition. During the 2,500 to 3,000 years of the lake's existence, varves were deposited (Antevs 1928), comprising alternating silty layers in summer and in winter the clayey layers that settled more slowly beneath the ice. Beds of clay and silt (nominally clay) up to 30 m thick, overlain by stratified outwash sand and gravel that reach a thickness of approximately 3 m, underlie the wetland sediments (Reeds 1933, Stone et al. 2002). Geologists were able to reconstruct a 2,550-year record by correlating thicknesses of varves from five clay pits and estimated that the glacial recession began about 15,000 years ago and proceeded at about 30 m/year (Schuberth 1968). Sandy or coarser-textured beaches also formed where currents and waves distributed materials along the shores of Lake Hackensack, and dune formation sometimes occurred when unvegetated sands were blown by strong winds.

It was postulated that Lake Hackensack drained as a result of breaching of the barrier dam (terminal moraine) on the south from gradual uplifting of the lake bottom as the land adjusted to decreased pressure from the retreating glacier (Heusser 1963). Draining of Lake Hackensack coincided with a postglacial rise in sea level of about 3 m in the last 2,000 years, which allowed encroachment of sea water and eventually the formation of tidal marsh as the estuarine tides flooded and ebbed on and off the wetland surfaces (Heusser 1963).

Soils in the Meadowlands are predominantly *Sulfihemists and Sulfaquents, frequently flooded* (tidal marsh soils) and anthropogenic cut-and-fill soils (*Udorthents*) (Goodman 1995). Natural mineral soils occur in limited upland areas (Sipple 1972). For scale, the Sports Complex, including stormwater ponds, occupies about 250 ha of fill. Only 1 ha remains of the greenspace at Granton Quarry and the unnamed greenspace 275 m to the northeast covers ca. 4 ha.

Wetland Soils

Vermeule (1897) described surficial deposits in the Meadowlands as consisting of "blue mud" or clay with portions covered by peaty soils. Sipple (1972) described soil in the Meadowlands as predominantly peat or muck with mineral material overlying the glaciolacustrine clays. The organic soils (*peat, muck*) contain the remains of aquatic plants, including fibers, logs, roots, and stumps, that accumulated in the wetlands during thousands of years. Peat soils contain more than 50% organic matter; muck soils 20–50% organic matter (Brady and Weil 1996). Peats formed either in freshwater wetlands (e.g., Atlantic white cedar swamps) or brackish marshes, and in places the latter overlie the former (Tedrow 1986, 208–211) as a result of sea level rise.

In the eastern portion of the Meadowlands Sports Complex, thicknesses of soil strata from top to bottom were: fill, 1.2–4.9 m; dark brown peat, *meadow mat*, or organic silt 1.2–3.0 m; fine sand with some silt, 0.3–0.8 m; varved clay and silt with some sand, 4.0–7.2 m; and till, 0.3–4.6 m. Decomposed shale occurred at 8.2–13.3 m, maximum 19.8 m below Mean Sea Level (MSW; McCormick & Associates 1978). A set of borings from the Sports Complex site was diagrammed in Agron (1980).

In the Bergen County soil survey (Goodman 1995), the wetland Sulfihemists (organic soils) and Sulfaquents (mineral soils) were mapped as one unit, and described as very deep, level or nearly level, very poorly drained soils. Sulfihemists in the Meadowlands have at least 41 cm, and usually more than 130 cm, of organic material overlying mineral material. The top 30 cm of organic material are dark colored (gray, reddish brown, brown, or black) and highly decomposed. Below 30 cm the organic material is less decomposed but may be interbedded with highly decomposed material. The underlying mineral material ranges in texture from very gravelly sand to varved silt and clay. Depth to bedrock typically exceeds 3 m in the Sulfihemists.

The Sulfaquents of the Meadowlands have either a surface layer of organic material less than 41 cm thick (dark reddish brown or black, and highly decomposed) or mineral material 10–30 cm thick (dark reddish brown, very dark brown, or black silt loam or fine sandy loam with organic content < 20%). The underlying mineral material has a wide range of color and has textures in the same range as those of the material underlying the Sulfihemists. Depth to bedrock exceeds 1.8 m in the Sulfaquents.

Both Sulfihemists and Sulfaquents have high available water capacity, and both soils have very slow runoff. The Sulfihemists and Sulfaquents have severe limitations (wetness, flooding) for building site development and sanitary facilities (sewage treatment systems and sanitary landfills; Goodman [1995]).

Both Sulfihemists and Sulfaquents are neutral or slightly acidic throughout when moist; however, when dried these soils become strongly acidic or very strongly acidic (Goodman 1995). The soils contain *pyrite* (iron sulfide). When the soils are exposed to air as a result of excavation or drainage, the sulfide oxidizes to sulfate which is very acidic. The generic name for soils that undergo this process is *acid sulfate soils*, and they occur widely in tropical and temperate tidal wetlands and occasionally in nontidal wetlands. Acid sulfate soils are very difficult to manage for cultivation as they must either be kept permanently wet (as in rice paddies), thoroughly leached, or treated with large amounts of lime for several years (Miller and Donahue 1995, Brady and Weil 1996). Sulfate acidification probably occurred where marsh substrate was excavated and deposited to create *wildlife islands* in wetland management projects. Acid sulfate soils might also have been a factor in the historic failures of drainage agriculture in the Meadowlands.

Urban Soils and Deposited Materials

Urban land in the Meadowlands was described as Udorthents with either a loamy substratum or a wet substratum and highly variable (Goodman 1995). Rock and soil material originating at other locations has been used intentionally as fill, or just dumped, at many locations in the Meadowlands, and it is likely that some of these materials have peculiar physical and chemical properties that created unusual habitats. One such site is Harrier Meadow (USACE 2004), where the marsh developed on rock rubble from highway construction outside the Meadowlands. Another may be where the western edge of the Core Meadowlands was affected by mine dumps and drains from the Schuyler copper mine in North Arlington (Puffer 1984).

Although serpentinite has not been found as a component of the Meadowlands bedrock, this unusual material occurs in Hoboken at Castle Point on the west shore of the Hudson River. Serpentinite is a metamorphic rock that tends to have an abundance of magnesium, iron, and chromium, and a scarcity of calcium and sodium. Ca. 2002, serpentinite excavated from a construction site in Hoboken was dumped in the Kane Natural Area in Carlstadt on an old fill deposit and in the edges of the surrounding wetland. We do not yet know if the serpentinite has created habitats for species (e.g., of lichens, mosses, or seed plants) that do not otherwise occur in the Meadowlands. Natural serpentinite outcrops typically support unusual assemblages of organisms.

Germine and Puffer (1981) stated that bedrock at Laurel Hill and the serpentinite bedrock at Castle Point are rich in asbestos. This may create an inhalation hazard for humans and wildlife, especially at Laurel Hill where dust is mobilized from the quarry floor by ORV (off-road vehicle) riding and other human activities.

Materials used to fill wetlands, build causeways, and cover garbage vary greatly. Native Americans disposed of food wastes and other materials in middens which likely resulted in small areas enriched with calcium from mollusk shells. European-Americans used logs of Atlantic white cedar (*Chamaecyparis thyoides*) and probably other trees to build early *corduroy* roads across wetland, memorialized in the name Paterson Plank Road. Rocks up to boulder size were dumped, pushed aside, or placed at Harrier Meadow, Laurel Hill, Bellman's Creek, and Kane Natural Area. Wastes dumped in the Meadowlands were almost limitless in variety (Sullivan 1998a, b). Dredged material (*spoil*) from Berry's Creek Canal, the Hackensack River, and other waterways was deposited in wetlands such as Oritani Marsh. *Overburden* and *tailings* (unwanted materials from mining) were dumped near mine sites (e.g., the Schuyler copper mine in North Arlington [Puffer 1984]), and brick manufacturing waste was dumped at Mehrhof Pond, Penhorn Creek, and Bellman's Creek. Unidentified sandy waste was dumped at Meadowlark Marsh.

Waste materials, because they are often different physically and chemically from the surrounding environment, may support unusual biota, and different materials provide habitats for different organisms. Pixie lichens (*Cladonia* spp.) were common on weathered rail ties in a dump at the west edge of the Kingsland Landfill, and on sandy-gravelly material that was probably old mine tailings northeast of Laurel Hill. Diverse moss species occurred on weathered artificial substrates such as concrete (see Chapter 5). Weak rush (*Juncus debilis*) has been found only at Meadowlark Marsh at the edge of the waste deposit. Clam shrimps occurred in rain pools on the fill of dirt roads in a few places but apparently not many others (Chapter 10).

Upland Soils and Talus

Natural upland (non-wetland) soils that have not been cut or filled are restricted in occurrence in the Meadowlands. Small areas at Little Snake Hill and Laurel Hill are the best examples; such soils also occur in outlying areas like Borg's Woods and Ridgefield Nature Center. Although mapped as udorthents, there may also be natural upland soils in the woodland at Mehrhof Pond and adjacent Losen Slote Park, though these areas are predominantly drained wetland. An unnamed hill 400 m northeast of Granton Quarry seems to have minimally disturbed upland soil. Because many organisms are burrowers in soil or leaf litter, the shortage of natural uplands likely plays a role in the absence or scarcity of certain *taxa* (related groups of organisms) such as amphibians and land snails.

Abundant clay and sand near the ground surface in Little Ferry and South Hackensack provided the natural resources for the brick industry, and that industry in turn left several large surface mine pits. Three water-filled pits remain as (from south to north) Mehrhof Pond, Willow Lake, and Indian Lake. Two additional small basins containing ponds were shown on the U.S. Geological Survey topographic map, east and northeast of Maple Grove Cemetery in South Hackensack, and a larger basin containing a pond east of Mehrhof Road in Little Ferry. These two basins were presumably mine pits that were filled. In other mined, and perhaps some non-mined, areas, clay is exposed but not flooded. These are the clay flats or wet clay meadows on the western portion of the Bergen County Utilities Authority (BCUA) property, and probably extending southward across Losen Slote and east of the Paterson Lateral Gas Pipeline (the last area now part of the Kane Natural Area Wetland Mitigation Bank). Soils were mapped by Goodman (1995) as Udorthents with wet substratum around Mehrhof Pond, and Sulfihemists and Sulfaquents frequently flooded between Losen Slote and the pipeline. These areas may have been mined for clay and sand for the brick industry or filled with *overburden* (non-usable material that was removed above the usable sand or clay).

Sandy and gravelly soils overlooking the Meadowlands on the east and west are occupied by at least twelve cemeteries (USGS 1967; also visible on Google Earth satellite imagery) collectively containing ca. 250 ha of greenspace. Bergen County soil maps (Goodman 1995) show that the cemeteries are associated with Boonton gravelly loam (till), Boonton-Urban land complex, and Riverhead sandy loam (outwash). This is a good example of environmental geology: elevated, well-drained, coarse-textured soils such as these are relatively easy to dig, even in winter, and therefore suitable for burials at almost any time.

Laurel Hill and Little Snake Hill have areas of talus (steeply sloping accumulations of rock fragments, in these cases boulders). The natural talus slopes on the north and east sides of Little Snake Hill include many large boulders with much bare rock surface exposed. On the south side of Laurel Hill there is a natural area of forested talus with smaller blocks and more accumulated soil and tree litter, and a small area of talus with larger blocks and less soil at the east end of the hill. At the western end of Laurel Hill there is anthropogenic talus of rock rubble from quarrying. This talus is mostly unvegetated and the rocks are little-weathered compared to the natural talus slopes. Both hills also have extensive exposed bedrock ledges. Much of Laurel Hill was mined (the northern and western portions), and the abandoned mine faces now comprise cliff-like features up to 30+ m high that support cliff-using biota such as blunt-lobed cliff fern (*Woodsia obtusa*) and nesting common raven (*Corvus corax*). Both hills, especially Laurel, have been used by ORV riders post-mining and this has contributed to further soil loss and exposure of bedrock slabs; Turnpike construction and improvement also resulted in alteration of bedrock and soil at Laurel Hill. Little Snake Hill has been difficult to access since construction of the Secaucus rail station in 2003; ORV use and graffiti-painting ceased earlier.

Wetland Sediment Accretion

In addition to its hydrological role in impounding the Oradell Reservoir for water withdrawal that greatly reduced freshwater discharge through the Hackensack River estuary, the Oradell Dam presumably traps much of the sediment carried into the reservoir from the upper Hackensack River watershed. Prior to damming, this sediment would have been transported down the estuary at high freshwater flows and partly deposited on the soils of the tidal wetlands in the Meadowlands. Reduced sediment supply may have limited accretion in the marshes. Sediment is also transported into the Hackensack River estuary from the south with the more-brackish water along the channel bottom (Suszkowski 1978). Minor sediment sources are sewage treatment plants (Shrestha et al. 2014), sediment from the surrounding landscape including remediation sites (E. Kiviat, pers. obs.), and deposition of airborne materials.

Most large deltas worldwide are sinking as a result of human activities (Syvitski et al. 2009). The fate of tidal wetlands in the Meadowlands and other coastal cities during this century will be influenced by the accelerating rate of sea level rise, the deposition, subsidence, compaction, and erosion of wetland sediments, the forces of more frequent and more intense storms, the function of marsh plants such as common reed and smooth cordgrass in accreting and stabilizing sediments, the shelter formed by existing fill, and improvements to engineering features that include *dikes* (berms or artificial levees), tide gates, and pumps. The magnitude and effects of all of these factors are controversial and challenging to predict. Walsh and Miskewitz (2013) opined that, if sea level rises 1 m during the present century, as has been predicted in a more aggressive sea level scenario of Gornitz et al. (2001), large areas of the Meadowlands District will be continuously flooded. The extensive Saw Mill Creek mudflats, where impoundment dikes were breached in a

1950 storm, appear to be eroding, and sediments have been dredged from the
marsh surfaces for reed control in many of the completed wetland management projects in the Meadowlands. Tiner et al. (2002) found that from 1889
to 1995, three-fourths of the wetlands in the District were lost, and deepwater
habitats increased by half; wetland loss occurred by a combination of filling
(mostly) and submergence. We could debate the details and magnitudes of
wetland sedimentation processes (e.g., see Kirwan et al. 2016), but it would
be cautious to assume a greater rather than lesser impact of sea level rise on
the Meadowlands. This approach requires anticipating ecological upheaval
affecting all habitats and organisms including people.

HYDROLOGY

A maze of waterways, mostly altered or wholly artificial, permeates the
Meadowlands; many of the larger ones are shown in Figures 1.8 and 1.9.
Water enters the Meadowlands from direct precipitation, wastewater treatment plant and combined sewer effluents, the outflow from the Oradell
Reservoir at the *head of tide* (greatest inland penetration of the tide) on the
upper Hackensack River estuary, freshwater discharge from the Passaic
River, outflows from smaller nontidal tributaries, diffuse or channelized
surface runoff from surrounding urban areas, groundwater discharging as
springs or seeps, and, of course, the tides. Surface water flows in streams and
estuaries (the Hackensack and Passaic river estuaries, tidal lower reaches of
smaller tributaries, and Newark Bay) and collects in a variety of pools, ponds,

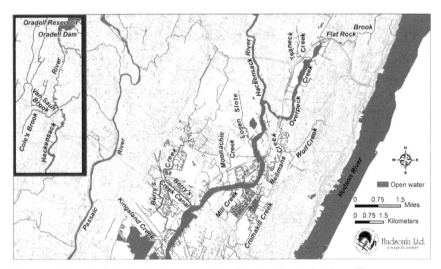

Figure 1.8 Greater Meadowlands streams (north). Map by Hudsonia staff.

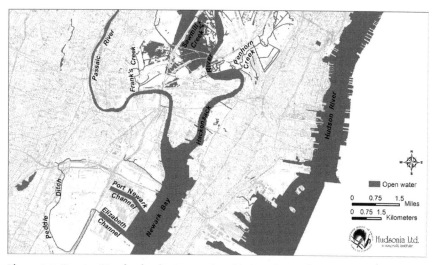

Figure 1.9 Greater Meadowlands streams (south). Map by Hudsonia staff.

and lakes including the clay pits, diverse intermittent and ephemeral pools, and a variety of stormwater ponds, waste treatment ponds, and ornamental ponds. How water moves and stands in the system has large influences on the structures, processes, and biota of the Meadowlands landscape. Paramount is the circulation of tidal and nontidal waters, their salinity variation in space and time, and the features of the natural and built landscape that accelerate or impede movements of water and salinity. Salinity at any point varies depending on the tide, recent rainfall in the watershed, and wind; salinity generally increases upriver to downriver.

Moving water carries materials that include natural salts, plant macronutrients, trace elements, living and dead organisms and their fragments, dissolved natural organic compounds, and pollutants. The Meadowlands have a *watershed* and a *seashed*. The Hackensack River component of the Meadowlands watershed (drainage basin or the region that contributes runoff water into the river) covers 523 km² from Haverstraw, New York, to the upper end of Newark Bay (Hales et al. 2007). This does not include occasional pumped interbasin transfers from Sparkill Creek in New York for the purpose of augmenting flow into Oradell Reservoir (*ibid.*). The Passaic River watershed and the watersheds of smaller streams draining directly into Newark Bay also contribute to the Meadowlands. The watersheds contribute the *freshwater discharge* to the Meadowlands, that is, the freshwater moving downriver independent of the reversing tidal currents. Effluent from publicly-owned sewage treatment plants accounts for most of the freshwater flow in the Meadowlands, except during storm events (USFWS 2007).

Raritan Bay, Arthur Kill, Kill van Kull, Upper New York Bay (New York Harbor), Lower New York Bay, and the New York Bight are the *seashed* of the Meadowlands. These areas contribute materials via up-estuary movement with the flood tide component of estuarine circulation or the density current of saltier water on the bottom of the channel. All the various sources of freshwater and brackish water make the Meadowlands a complex hub of estuarine flows.

More than any other habitat, wetlands characterize the Meadowlands. In 1889, there were about 8,115 ha of wetlands in an area slightly larger than the Meadowlands District (Tiner et al. 2002). In 1953–1954 there were more than 5,425 ha of wetlands (*ibid.*). In 1995, 2,243 ha of wetlands remained. The analysis of Tiner et al. did not include the Newark-Elizabeth Meadows and the upper Hackensack River estuary wetlands that covered perhaps another 3,000 hectares, now nearly all filled. (There are somewhat higher estimates of the area of wetlands in the Meadowlands District in other literature such as Obropta et al. [2007]).

Mean tide range (the average vertical distance between Mean High Water [MHW] or high tide level and Mean Low Water [MLW] or low tide level) for the Meadowlands as a whole is about 1.5 m. In the Hackensack River at Kearny, the mean tide range (computed from ca. 30 years of data) was 1.53 m, and at the mouth of Mill Creek in the Hackensack River, the farthest upstream site at which MERI (the Meadowlands Environmental Research Institute) maintained a tide gauge, 1.66 m (MERI 2008).

As generally along the East Coast of the United States, there are two high tides alternating with two low tides each approximately a 25 hour period, or a full tidal cycle (high tide to high tide) every approximately twelve hours and twenty-five minutes. However, because of the distance from the outer coast and the topographic constrictions caused by Staten Island and Newark Bay, the time and height of high and low tides are more variable. The amount of upland runoff entering the estuary, the wind direction, speed, and duration, and the phase of the moon can have advancing or retarding effects on tide timing and height. For example, actual high tide can be later and lower than the predicted high tide if there is a strong north wind; both high and low tides can be higher and the high tide can last longer if there is a large amount of upland runoff entering the estuary after a big storm or snowmelt. Tide range is greater around full moon and new moon, and less around half moon dates.

The tides move up the Hackensack River estuary as a wave with a long wavelength and a low wave height. On average, the wave height equals the mean tide range of about 1.5 m. High tide occurs in Newark Bay before it occurs up-estuary at the junction of Overpeck Creek and the Hackensack River. Before the Oradell Dam was completed in 1921, the maximum inland extent of tidal influence (the height of tide) was at approximately the location of the dam. Thus the construction of the dam did not greatly alter the inland penetration of the tides (but this might have changed with rising sea level). Presumably the dam was sited there because it would impound water that was ample in quantity and fresh enough to drink, as well as taking advantage of topography suitable for engineering. The Oradell Reservoir, and other reservoirs in the upper Hackensack River watershed, impounded virtually all the freshwater discharge of the river for water supply for surrounding communities (Hales et al. 2007). This removal of freshwater increased the salinity of the upper Hackensack River estuary and the Meadowlands (Quinn 1997), causing salinity to intrude farther into the Meadowlands (because the freshwater discharge would have pushed salinity farther downriver under any particular set of conditions).

In the late 1800s and early 1900s, Newark Bay and the lower Hackensack River and Passaic River estuaries were dredged to improve shipping channels (Marshall 2004). A 9 m deep shipping channel was dredged the length of Newark Bay and a 3 m deep channel from the mouth of the Hackensack River up to Lyndhurst (Crawford et al. 1994). Dredging, as well as relative sea level rise, increased salinities upriver and presumably increased the magnitude of storm surges. Increased salinities converted areas of freshwater tidal marsh into *oligohaline* (slightly brackish) marsh and oligohaline marsh into *mesohaline* (medium brackish) marsh (Marshall 2004). Increased salinities, along with logging and burning of the Atlantic white cedar swamps, caused the extirpation of this tree in the late 1800s and early 1900s (Heusser 1949); an undocumented and probably unusual biological community (Laderman 1989) was presumably lost with the cedars.

Tidal flows pump in and out of the Hackensack River—Newark Bay estuary and its tidal marshes from the seashed. There is a limited hydrological connection across the watershed boundary between the lower end of the Hackensack River estuary and the lower end of the Passaic River estuary via the Kearny Marshes. A damaged bulkhead at Frank's Creek, which usually drains water from the southwestern part of Kearny Marsh West into the Passaic River, actually reverses flow and feeds Kearny Marsh West when rainfall coincides with a high tide and the downstream tide gate closes (Obropta et al. 2008). Water also flows out of the eastern end of Kearny Marsh West toward the Hackensack River. Before hydrological alterations, we assume that much of the area of Kearny was wetland across which water moved in both directions between the two rivers. The up-estuary flows from Newark Bay presumably also mix waters from the Hackensack and Passaic rivers.

Most of the Meadowlands is at or close to sea level; 90% of the Meadowlands District lies within 0.6 m elevation above Mean High Water. The following information on flooding and flood control infrastructure was summarized by Obropta et al. (2007, and references therein). *Storm surges* (unusually high tides due to some combination of abundant storm runoff, higher-than-normal high tides, and onshore or up-estuary winds) may reach 1+ m above the MHW elevation of 0.6 m. Major flooding can occur in any season but is more likely in late summer and early fall. Annual precipitation has increased during the past century, and the frequency and intensity of *Nor'easters* (strong onshore storms) have increased (see Climate, below).

Historically, large areas of the Meadowlands have been built up slightly by means of low-lying wetland fill, or more greatly by means of garbage land-fills. Some of the filled areas are subsiding because fill was placed on deep wetland soils and because of compaction of the fill material itself. Some areas of the Meadowlands are protected to a degree by dikes, tide gates, and pumps; there are 25 tide gates and 9 pumping stations in the Meadowlands District and some have failed recently, e.g., during Hurricane Sandy. Extensive cover by impervious surfaces (pavement and roofs) increases runoff and exacerbates flooding. During the more intense flood events, many streets and highways are impassable. In October 2012, Hurricane Sandy caused the highest storm surge in recent years, flooding most of the Meadowlands with brackish water during a single high tide, to a depth of almost 3 m above the predicted high tide (Artigas et al. 2017, National Weather Service no date). The Meadowlands are experiencing a dilemma of planning and engineering common to low-lying coastal cities: try to keep the sea out or move buildings and infrastructure to higher ground (Nordenson et al. 2018).

Many Meadowlands wetlands were impounded or diked in attempts to grow crops or control mosquitoes (for historical information, see Quinn [1997] and Marshall [2004]). Formerly tidal wetlands were transformed into nontidal (or restricted-tidal) wetlands when the tides were excluded by dikes or tide gates. Impounded wetlands are fed by stormwater runoff from adjacent urban areas as well as by precipitation and in some cases tidal waters passing through culverts or seeping through the dikes. As is true of engineered flood control features anywhere, the dikes and tide gates in the Meadowlands are vulnerable to damage; for a cautionary tale see the Van Baars and Van Kempen (2009) analysis of the numerous dike failures in the Netherlands. In 1950, the dikes impounding the Saw Mill Creek marshes in the Meadowlands were breached by a storm. In 2007, the dikes protecting the Kane Natural Area marshes were breached by a storm, allowing brackish tidal water to flood a large marsh that had been fresh for many years. These dikes, bordering the Hackensack River mainstem, were in effect artificial levees. Many of the *diked* (empoldered, water kept out) or *impounded* (water kept in) marshes are dominated by common reed (*Phragmites australis*), and this has generally caused scientists and managers to consider these tide-restricted wetlands degraded and in need of reed removal or re-opening to the brackish tides. Yet some of the rare and uncommon species of animals found in these wetlands are closely and apparently positively associated with reedbeds or combinations of reedbeds and other habitats. Common reed also stabilizes and builds sediments in the marshes and provides other ecosystem services (Kiviat 2013).

Given the numerous commercial and residential buildings, highways, and other infrastructure built barely above sea level, subsidence of land and structures built on filled wetland, rising sea levels, and the increases in impervious surface coverage throughout the watershed, it is not surprising that there are flooding problems after heavy rains and during storm surges. Water control or avoidance of water is always an important environmental issue in wetland regions (Kiviat 1991, 2014). The Meadowlands have myriad drainage canals and ditches, tide gates, pumps, dikes, and ponds intended to keep storm surges out of developed areas and help stormwater drain off the land. These stormwater management systems are in many cases old, and were not necessarily designed for the modern situation, nor should they be expected to function well due to their location at or barely above sea level. The canals, ditches, and stormwater ponds we have seen appear *eutrophic* (rich in plant nutrients) and otherwise with poor water quality; nonetheless, some of these systems are important for biodiversity. A shallow stormwater pond at the northern base of the 1-E landfill in September 2006 supported foraging of several species of shorebirds and an array of flower-visiting and predatory insects. A stormwater pond in Moonachie supported a chorus of Atlantic Coast leopard frog in 2006 (Kiviat 2011) which was still present in 2012–2013. Agencies are promoting economic growth (Anonymous 2007) and development or redevelopment (see NJSEA 2019) of currently vacant lands in the Meadowlands despite the vulnerability of many areas and much infrastructure to flooding and the concomitant importance of those areas to birds and other native biota. For example, a large school, completed in 2018 on ca. 7 ha of prior greenspace at the north end of Laurel Hill Park, has a ground floor elevation of ca. 2–3 m, and some new or proposed commercial and government buildings elsewhere are even lower.

The ecology of the Meadowlands will continue to change due to human activities and other factors. Rising sea level and new patterns of coastal storms resulting from global climate change will increase the frequency and duration of flooding, as well as the frequency and intensity of salinity intrusions in portions of the Meadowlands. Probable establishment of a beaver (*Castor canadensis*) population in the freshwater and slightly brackish portions of the Meadowlands will substantially alter hydrology. Implementation of mitigation and restoration projects in wetlands is probably having a large effect. Proposed deepening and widening of the shipping channel in Newark Bay, and potentially in the lower end of the Hackensack River estuary, will affect hydrology in the Meadowlands by increasing tidal range and salinity.

Proposed redevelopment and flood control projects would also have profound effects on hydrology, habitats, and species.

WATER QUALITY

In general, the quality of urban surface waters is poor by standards of undeveloped areas, and this is certainly true of the Meadowlands. Good water quality is characterized by generally high levels of dissolved oxygen (DO) and low levels of phosphorus and nitrogen compounds (including phosphates, nitrates, and ammonium), suspended and dissolved solids, and fecal coliforms and other potentially harmful bacteria. Throughout the 1800s, portions of the Meadowlands experienced substantial sewage and industrial pollution and reduced freshwater discharge (due to water withdrawals) from the Passaic River (Marshall 2004). The City of Newark suffered flooding and epidemics due partly to its location adjoining and in wetlands (Galishoff 1988, Iannuzzi et al. 2002). Sewage and industrial effluents were released untreated into waterways through the mid-to-late 1900s (Crawford et al. 1994). All (dry weather) sewage was treated by 1995, but during wet weather, combined sewer outflows (CSOs) continue to release untreated sewage into waterways (Stinnette et al. 2018).

Since 1993, the Meadowlands Environmental Research Institute (MERI) has monitored approximately 25 water quality variables quarterly at 14 stations on the Hackensack River estuary mainstem, tidal tributaries, and impoundments. We examined the data from 6 of those stations selected to represent better and worse conditions: the Hackensack River just south of Route 46 (near the confluence with Overpeck Creek; MERI station 1), Saw Mill Creek (MERI 7), Berry's Creek (MERI 8), upper Cromakill Creek (MERI 11), Kearny Marsh (MERI 13), and the 1-A Landfill (MERI 14; see Figures 1.4 and 1.5; data available from MERI 2019). All stations were slightly to moderately brackish, with mean salinities ranging from 9.9 ppt (parts per thousand) at Saw Mill Creek to 1.5 ppt at upper Cromakill Creek (compared to about 35 ppt salinity in the Atlantic Ocean). Mean pH was slightly alkaline (means for the 6 stations were 7.3–7.7); values ranged from 6.0 to 9.1 standard units.

Low DO, or *hypoxia*, occurs when nutrient enrichment promotes an abundance of organic matter, for example, from algal blooms, and the decomposition of this material or other organic matter (e.g., sewage) then robs the water of oxygen. US EPA criteria for the protection of saltwater life are 4.8 mg/L (milligrams per liter DO) for chronic hypoxia and 2.3 mg/L for acute hypoxia. Below approximately 5 mg/L DO, many fishes are physiologically stressed. Low DO was a persistent problem in Newark Bay through much of the 1900s with some improvement in the 20 years preceding publication of Crawford et al. (1994). The larger New York–New Jersey estuary region has experienced decadal improvements in DO, but hypoxia remains a problem: during the 1970s, chronic hypoxia was exceeded in 50–70% of samples in most years, compared to approximately 20% of samples in the 2008–2018 period (Stinnette et al. 2018).

No trend in DO (1993–2018) was apparent when data from the 6 Meadowlands (mostly off-mainstem) stations we examined were combined, and all sites had some measurements below 4.0 mg/L in all decades. Cromakill Creek, which is influenced by both sewage treatment plant effluent and CSOs, experienced worsening DO levels during this period. Median DO for 1993–2018 was only 4.1 mg/L at Cromakill Creek, but over 7 mg/L at the other stations. However, DO levels in the Meadowlands portion of the Hackensack River mainstem showed improvement over this time period (Artigas et al. 2016), perhaps due to more tidal mixing here than in the wetlands and tributaries.

Many aspects of the distributions of organisms are shaped by environmental extremes rather than means; single point (one station at one time) DO measurements may be more useful in understanding the adequacy of DO to support particular species of animals, although it is not always known how long a measured level persists. Even in the 2014–2018 period, warm-season DO measurements fell below 4.0 milligrams per liter at 5 stations (excluding the 1-A Landfill), and there were several measurements below 1 mg/L at Cromakill Creek. Given the highly urban character of the surrounding landscape, and the intensity to which the estuary has been subjected to organic wastes including sewage and landfill leachate, it is not surprising that DO is frequently low in hot weather and has improved little in recent years. Notwithstanding, the estuary supports moderately diverse fish and benthic invertebrate communities (Chapters 9–10). Areas with contamination sources and little water circulation may reach extreme values. For example, surface water pH ranged from 7.0 to 9.5 (neutral to very alkaline) in channels bordering the Keegan Landfall in Kearny Marsh West (Mansoor et al. 2006); many organisms might not survive at the high end of that range.

The nutrients that contribute to hypoxia are primarily nitrogen and phosphorus compounds, with nitrogen generally the more important (limiting) nutrient in marine and brackish water systems and phosphorus the limiting nutrient in freshwater systems. Total nitrogen levels less than 0.45 mg/L are considered supportive of marine health. The New Jersey criteria for un-ionized ammonia in estuary waters are 0.115 mg/L (3-h mean) and 0.030 mg/L (30-d mean). The Meadowlands sites are consistently well above these levels: mean ammonia for the 1993–2018 period was lowest at Kearny Marsh (0.97 mg/L) and highest at Cromakill Creek (8.3 mg/L). A modest decline in ammonia occurred over this period, due mainly to fewer extreme values in recent years. Hackensack River nitrate, ammonia, and phosphate levels measured during non-storm periods in 2015–2016 were significantly higher than levels in Newark Bay or the Passaic River (Jung 2017). The major source of ammonia to estuaries is sewage.

New Jersey water quality standards for fecal coliform counts in estuarine waters not suitable for swimming or shellfish harvesting are 70–1,500/100 mL (mean), and *Enterococcus* standards are 35/100 mL (mean) and 104/100 mL (single sample maximum) for estuaries. Between 2007 and 2016, the mean concentration of *Enterococcus* in the Hackensack-Passaic-Newark Bay area was above the acceptable contamination level in every year (Stinnette et al. 2018). Donovan et al. (2008) analyzed levels of fecal coliform, total coliform, fecal *Streptococcus*, fecal *Enterococcus*, *Giardia lamblia*, *Cryptosporidium parvum*, and several viruses sampled in August and September in the lower Passaic River, including locations at combined sewer outflows (CSOs) during a rain event. Fecal coliform mean was 2,980/100 mL, with samples near the CSO of > 30,000/100 mL. Mean *Enterococcus* was 800/100 mL, with individual samples exceeding 30,000/100 mL. They estimated that the risk of contracting gastrointestinal illness from accidental ingestion of the water was 14% for anglers or boaters, 68% for swimmers, and 88–100% for homeless people using the river for washing or bathing (Donovan et al. 2008). Meadowlands coliform counts remained comparable from 1993–2018, except for an increase in extremely high individual counts in the most recent six years, primarily at Cromakill Creek.

Water clarity is important for photosynthesis of submergent plants and helps determine the depth and extent to which submergents can grow, which in turn affects the overall productivity of an estuarine system. High turbidity or high levels of total suspended solids can cause aquatic plant dieback and interfere with movement, respiration, and feeding of aquatic invertebrates and fishes. Water clarity can vary seasonally in unpredictable ways and is highly affected by storm events. Meadowlands water clarity was least in winter and greatest in summer in 1987, but greatest in winter and least in summer and fall in 2002 (a drought year; Bragin et al. 2009). Between 1993 and 2018, mean turbidity was similar among Meadowlands sites, with the most extreme values seen in the Hackensack River mainstem. From 2006–2018, many more high turbidity (low water clarity) values were recorded than previously. High turbidity, along with other factors, may explain the paucity of submergent plants in Meadowlands water bodies.

Water temperature affects other measures of water quality, especially DO, because warmer water holds less oxygen than cooler water. In the New York–New Jersey harbors, mean winter temperatures at the bottom experienced a significant warming trend between 1987 and 2017 (Stinnette et al. 2018).

Summer water temperatures are more limiting to aquatic life; these have been stable but will also rise as regional climate continues to warm.

Toxic Contaminants

A wide variety of toxic contaminants also affects water quality in the Meadowlands. A long industrial history, including the manufacture of paints, chemicals, pharmaceuticals, and plastics and the refining of copper and petroleum, have left substantial contamination in soils and sediments. Present contaminant sources include power plants, sewage and stormwater discharges, petroleum refineries, and landfill leachate (Crawford et al. 1994). Contaminants that occur at elevated levels in the Meadowlands include the metalloid arsenic and metals such as cadmium, chromium, copper, lead, mercury, nickel, and zinc; a large number of different PCB compounds; polycyclic aromatic hydrocarbons (PAHs) and other petroleum hydrocarbons; and many organochlorine pesticides, dioxins, related organic compounds, and other endocrine-disrupting chemicals (USFWS 2007).

Cadmium, chromium, lead, nickel, and zinc declined in Meadowlands water samples between 1993 and 2018. Copper and manganese remained at similar levels, and iron increased. Of these, copper, lead, iron, and zinc regularly exceeded EPA water quality criteria in recent years. Dioxins were present in Newark Bay waters in 2008–2009 but at levels three times or more below the criterion values (Friedman et al. 2012). The concentration of a contaminant in water is a better indicator of its bioavailability than sediment concentration (Friedman et al. 2012), but most measurements we have are from sediments.

Sediment levels of mercury, dioxins, PCBs, and DDD (the last is a breakdown product of DDT) in Newark Bay have all declined substantially from 1950s or 1960s values, but all continue to be above threshold values for adverse effects to aquatic life (Steinberg et al. 2004, Lodge et al. 2015, Stinnette et al. 2018). Sediment levels of arsenic, chromium, copper, lead, and zinc in the Hackensack River are also above threshold values (during non-storm periods in 2015–2016; Jung [2017]). Other contaminants present in high levels in different parts of the Meadowlands include PAHs, volatile organic compounds (VOCs), phenol, petroleum hydrocarbons, and the insecticide dieldrin (USFWS 2007). Contaminants of emerging concern in the New York–New Jersey Harbor Estuary include microplastic particles, many different pharmaceuticals and pesticides, caffeine, sucralose, nicotine, the insect repellent DEET, and the industrial chemical methyl benzotriazole (Stinnette et al. 2018).

When present and bioavailable, contaminants can have a range of effects on organisms including neurotoxicity, altered metabolism, endocrine disruption, developmental toxicity, immunotoxicity, and impaired photosynthesis (e.g., Tsipoura et al. 2011, Köhler and Triebskorn 2013), with cascading effects on biodiversity, nutrient cycling, and other ecosystem processes (although research on higher-order effects is scant; Köhler and Triebskorn 2013). Metals, organic contaminants, and pesticides can compromise higher plant or phytoplankton growth, changing plant community composition (Gallagher et al. 2008, Watts et al. 2008, Köhler and Triebskorn 2013). Heavy metal concentrations in Hackensack River water increase downstream through the Meadowlands District (Shin et al. 2013). Mercury and possibly other metals have been found to affect the community structure of benthic macroinvertebrates in some locations in the Meadowlands, resulting in lower diversity and more pollution-tolerant species inhabiting highly contaminated areas (Weis et al. 2004, Goto and Wallace 2009). Metals affect feeding behavior, growth, and survival of invertebrates and small fish in the Meadowlands (e.g., Zhou et al. 2000, Reichmuth et al. 2009). Many sublethal but damaging behavioral and physiological effects on mummichogs (*Fundulus heteroclitus*; an abundant small fish important in estuaries) have been demonstrated, caused by mercury and the mixture of pollutants found in a tributary of the Arthur Kill downstream of the Meadowlands (Weis and Candelmo 2012). In one case, altered mummichog feeding behavior caused changes in the population density of its prey, grass shrimp. Such indirect effects of contaminants are probably common.

Water Quality Amelioration by Wetlands

Vegetated subtidal shallows and intertidal and nontidal marshes tend to improve the quality of surface water flowing through them. Marshes remove suspended sediments, plant nutrients (especially nitrogen and phosphorus), and metals from the water (Ehrenfeld et al. 2011). Marsh plant uptake of metals is an important phenomenon in the Meadowlands (Burke et al. 2000, Windham et al. 2001a, b). Portions of these materials are incorporated into plant biomass and eventually the marsh soils, while other portions are released to the water column. Other materials, such as nitrate, organic matter, and organic contaminants may be partly broken down in the marsh, often into innocuous compounds. Floating debris of natural and human origins is often physically trapped by the marsh. Some marsh uptake of materials is temporary in that death of aboveground plant parts and violent storms return some materials to the surface waters, but large quantities of materials are typically retained in the marsh soils indefinitely. For example, Meadowlands marsh sediments appear to be an effective long-term carbon sink, with an estimated 78% of fixed carbon persisting for 130 years, and 50% of the original fixed carbon persisting for > 645 years (Artigas et al. 2015).

The water quality amelioration function of marshes has been little studied in the Meadowlands. A report on the nitrogen budget of Mill Creek before the mitigation projects were built (Trattner and Mattson 1976) estimated a 5.45% removal of nitrogen between flood tide and ebb tide during a single daytime tidal cycle, despite a sewage treatment plant discharge contributing nitrogen to Mill Creek and a massive tidal input of nitrogen from the Hackensack River. Nitrogen processing in urban tidal marshes is likely to vary greatly in space and time and estimates such as these are subject to considerable error. Undoubtedly, however, the great extent of marshes in the Meadowlands provides a significant degree of processing of nitrogen and other materials in stormwater and discharged wastewater, and the Hackensack River and the greater New York–New Jersey estuarine delta would be much worse off without these wetlands. Ravit (2008) suggested constructing wetlands for tertiary sewage treatment; one such project is under consideration. It is unfortunate there were no studies of the effects of wetland mitigation on the nitrogen processing in Mill Creek; replacement of common reed stands with shallow open water and smooth cordgrass likely reduced marsh removal of nitrogen because there was less biomass of marsh vegetation after mitigation, because cordgrass is probably less productive than common reed, and because waters presumably move through the site more quickly than they did when it was dominated by reedbeds.

It is the stated goal of the federal Clean Water Act to restore and maintain water quality for "the protection and propagation of fish, shellfish, and wildlife and . . . recreation in and on the water" (Federal Water Pollution Control Act, 33 U.S.C. 1251 et seq.). On a more straightforward, local level, Hackensack Riverkeeper is "dedicated to restoring the river so that humans can safely swim and fish, and so that fish and other aquatic organisms can lead normal lives as part of a healthy ecosystem" (Len 2015). These ideals are certainly worth pursuing but given the history and condition of the Meadowlands as an urban-industrial wetland complex and dumping ground, safe fish consumption and swimming may be elusive goals. It is, however, important to protect people and other organisms from gross threats to health and overtly unpleasant conditions. Because management actions must be designed and implemented in the context of existing conditions, scientific understanding, and environmental, human, and fiscal constraints, it may be wise to consider the benefits of leaving sediments and invasive plants in place. The nonnative form of common reed, although controversial, may be the best and cheapest ally in the battle for better water conditions and protection from rising sea level.

CLIMATE AND WEATHER

Temperature, precipitation, wind, and other elements of the weather and climate strongly influence the distribution patterns of organisms. The Meadowlands are in a coastal location although somewhat sheltered from maritime influences by Staten Island and the Palisades Ridge on the south and east. The *microclimates* (climates near the ground) of this region are also influenced by extensive urbanization.

Our climate data, unless cited otherwise, are from the Office of the New Jersey State Climatologist (ONJSC 2010). Annual precipitation (the total liquid equivalent of rain, snow, and other forms of precipitation), based on means from about ten weather stations in northern New Jersey, has increased since records have been kept. Annual precipitation has been divided into three periods: 1895–1970, when precipitation was 1,132 mm per year; 1971–2000, 1,265 mm; and 2001–2009, 1,311 mm. This is an increase of 16%. The smallest amount of precipitation was about 762 mm in 1965, a year of intense drought. The largest amount was about 1,626 mm in 1983.

Mean annual temperatures, for the same region and time periods, were: 10.3 C; 10.5 C; and 11.4 C. The lowest mean annual temperature was 8.5 C in 1904, and the highest was 12.4 C in 2006. The growing season (freeze-free period) in the region that includes the Meadowlands is about 179 days. The Meadowlands are in the USDA Plant Hardiness Zone 7a, with a mean annual extreme minimum temperature of -17.8 to 15 C.

Prevailing winds in the Meadowlands are from the west; however, winds blow from all directions, including east (Willis et al. 1973). Fragmentary wind data are available from MERI (2019), from the Kingsland Impoundment weather station. Prevailing winds for the year 2007 were north-northwest to south-southwest, 50% of the time wind speeds were 0.9–2.2 m/s (meters per second), and speeds rarely exceeded 4.5 m/s.

Urban areas generally have microclimates characterized as *urban heat islands* (UHI). This is because the concrete, brick, stone, blacktop, roofing, and metal surfaces so abundant in most cities absorb more solar heat than natural soil and vegetation. Thus, the cities become warmer than their non-urban surroundings. The UHI associated with the New York City area is active from late afternoon until just after dawn in all seasons; it increases air temperatures an average of 4 C in summer and fall, and 3 C in winter and spring, but can reach extremes greater than 8 C. Sea breezes often shift New York City's UHI westward to the Meadowlands during the early hours of the night (Gedzelman et al. 2003). Rosenzweig et al. (2005) documented an urban heat island at Newark with a smaller temperature increase.

AIR QUALITY

Air pollutants can affect the health and survival of cryptogams, seed plants, and animals (Barker and Tingey 1992) as well as humans. Lichens and mosses are especially sensitive to air pollutants including sulfur dioxide, fluorine, nitrogen oxides, ozone, and some of the heavy metals. Northeastern lichen communities lose many species as sulfur and nitrogen oxide concentrations in the air increase (Will-Wolf et al. 2015). When levels of pollutants exceed National Ambient Air Quality Standards, human health is likely to be affected, particularly in members of high-risk groups; critical levels for non-human organisms have not been established, and for many species critical levels may well be much lower.

New Jersey data indicate a significant statewide improvement in air quality between the 1970s and 1990s, and a modest improvement between 1997 and 2017, in measured concentrations of nitrogen dioxide, ozone, sulfur dioxide, carbon monoxide, and fine particulates (Lioy and Georgopolous 2011, NJDEP 2018b). In recent years, ozone has been the only monitored pollutant that regularly exceeds the critical value (e.g., 14 days in 2017), generally during hot summer conditions. Toxic air pollutants such as benzene, formaldehyde, and chloromethane have shown modest declines in recent decades, but are still far above benchmarks for elevated human cancer risk (Lioy and Georgopolous 2011, NJDEP 2018b).

Data from the monitoring stations closest to the Meadowlands indicate that Meadowlands air quality has generally followed statewide trends: much improved since the 1970s but still problematic for human—and presumably wildlife—health, and above the tolerance level for many lichens (and possibly other organisms). For plants, some species are likely to be more sensitive to short-term and others more sensitive to long-term concentrations of air pollutants, and combinations of pollutants may be more toxic than the individual constituents (Wolfenden et al. 1992).

Nitrogen dioxide air quality standards are 53 ppb (parts per billion; annual mean) and 100 ppb (1-h mean). Meadowlands nitrogen oxides had an annual average of 113 ppb in 1970 (Willis et al. 1973); in 2007–2008 concentrations ranged from 17 to 29 ppb, with an annual mean of 23 ppb (Gao 2009). Nitrogen dioxide measured in the Meadowlands in 2007–2008, although it rarely exceeded air quality standards, was strongly correlated with respiratory-related hospital admissions in surrounding areas (Roberts-Semple and Gao 2017). Tree trunk lichen communities in North Carolina declined in diversity with increasing nitrogen dioxide concentrations, even at levels as low as 1 ppb (3-d mean); moss and liverwort frequency declined as well. The number of lichen species fell from 15–33 species at the lowest levels of nitrogen dioxide to 2–3 species at the highest levels (which were much lower than Meadowlands concentrations, between 3 and 4 ppb; Perlmutter et al. [2018]).

Sulfur dioxide air quality standards are 30 ppb (annual mean) and 75 ppb (1-h mean). Sulfur dioxide in the Meadowlands had an annual mean of 43 ppb in 1970; annual means for the Meadowlands area (three nearest stations) were 7 ppb in 1997, 5 ppb in 2007, and 0.3 ppb in 2017. This 20-fold decrease was also noted statewide (NJDEP 1998, 2008, 2018b).

Ground-level ozone is a pollutant formed when nitrogen oxides and volatile organic chemicals react in the presence of sunlight; it can damage respiratory and immune systems in humans and analogous systems in plants (affecting photosynthesis, respiration, growth, and disease resistance). Air quality standards (for both human health and protection of crops and trees) are 70 ppb (4th highest daily maximum 8-h mean) and 120 ppb (1-h mean). In 1997, monitoring stations near the Meadowlands exceeded the 70 ppb standard, with an 8-h mean of 100 ppb; in 2007, the mean was 92 ppb, and in 2017 it was 68 ppb (NJDEP 1998, 2008, 2018b). Increasing ozone over approximately 50 ppb (8-h mean) corresponded with increasing numbers of pediatric asthma emergency department visits in Newark (Gleason and Fagliano 2015), suggesting that the current standard should be lower to protect children's health.

Other combustion products that characterize urban air include PAHs, volatile organic compounds (VOCs), elemental and organic carbon, hexavalent chromium, and other particulate metals (Yu et al. 2014, NJDEP 2018b). Pollutants in sediments and waters can also volatilize and contaminate the air: in one study, almost half the amount of a dioxin dissolved in Newark Bay was volatilized (Friedman et al. 2012). Many VOCs (including formaldehyde, chloromethane, and benzene) and some toxic elements (hexavalent chromium, arsenic, and cadmium) exceeded health benchmarks in the Meadowlands or at nearby stations in recent years (Yu et al. 2014, NJDEP 2018b). Airports contribute particulates, aldehydes, other hydrocarbons, ozone, carbon monoxide, and nitrogen oxides to air pollution (Milford et al. 1970, Penn et al. 2017); we have found no information about the pollution impacts of Newark and Teterboro airports on surrounding areas.

Air quality is directly linked to levels of atmospheric deposition of pollutants to aquatic and terrestrial systems, both through precipitation (wet deposition) and particulate matter (dry deposition). Estimated inorganic nitrogen wet deposition for the nearby lower Hudson–Long Island region was 6.9 kg/ha in 2002–2004 (Golden and Boyer 2009), compared to estimates from the 1960s of 3–5 kg/ha and from the 1980s of 4–8 kg/ha (Gronberg et al. 2014). Degraded lichen communities were observed at sites with annual nitrogen deposition values from 2.3 to 9.5 kg/ha in Pacific Coast states and at 4.0

kg/ha in the Rocky Mountains at high elevations (McMurray et al. 2013). Critical nitrogen deposition values for lichens have not been established for the Northeast. Fortunately, in recent years, total inorganic nitrogen deposition has decreased substantially in the Northeast, although there has been a partial shift from nitrate to ammonium deposition (Li et al. 2016). In 2019, the Keegan Landfill (Kearny) undergoing remediation emitted toxic levels of hydrogen sulfide.

PALEOECOLOGY AND ENVIRONMENTAL HISTORY

Fossils and other evidence preserved in environmental media, including sediments and rocks, can be used to reconstruct the biota and environmental conditions of the past. Fossil pollen in sediment cores is a well-known source of paleoecological data. Organic wetland sediments, because they are perennially wet and anaerobic, are especially good repositories of materials such as pollen, microscopic animals and plants, bones, wood, seed plant parts, and artifacts. The few paleoecological studies in the Meadowlands have yielded intriguing information.

Much evidence for the development of plant communities has been gained through studies of pollen, spores, other plant remains, and Foraminifera in peat core samples, as well as historical accounts of the vegetation. There is significant evidence that the wetlands that developed on the glacial lake sediments were dominated by freshwater plant associations for millennia. Analysis of pollen and other plant remains in peat core samples from Secaucus demonstrated a progression of plant communities over several thousand years, reflecting climatic and salinity changes in the Meadowlands (Heusser 1963, Harmon and Tedrow 1969). The oldest peat samples taken from Secaucus were dated at 2,025 ± 300 YBP (Years Before Present; Heusser 1963). Peat cores in the Meadowlands indicated that the organic material was deposited by freshwater plants (Heusser 1963, Harmon and Tedrow 1969). Salt marsh peat later developed over these freshwater peats (Tedrow 1986).

The first postglacial wetland community in the area was dominated by black ash (*Fraxinus nigra*) and then a mixture of black ash and northern peatland species including tamarack (*Larix laricina*) and black spruce (*Picea mariana*). More than 500 YBP, Atlantic white cedar moved into the area. Atlantic white cedar had been established in more southern areas of New Jersey for thousands of years (e.g., Rosenwinkel 1964). Finally, the peat cores show that the Atlantic white cedar wetlands were encroached upon from the periphery by marsh composed of Olney three-square (*Scirpus americanus*), black rush (*Juncus gerardii*), and narrow-leaved cattail (*Typha angustifolia*), all either salt or brackish marsh species, according to Heusser (1963), although narrow-leaved cattail also grows in fresh water. A core taken just west of the Hackensack River on the north side of Route 3 showed a discontinuity between a basal clay and the oldest organic deposit radiocarbon-dated at 2,610 ± 130 YBP (Carmichael 1980). Above the clay were 100 cm of alder (*Alnus*) peat, then 160 cm of sedge (Cyperaceae) peat, and finally 120 cm of silty reed (*Phragmites*) muck. The alder-sedge transition was dated 2,060 ± 120 YBP, and the sedge-reed transition between 810 ± 110 and 240 ± 110 YBP. The reed muck contained various weedy plants associated with human disturbance (Carmichael 1980). Rue and Traverse (1997) reported concentrations of cattail, grass family (Poaceae), pigweed family (Chenopodiaceae), and amaranth family (Amaranthaceae) pollen beginning 3,000 YBP that they interpreted as a shift to brackish marsh conditions. These groups of plants could also indicate increasing human (Native American) influences such as vegetation clearing and nutrient inputs to the estuary. However, Rue and Traverse obtained a radiocarbon date of 2,050 YBP 1 m below the surface of the organic tidal marsh soil indicating a mean sediment accretion rate of only 0.5 mm per year. This low rate of sediment accretion does not seem consistent with intense human activity although undoubtedly rates increased with modern urbanization.

Historic Vegetation

The harvest of *salt hay* (primarily saltmeadow cordgrass [*Spartina patens*]) in the Meadowlands began no later than 1697 (Quinn 1997), and old maps indicated that salt marsh occupied fairly extensive areas of the Meadowlands at that time. However, surveys conducted in the early 1800s (Torrey 1819) did not report significant coverage by brackish marsh species in the Meadowlands. Common reed was not reported by Torrey in 1819, though it may have been present as indicated by its occurrence in nearby Elizabethtown (now Elizabeth; Sipple 1972). In an 1877 list of New Jersey species, common reed was included as occurring in the Hackensack Meadows (Willis 1877). Interestingly, peat profiles published by Waksman (1942–43) showed that common reed occurred widely in lowland peat deposits in New Jersey, including at least one site in the Meadowlands, at depths where the reed materials must have been deposited in pre-Columbian times (this would have been native common reed, *Phragmites australis* subspecies *americanus*). Harshberger and Burns (1919) described salt marsh vegetation along the creeks and the river. In addition, they reported extensive areas of brackish marsh covered by common reed and less brackish areas dominated by cattail (*Typha*). Vermeule's (1897) maps indicated that islands of Atlantic white cedar swamp still existed between lower Berry's Creek and the Hackensack River and in areas of Carlstadt, East Rutherford, and Ridgefield. Therefore, evidence suggests that the widespread brackish and salt marsh communities of the Meadowlands of today and the now-ubiquitous common reed were present before European settlement but it is unclear how extensive different plant communities were at various times. The apparent contradictions among different sets of paleoecologic and historic evidence concerning Meadowlands vegetation may be due to the temporal and spatial variation in plant community coverage as well as to the inherently coarse scale and potential sources of error in the methods of paleoecology and historical ecology (Egan and Howell 2005, du Châtelet et al. 2018). Few peat cores have been analyzed from the Meadowlands and may not be representative of spatiotemporal variation in the juxtaposition of glacial, lacustrine, riverine, estuarine, and human effects. Fortunately, the most recently extant cedar swamps left conspicuous macroscopic remains at locations such as Mill Creek, Saw Mill Creek, and Upper Penhorn Marsh. Cedar stumps and logs have been found in many areas of the Greater Meadowlands (Anonymous 1937) but these trees were almost certainly not all alive during historic times. Atlantic white cedar tolerates very little salinity. Most plants of nontidal fens and bogs also do not tolerate salinity. Most species of northeastern tidal swamp trees and shrubs apparently tolerate little salinity, although a few species occur well down the Hudson River estuary where local maximum salinities may be in the range of 1–15 ppt. Cedar swamps in the Meadowlands would have been freshwater or very close to fresh, and nontidal or flooded only a little by tides. If it existed, tidal hardwood swamp would have occurred in freshwater tidal to oligohaline conditions. Salinity intrusion due to the Oradell Dam and other hydrological alterations must have played a role in the demise of these plant communities in the Meadowlands. Therefore, a vegetation mosaic that included cedar swamps as well as brackish and saline tidal marshes would have occurred along a gradient from nontidal freshwater to quite brackish tidal water. Likely saline marshes were extensive along the western side of Newark Bay which is much saltier than the Hackensack River; Urner (1923) described large areas of salt hay at Elizabeth which would have been high salt marsh dominated by saltmeadow cordgrass.

FIRE

Fire almost certainly affected the development of Meadowlands plant communities (Figure 1.10). As organic materials from plants accumulate in marshes and peatlands, surface elevation gradually increases above the water table and moisture decreases (Harmon and Tedrow 1969). After this dried peat burns, the surface elevation is substantially decreased. Harmon and

Figure 1.10 Soon after an early spring fire in the Kane Natural Area. Photo 15 April 2012 by Erik Kiviat.

Tedrow (1969) found buried ash in soil cores suggesting that ancient fires were fairly common in the Meadowlands.

Vegetation fires were reportedly frequent in the Meadowlands marshes in the early 1900s (Anonymous 1937, White 2005) and are still common. They are started intentionally by vandals and probably accidentally from discarded cigarettes and the hot engines of ORVs (Burkhardt 2014), among other sources of ignition. Fires are common during dry periods in early spring (as is generally true in the northeastern states), and also occur in other seasons. The standing and fallen litter of dense reedbeds that lack surface water (or above the water) is especially combustible (see Kiviat 2010 for a summary of fire ecology in reedbeds). Natural methane emitted from organic wetland soils is also highly combustible and may potentiate reed fires (Hartig and Rogers 1984). Garbage landfills were frequently burned intentionally and accidentally in the early to mid-1900s (White 2005), and methane emissions from these landfills presumably increased their combustibility. Landfill fires still occur occasionally; the 1-D Landfill in Kearny burned in March 2006 (Brett Bragin, NJSEA, pers. comm.). Fires in garbage and reedbeds both produce thick dark smoke that can be a hazard to highway and rail traffic (e.g., Wakin 2001) and presumably to boat and aircraft activities as well. The smoke degrades air quality and may be an inhalation hazard to people and wildlife. Chemical foam containing PFAS (per-and polyfluoroalkyl substances) is used to suppress fires and may also enter aquatic ecosystems in landfill leachate and via other sources; we have been unable to confirm the extent of this substance in the Meadowlands.

Fire affects vegetation. When the soil is wet in reedbeds (common reed stands), fire burns off the litter and spares the rhizomes allowing rapid regrowth of reed. In a dry summer or drought, when soils are dry, reed fires can burn into organic soils, kill reed rhizomes, and create shallow pools (see Kiviat 2010). Reed fires readily kill tree-of-heaven (*Ailanthus altissima*) and probably many other woody plants, helping to maintain the high dominance of reed in some Meadowlands marshes.

CONCLUSIONS

The environmental setting of the Meadowlands is complex at many levels. Juxtaposition of bedrock, glacial deposits, patterns of water flow, vegetation, natural disturbances, and centuries of intensive human activities has created a palimpsest of habitats on the landscape. The Meadowlands have never been static, and continue to change in response to natural processes, human use, sea level rise, and accelerated climate change. This spatial and temporal complexity creates an abiotic environment that is variably suitable for diverse and shifting assemblages of organisms and provides the context for appreciating and managing nature. People have used, and continue to use, the Meadowlands region for a large variety of activities (Kiviat 2020), including some that are influenced by the environment and some that have strong effects on the environment. The juxtaposition of the built environment and the rest of nature, and the way they are exploited commercially, was expressed by the Meadowlands Regional Chamber (2019): "The Meadowlands proudly hosts not just an ecosystem of rebounding, flourishing nature but also a surging ecosystem of profound economic opportunity and boundless connections . . . We invite you to discover how your business can grow with us here in the Meadowlands." In the rest of this book we address the structure and formation of Meadowlands habitats, the organisms that use them, and the stresses they are subject to. The material above illustrates the challenges and opportunities confronting organisms in the Meadowlands.

NOTES

1. Kristen Bell Travis (Hudsonia) prepared the water quality and air quality material and drafted the maps for this chapter. Maps were edited by Lauren Bell and Lea Stickle (Hudsonia).

Chapter 2

Habitats

Marshes, Ponds, and Channels

Nearly half of the Meadowlands District is water (ponds, impoundments and streams) and herbaceous wetlands (NJDEP 2015). Herbaceous wetlands and deepwater habitats dominate the natural landscape of the Meadowlands, and support many of the region's spectacular organisms such as the swamp rose mallow (*Hibiscus moscheutos*) and great egret (*Ardea alba*). These open, wet habitats also occupy a prominent position in the politics, economics, and aesthetics of the Meadowlands: the business community and educational sector do not try to attract people to the Meadowlands by showing photographs of thickets, dumps, and swamps! Nonetheless, even the tidal marshes have not been described in detail in the regional literature. Many examples of the habitats described in this chapter were profiled briefly by Kane and Githens (1997) and U.S. Army Engineers (USACE 2004).

DEEP CHANNELS

There are about 266 km of river, stream and channel in the Meadowlands District alone (NJDEP 2015). Deep channels, as far as we know, lack rooted vegetation except at their upper intertidal zone edges (and as floating fragments). However, these habitats in the mainstem of the Hackensack River estuary and its major tidal tributaries are the homes, movement corridors, or foraging habitats of most of the fishes, shrimps, crabs, swimming and wading birds, and marine mammals of the Meadowlands. The channels also support much of the boating and fishing, and a portion of the ecotourism, that take place.

Because water in the unconfined channels is usually very turbid (murky), due to a combination of natural estuarine processes and urban pollution, light cannot penetrate far and submergent plants are rare or absent. We assume that this ecosystem is *heterotrophic*, that is, particulate and dissolved organic matter from wetland and upland plants, rather than living algae, is the principal base of the food web. Yet *phytoplankton* (microscopic drifting algae), and mud algae (also called *microphytobenthos*, the algae that live on the sediment surface) in the intertidal zone, surely contribute (Foote 1983). The dead organic matter originates in the tidal marshes and on land where tree and grass leaves and other dead tissues of plants are washed into the estuary and form food for microorganisms, invertebrates, and small fishes such as mummichog.

The channel beds are typically composed of fine sediments, although sands and coarser materials also occur. In portions of the uppermost reach of the Hackensack River estuary the bottom is partly rocky. And of course, demolition rubble and other anthropogenic materials are common. There must also be shelly areas because oyster and hard clam shells are abundant in dredge spoil from Berry's Creek Canal that was deposited on portions of Oritani Marsh in 1911.

VEGETATED SHALLOWS

Submergent vegetation comprises plants adapted to life entirely or largely beneath the water in either nontidal or tidal habitats. This plant community is also called *submerged aquatic vegetation (SAV)*, *submerged aquatics*, or (in estuaries) *subtidal vegetation*. In low salinity tidal habitats these plants (in the Hudson River, common coontail [*Ceratophyllum demersum*], wild-celery [*Vallisneria americana*], pondweeds [*Potamogeton* spp.], naiads (*Najas* spp.), and Eurasian watermilfoil [*Myriophyllum spicatum*]) may be exposed on mudflats in the lower intertidal zone at low tide, and most species have flowering structures that are prominently emersed above the water surface for aerial pollination. We are also including in this plant assemblage certain free-floating or floating-leaved plants that occur with the submergent species. Submergent vegetation in estuaries, lakes, and ponds is generally important for oxygenating the water and providing crucial habitat structure or food for many invertebrates, fishes, amphibians, turtles, birds, and the muskrat.

In general in the northeastern states, common submergent plant species in fresh and brackish waters include the pondweeds (*Potamogeton*, many species), naiads (*Najas*, several species), coontails (*Ceratophyllum*, 2 species), watermilfoils (*Myriophyllum*, several species), wild-celery, waterweeds (*Elodea*, 2 species), horned-pondweed (*Zannichellia palustris*), wigeon-grass (*Ruppia maritima*), and eelgrass (*Zostera marina*). (The stoneworts, algae that superficially resemble submergent flowering plants, and the quillworts are discussed in Chapter 5.) The free-floating flowering plants are principally duckweeds (several species of Lemnaceae). The floating-leaved plants are the water-lilies (*Nymphaea* and *Nuphar*) and water-chestnut (*Trapa natans*). (Not all are found in the Meadowlands.)

The paucity of this plant assemblage in the Meadowlands is striking: the submergent flora is species-poor, spatially spotty, and virtually unstudied. In both the tidal and nontidal habitats, we have found only leafy pondweed (*Potamogeton foliosus*), curly pondweed (*Potamogeton crispus*), Nuttall's waterweed (*Elodea nuttallii*), horned-pondweed (*Zannichellia palustris*), and common coontail (*Ceratophyllum demersum*). Additionally, wigeon-grass (*Ruppia maritima*) occurs sparsely as small, inconspicuous tufts in the intertidal zone such as at Kingsland tidal marsh and Skeetkill Marsh (DiBona 2007; Kiviat, pers. obs.). More species of pondweeds (*Potamogeton* spp.) were reported historically (Foote and Loveland 1982). Additional survey work, for example, by means of grappling hook or rake in the ponds, lakes, and low-energy tidal creeks may well discover some submergent species that are not visible from the shorelines.

Free-floating plants on quiet Meadowlands waters were common duckweed (*Lemna minor*), great duckweed (*Spirodela polyrrhiza*) and watermeal (*Wolffia brasiliensis* and *Wolffia columbiana*). These tiny plants occur among the floating canopy of submergent species such as the pondweeds, and also in the smaller, quieter ponds, as free-floating individual thalli (plant bodies) or mats of densely packed thalli, often adjoining shorelines or among the stems of emergent or shoreline plants intersecting the water surface.

Submergent plants, in general, require waters that are not too high-energy (waves and currents), nor too turbid. The submergent and free-floating plants in the Meadowlands are probably all species that are relatively tolerant of turbidity and high nutrient levels. They tolerate some salinity, and wigeon-grass is notably salinity-tolerant. Most of these plants are important waterfowl foods. The edges of Overpeck Creek, which have low salinity and relatively low turbidity, as well as small reedbeds that provide shelter from waves and currents, had several submergent species. Horned-pondweed was abundant in some years in the low energy brackish waters of Kingsland Impoundment near the NJSEA buildings (Kyle Spendiff, formerly NJMC, pers. comm.). Lower Muddabach Creek, in the Kane Natural Area, has a tide gate at its mouth at the Hackensack River, and prior to the 2010 creation of the mitigation bank, was densely grown with curly pondweed. Leafy pondweed was prominent in a small, rocky, quarry pool in North Arlington. We were not able to explore the clay pit lakes, or many other potential habitats, for submergent plants.

In 2019, hydrilla (*Hydrilla verticillata*) was reported and verified for the first time in the Meadowlands, growing densely in the Leonia portion of Overpeck Creek (Fallon 2019a). Hydrilla is a nonnative submergent plant that is usually considered a pest, although its effects on the biota are mixed (see Kiviat 2009).

TIDAL MARSHES

By way of introduction to this habitat type, it is important to note that many Meadowlands tidal or formerly tidal marshes have been subjected to *mitigation*. Wetland mitigation is the creation, enhancement, or restoration of wetland to replace permitted destruction of other wetland. A mitigation bank is a large mitigation project from which credits are sold to compensate for other wetland destruction instead of individual mitigation projects. Mitigation, restoration, or enhancement in the Meadowlands typically constitutes spraying glyphosate or another herbicide, usually multiple times, to kill common reed stands, removing the top 30 cm of soil with reed rhizomes, and planting cordgrasses (*Spartina*) or other native plants. In some marshes, dikes are breached to reconnect the impounded marsh to brackish tides, tidal creeks and pools may be sculpted into the marsh sediments, and dredged sediment may be used to build islands. Several hundred hectares of marshes have been managed to date in the Greater Meadowlands, mostly in the Meadowlands District.

The tidal marsh habitat type is extensive in the Meadowlands. Tidal marshes occur on a gradient from quite brackish to nearly fresh. Riverbend Marsh, south of Laurel Hill at about river km 5.0 (measured along the curve of the Hackensack River channel) has mean salinity 16 ppt in the open tidal waters. That is almost one-half seawater salinity, and salinity is presumably higher where water evaporates at low tide on the high marsh. Of course, salinity varies around that mean, being lower during periods of high freshwater discharge in spring or after rainstorms, and higher during periods of low freshwater discharge. Riverbend Marsh has a *high salt marsh* community formerly dominated by saltmeadow cordgrass (*Spartina patens*) in the supratidal zone just above MHW (Mean High Water or average high tide level), and a *low salt marsh* community of smooth cordgrass (*Spartina alterniflora*) between MHW and Mean Sea Level (MSL). The cordgrasses are superbly adapted to relatively high salinities and tidal fluctuation. In the Meadowlands District there are about 583 and 145 acres of low and high salt marsh, respectively (NJDEP 2015). Saltmeadow cordgrass has been declining at Riverbend; ca. 2010, the high marsh was about 80% covered by common reed (Claus Holzapfel, pers. comm.). The nonnative subspecies of reed is potentially vigorous at somewhat lower salinities but can thrive in such higher salinity marshes where fresh surface water or groundwater enters the marsh or where fill has been placed on the marsh surface. Lack of oxygen and associated increase in sulfide seem to inhibit common reed (C. Holzapfel, Rutgers University, pers. comm.).

On the northeastern coast, intact high salt marsh assemblages, in addition to saltmeadow cordgrass, also have salt grass (*Distichlis spicata*), black rush (*Juncus gerardii*), and sometimes other species such as marsh straw sedge (*Carex hormathodes*). Sea-lavender (*Limonium carolinianum*) occurs rarely; three stands are known (C. Holzapfel, pers. comm.). Glasswort (*Salicornia*) occurs, especially in pannes (shallow depressions that accumulate salts from evaporation of the irregularly-flooding tidal waters). Marsh straw sedge was abundant in the 1980s on the high marsh of Piermont Marsh, the most brackish major marsh on the Hudson River, but has not been reported in the Meadowlands since 1895 (Appendix 1). Black rush and salt grass are members of the remnant high salt marsh patches in Riverbend Marsh (C. Holzapfel, pers. comm.). Plant species diversity in the high salt marsh may be a good indicator of ecological integrity or a high quality habitat in the more brackish portions of the estuary. In addition to supporting the salt hay industry that was so important socioeconomically and culturally in the Meadowlands (see Quinn 1997), the short *graminoid* (grass-like) plant communities of high salt marshes were the habitat for several breeding birds of conservation concern, notably saltmarsh sparrow and seaside sparrow.

In general, in the Meadowlands the cordgrass communities have been, or are being, replaced by common reed, where salinities are moderate to low, and this process is facilitated by alterations to the hydrology, water quality, and soils of the marshes. A brackish lagoon and marsh at the western edge of Newark Bay supported smooth cordgrass and common reed when visited in 2019 (Figure 2.1). Reed tends to replace smooth cordgrass in the higher portions of the intertidal zone, and saltmeadow cordgrass in the supratidal zone. Fill, mosquito ditching, impoundment for salt hay production or mosquito control, and tidegating for flood control were the most common alterations. Reed colonies often initiate from the burial of large fragments of reed rhizome in well-drained, low salinity microhabitats (Bart et al. 2006). Shoreline or watershed development, and nutrient loading from human activities, also favor reed over cordgrasses (Silliman and Bertness 2004, McCormick et al. 2010); heavy metals can reduce germination and seedling survival of smooth cordgrass as well (Waddell and Kraus 1990). Cordgrass communities are generally considered high quality native saltmarsh and are often the elusive goal of marsh management. Because of the stresses of urbanization and other human activities, once the cordgrass stands are gone they are difficult to reinstate sustainably. After active maintenance of cordgrasses ceases in mitigated marshes, the managed areas are commonly either re-colonized by common reed or degrade to eroding mudflats.

Figure 2.1 Brackish lagoon and marsh at the western edge of Newark Bay in 2019. Photo by Erik Kiviat.

The Saw Mill Creek tidal marsh was diked for mosquito control in the early 1900s (USACE 2004). Within memory this marsh was evidently reed-dominated as is usual for tide-restricted cordgrass marshes that are not actively managed for salt hay. A 1950 storm breached the dikes and the Saw Mill Creek marsh became tidal again. The majority of this marsh is now intertidal mudflats that have not been colonized by seed plants. There are, however, extensive patches of smooth cordgrass and common reed, especially along Saw Mill Creek (we suspect the marsh surface elevations are a little higher or more stable along the creek banks, because a *natural levee* or slightly raised area typically forms along a major tidal creek). These plants, however, have apparently not been able to colonize or re-colonize the areas that are now mudflats. The mudflats appear to be eroding, although we do not know of any measurements of sedimentation at this site.

Just north of the Saw Mill Creek tidal marsh is the Kingsland Impoundment, a tide-restricted brackish marsh, in Lyndhurst adjoining the NJSEA headquarters. We have made numerous visits to Kingsland Impoundment, with its pleasant boardwalk and perimeter access on trails and roads. Extensive reedbeds were interspersed with large areas of shallow open water, at least some of which supported submergent plants. Although vegetation in the impoundment was not diverse, this area was high quality habitat for marsh, water, and shore birds. A rare plant, floating marsh pennywort (*Hydrocotyle ranunculoides*), was present in the impoundment but much less abundant than in the sheltered, apparently nontidal, portions of Kingsland Creek north and west of the nearby Kingsland Landfill and upstream of the impoundment.

Some tidal marsh reedbeds in the Meadowlands are stippled with small pools of unknown origin. For example, this pattern is well developed in Walden Marsh west of the Sports Complex. The pools are used by birds such as sora and least sandpiper (Christopher Graham, Hudsonia, pers. comm.), and contribute to the biological diversity of the marshes.

Mean and maximum salinities decline in an up-estuary direction. There is generally less of the cordgrass-dominated communities and more reedbeds or mixed brackish marsh communities. At intermediate salinities, we can reasonably deduce based on tidal marshes in nearby estuaries, that plants such as narrow-leaved cattail (*Typha angustifolia*) and big cordgrass (*Spartina cynosuroides*) were abundant before large-scale colonization and dominance of the marshes by common reed.

The species diversity of flowering plants increases from higher to lower salinities. Slightly brackish marshes and freshwater tidal marshes may have a rich flora (as is evident on the Hudson River). The Meadowlands must have once supported substantial areas of freshwater tidal marsh before construction of the Oradell Dam and other factors increased the intrusion of saline waters. The wild-rice (*Zizania*) stands that were reportedly extensive at one time would have been in these fresh and slightly brackish tidal marshes with maximum salinities about 2 ppt. Freshwater tidal marshes have been well-studied in the Hudson River (see Kiviat et al. 2006) and elsewhere (Barendregt et al. 2009) and are important for biodiversity.

NONTIDAL MARSHES

Most of the extant nontidal marshes of the Meadowlands were formerly tidal and were diked for cultivation of salt hay or mosquito control, or inadvertently as tidal marshes were partly filled for infrastructure development. This contributed to common reed dominance in many areas. We are most familiar with the formerly diked and reed-dominated marsh of the Richard P. Kane Natural Area in South Hackensack and Carlstadt. This marsh was mostly reed-dominated from 1999 when we first saw it to 2010 when large areas were managed to create a mitigation bank. In 2007, a storm breached the dikes and at least portions of the Kane area became brackish-tidal prior to creation of the mitigation bank. In 2010, a third of the marsh was managed to transform weakly tidal common reed marsh to strongly tidal cordgrass marsh and in a portion freshwater swamp.

Despite the appearance of solid reedbeds, there was much subtle diversity at Kane (Kiviat and MacDonald 2002; Kiviat, pers. obs.). Certain features were noteworthy. Areas with fill, such as along the major creeks and ditches, had patches of supratidal vegetation with a diverse mixture of native and non-native herbaceous and woody plants. An area of perhaps 0.4 ha on the north side of Paterson Plank Road, that had been cut for reed harvest annually, had an understory of mixed native and nonnative herbs. Those two features may persist in the portions of the marsh not converted to tidal marsh for the mitigation bank. Most interesting, a strip of about 10 ha of marsh, 100 m wide, on the west side of the Paterson Lateral Gas Pipeline, had many small patches of native bluejoint grass (*Calamagrostis canadensis*)-dominated wet meadow in a reedbed matrix. The bluejoint meadows had an admixture of goldenrod (*Solidago rugosa* and probably other *Solidago* species), sow-thistle (*Sonchus*), blackberry (*Rubus ?pensilvanicus*), purple loosestrife (*Lythrum salicaria*), steeplebush (*Spiraea tomentosa*), and other species. Bluejoint communities were mapped in the Meadowlands by Sipple (1972) although not at this location. Although individually small, the bluejoint meadows contributed to the habitat complex for a variety of interesting wildlife. However, this vegetation was eliminated for development of the Kane Natural Area Mitigation Bank.

Bluejoint is a freshwater grass of somewhat northern affinities and is rather fire-tolerant. It is possible that bluejoint meadows persisted close to the pipeline as a result of fires started accidentally or intentionally by ORV riders and other users of the pipeline road (ORVs have exhaust systems that can become hot enough to ignite dry plant material [Baxter 2002]). It is also possible that climate warming had been a stress on the bluejoint. In any case, the reflooding by brackish tidal waters and the mitigation banking project eliminated this regionally scarce community.

Kearny Marsh West (Kearny Freshwater Marsh) is another noteworthy, nearly freshwater marsh. The marsh was accidentally impounded by railroads, highways, and a garbage landfill, and became a deepwater nontidal marsh sometime in the first half of the 1900s. Reedbeds dominated Kearny West during our Meadowlands studies. These reedbeds were interspersed with small and large pools of shallow open water. The sediments are peaty. The reedbeds appeared to be stressed, as evidenced by the breaking-up of some extensive reedbeds into small patches and clumps of reed, and also by the appearance of flotant (see below). The cause of stress was evidently deep water combined with the grazing activities of muskrats (*Ondatra zibethicus*) and common carp (*Cyprinus carpio*) on the rhizomes and *culm* (aerial stem) bases of the reeds, and perhaps also leachate pollution from the unremediated Keegan Landfill in the marsh. Reedbeds, open water, and flotant combined to form a diverse and highly interspersed marsh mosaic providing high quality habitat for marsh birds. In 2013 we observed very low water in Kearny West with exposed peat at reedbed margins colonized by diverse, mostly low-growing, seed plants. Bryophytes were just beginning to colonize, presumably after death from brackish water flooding. Peat masses had also been moved around by the Hurricane Sandy storm surge of 2012 and probably also the Hurricane Irene flooding of 2011.

The ca. 66 ha Upper Penhorn Marsh is isolated from normal high tides by a tide gate on lower Penhorn Creek. Old canals, evidently dredged for logging, contain Atlantic white cedar (*Chamaecyparis thyoides*) stumps. Much of this marsh has been reedbed-dominated during our studies. However, in 2015 we discovered that part of the marsh had an assemblage of smaller, sparser *Phragmites* mixed with diverse other seed plants and a patchy bryophyte layer. The seed plants included marsh St. John's-wort (*Triadenum virginicum*), false-nettle (*Boehmeria cylindrica*), goldenrod (*Solidago ?rugosa*), and flat-sedges (*Cyperus* spp.). Surprisingly, the moss layer had abundant peat moss (*Sphagnum fimbriatum* and *Sphagnum palustre*), along with other mosses and liverworts. No similar assemblage is known elsewhere in the region. Upper Penhorn Marsh also has a large breeding population of Atlantic Coast leopard frog (*Lithobates [Rana] kauffeldi*), as well as least bittern, Virginia rail, and various ducks. The marsh is surrounded by a busy railroad, New Jersey Turnpike ramps, commercial development, and old fill which is slated for a hotel. Several years ago, radio broadcast towers were installed in the marsh on NJSEA property. In 2019, fresh, non-permitted fill was emplaced in the marsh on private property some distance from the proposed hotel site.

FLOTANT

This small habitat type was important for biodiversity in Kearny Marsh West. Patches of common reed had died due to the stresses mentioned above. The reed rhizomes and associated peat had floated to the surface, and these floating rafts of soil, at first bare of live plants, had become vegetated. The colonizing plants probably grew from seeds and other propagules dormant in the soil (the *seed bank* or *propagule bank*), along with seeds, spores, and vegetative fragments that drifted on the water or blew in the wind.

Sparsely vegetated, relatively recently-floated rafts had saltmarsh-fleabane (*Pluchea odorata*), purple loosestrife, common reed, and a few other plants, with ample space between the living shoots. This bare soil provided a substrate for the growth of mosses and pellia (*Pellia*), a large thallose liverwort that is rare in the Meadowlands. Sparsely vegetated peat rafts were probably used by waterbirds for loafing, shorebirds for resting and foraging, turtles for basking, and muskrats for foraging and building lodges, as we have observed on flotant in other nontidal marsh systems in the eastern United States. We also saw several species of butterflies and signs of meadow vole (*Microtus pennsylvanicus*) on the Kearny flotant.

At least some of the peat rafts eventually became densely vegetated again with common reed. We do not know if this happens to all the rafts, or whether the origin and development of the rafts is cyclic or episodic.

WET MEADOWS

The common professional usage of marsh and wet meadow in North America should not be confused with the historic names such as Hackensack Meadows and Newark Marshes which were less specific biologically. Wet meadows are rarely flooded, or flooded only for short periods, but the soil is saturated near the surface long enough to deplete oxygen and create wetland conditions. It may be difficult to distinguish wet meadows from marshes, and a single habitat unit may contain both. We found nontidal wet meadows at only a few locations. The most extensive occurrence was on the Bergen County Utilities Authority (BCUA) property southwest of Mehrhof Pond in Little Ferry, and in the adjoining Losen Slote Park. The meadows were apparently underlain by clay in an area that may have been strip-mined for clay or sand for brick-making. Another wet meadow was between the quarried areas at the western edge of the Meadowlands in North Arlington and the abandoned railroad just west of Harrier Meadow and the 1-E Landfill. A third area, which we have not surveyed in detail, was on the east side of the Hackensack River just south of New Bridge Crossing in Teaneck. Some of the characteristic plants of the wet meadows were sedges (Cyperaceae) and purple loosestrife. We found Canadian burnet (*Sanguisorba canadensis*), and swamp lily (*Lilium superbum*), both rare in the Meadowlands and perhaps in the larger region of northeastern New Jersey. The reed—peat moss area of Upper Penhorn Marsh (see above) could be called wet meadow.

SPRINGS

Springs (and their more diffuse relatives, seeps) are of ecological interest because the groundwater discharged into pools, streams, or currents is chemically and physically different from nearby surface waters. Groundwater discharges at mean annual temperature, about 13 C in northern New Jersey, making springs and their immediate surroundings warmer in winter and cooler in summer than the surrounding area. Thus, in general, springs can serve as refuges for animals needing warmth in the winter, such as an occasional overwintering Virginia rail. Springs can also potentially serve as summer refuges for species of northern affinities. Groundwaters in the northeastern states may acquire calcium and other minerals from deep bedrock or surficial materials that are chemically different from materials at the ground surface, making springs sites that potentially support unusual plants.

Few springs have been documented in the Meadowlands, and, although disproportionately important to biodiversity, virtually nothing has been written about them. Groundwater that discharges in springs may originate in upland areas outside the Meadowlands-proper thus possibly has lower levels of the contaminants that are prevalent in the Meadowlands. For these reasons, springs in the Meadowlands may provide unique habitats or refuges for species that may not tolerate the conditions (warm, dry, or polluted) in other wetlands of the Meadowlands. One of few occurrences of crested liverwort (*Lophocolea heterophylla*) in the Meadowlands, the only place where we found two interesting seed plants, lizard's-tail (*Saururus cernuus*) and great blue lobelia (*Lobelia siphilitica*), and (at least formerly) a population of the northern red salamander, in a spring in Ridgefield Park, are consistent with this prediction. Other springs are in the Ridgefield Nature Center (Karen Riede, formerly Ridgefield Nature Preserve, pers. comm.), formerly the source for a commercial water bottling company, and the area just west of Harrier Meadow and the 1-E Landfill in North Arlington. Kearny Marsh West has large inputs of groundwater (Obropta et al. 2008), and star patterns in the winter ice reveal underwater springs. A rock pool on the abandoned quarry floor at Laurel Hill also appeared springfed.

PONDS AND LAKES

Ponds and lakes of the Meadowlands range from large lakes like Mehrhof Pond down to intermittent pools of a few square meters. (The largest lake is the Oradell Reservoir, but we do not include the reservoir in the Greater Meadowlands.) These water bodies have diverse origins, as mine pits, ornamental lakes and ponds, stormwater ponds, tidal power impoundments, ponds of uncertain origin that may simply have been left as wetlands and channels were filled, possibly natural pools associated with springs, shallow pools within marshes like Kingsland Impoundment and the Kearny Marshes, and small pools created inadvertently by motor vehicles.

There are three large, extant, clay pit lakes, from south to north, Mehrhof Pond, Willow Lake, and Indian Lake, all in Little Ferry. Mehrhof Pond has a surface area of ca. 16 ha and a shoreline perimeter of 1.7 km. The area of Willow Lake is 1.2 ha and perimeter 0.5 km. For Indian Lake, 2.2 ha area and 0.7 km perimeter. Little biological information is available on these lakes which were mostly open water when observed. Mehrhof Pond and Willow Lake were both bordered by woody vegetation on three sides, and this included well developed hardwood forest on the west and southwest of Mehrhof. Indian Lake is in a public park and was somewhat manicured, with small patches of yellow water-lily (*Nuphar*) in the pond shallows, and tufts of emergent plants at the shoreline including arrow arum (*Peltandra virginica*), purple loosestrife, and common reed. Mehrhof and Indian are reportedly subject to occasional influence of the brackish water from the Hackensack River. Andreas Pond (Teaneck) is the best example of an extant tidal pond that was created for tidal water power historically. A pair of culverts allows the Hackensack River tides to rush in and out of the pond. Andreas Pond had very little emergent vegetation, and most of the pond was surrounded by woodland.

Mehrhof Pond is an old mine pit, not an impoundment, and is apparently susceptible to occasional flooding from the estuary. Indian Lake was reported to have a tidal connection to the river via a culvert; a salinity of 0.6 ppt (parts-per-thousand) measured on 6 May 2004 (Konsevick 2004) supports a current or historic estuarine influence. The large, flooded marsh in Lyndhurst west of Polito Avenue and north of Valley Brook Road has or had a hydrological connection to Berry's Creek (Brett Bragin, NJSEA, pers. comm.) and this is visible on a 23 October 2020 Google Earth satellite image.

These larger ponds and lakes have apparently not been well-studied. They potentially support unusual plants such as quillworts (*Isoëtes*) and fishes or invertebrates that do not occur elsewhere in the Meadowlands. Mehrhof Pond is important nonbreeding habitat for diving ducks; for example, as many as several hundred ruddy ducks may be seen on a winter day (Wright 2018). Mehrhof also attracts other ducks, bald eagles, herons, and other birds.

A 1.6 ha freshwater pond at Meadowlark Marsh northwest of Bellman's Creek in Ridgefield had a large population of common stonewort (*Chara vulgaris*). This pond probably originated from tidal marsh isolated by a highway, a service road for the electric transmission lines, and a waste dump.

There is a variety of stormwater ponds in the Meadowlands. These are constructed ponds (either excavated or dammed) that are presumably intended to hold storm runoff from impervious surfaces (pavement, roofs), settle suspended sediments, and slow the movement of runoff to estuarine waters. The stormwater ponds are in the range of perhaps 15–40 m diameter. Most of the stormwater ponds we have seen are permanently flooded or nearly so, that is, they contained standing water whenever observed. There are two intermittently flooded ponds in Moonachie at the southeastern corner of the Teterboro Airport Woods. One is nearly permanent although it dried down during the summer 2010 drought; this pond supported a breeding population of the Atlantic Coast leopard frog (Kiviat 2011, reported erroneously as southern leopard frog prior to the formal description of the former species). The other pond was dry most of the time and supported extensive growth of low emergent plants. Another intermittent stormwater pond is at the north-western foot of the 1-E Landfill and when dried down in September 2006 supported low and tall emergent plants across much of the bottom. These, and most of the other stormwater ponds we have seen, had fringing emergent vegetation such as common reed, purple loosestrife, swamp rose mallow (*Hibiscus moscheutos*), sunflower (*Helianthus*), saltmarsh-fleabane (*Pluchea odorata*), shrubs, and trees. We observed mallard and other ducks, Canada goose, shorebirds, muskrat, red-eared slider, and a variety of dragonflies, damselflies, butterflies, moths, and other wildlife at stormwater ponds. There are also small ornamental ponds associated with buildings. These ponds tended to be manicured and apparently supported less wildlife use because of the lack of emergent vegetation and the nature of the surroundings (lawns, buildings, pavement).

There is a complex of about 10 small ponds, freshwater and apparently permanent, east of Laurel Hill, and west, north, and east of Little Snake Hill. These ponds, variably bordered by reedbeds, evidently originated when portions of the Riverbend tidal marshes were cut off by railroads and other fill. Although we were able to observe there infrequently, we found herons, ducks, shorebirds, winter wren (singing), painted turtle, Atlantic Coast leopard frog, and green frog. The marsh and pool complex supported a substantial chorusing population of the leopard frog in 2012–2013. A 1.1 ha pond farther north between the Secaucus rail station and Laurel Hill has had as many as 108 black-crowned night-herons roosting although recent maxima are ca. 50 (C. Holzapfel, pers. comm.). The reedbeds and the difficulty of human access to these ponds may screen them to some extent from visual and noise disturbance and provide a more secluded environment for wildlife. A group of ponds in Lincoln Park (Jersey City) may have originated as soil mines.

INTERMITTENT POOLS OTHER
THAN STORMWATER PONDS

In rural areas of northern New Jersey and southeastern New York, there are numerous intermittent pools (temporary pools or vernal pools), many of which provide important habitats for amphibians, reptiles, birds, invertebrates, and plants (see Kiviat and Stevens 2001, Colburn 2004). Intermittent pools in both rural and urban areas tend to be disregarded as habitats, despite considerable recent interest in woodland pool-breeding amphibians. We have seen few woodland pools in the Meadowlands, although other kinds of intermittent pools (other than stormwater ponds) were present. A 125 m long intermittent woodland pool in Borg's Woods, Hackensack, is a relatively natural pool of a type that is rare in the Greater Meadowlands (Figure 2.2). Land use intensification is adverse to intermittent pools and the surrounding habitats that protect the pools and their biota (Colburn 2004).

During our study, wood frogs bred in the Borg's Woods pool and there was a possibly rare liverwort (*Riccia sullivanti*) in the pool; a ditch has lowered the water levels. There is a large, springfed woodland pool, perhaps 25 m long, in the southern corner of the Ridgefield Nature Center in Ridgefield. However, in 2013 this pool was permanently dry as a result of water withdrawal (K. Riede, pers. comm.). At the western end of Laurel Hill on the abandoned quarry floor there were two small intermittent pools of different character but connected by surface water after heavy rain. We surmise these pools are partly fed by groundwater discharge from the remaining portions of the hill. The northern pool supported a carpet of opposite stonewort (*Chara contraria*) on its bottom (see Chapter 5). The pool was partly filled with soil during athletic field expansion between 2006 and 2008. Intermittent swamp and wet meadow pools occurred in several locations at Teterboro Airport, and there were small intermittent swamp pools in Schmidt's Woods and Losen Slote Creek Park. A 20 m diameter ephemeral pool in a mowed wet meadow southwest of the parking lot at Losen Slote Park supported a population of the citrine forktail damselfly (*Ischnura hastata*). There were two, ca 12 m long,

Figure 2.2 Large intermittent woodland pool in Borg's Woods. Photo by Erik Kiviat.

pools, either intermittent or permanent, in the abandoned quarry on the south side of Long Swamp in Palisades Park just outside the Core Meadowlands.

Some of the service roads associated with railroads and pipelines, other dirt roads, and ORV trails in the Meadowlands have small intermittent pools created and maintained by vehicle tires. For example, we saw pools on the road running west from the NJSEA buildings in Lyndhurst, the north-south road crossing the Bellemead mitigation site (Lyndhurst), the railroad service road north of the Kingsland Impoundment (Lyndhurst), and the service road of the abandoned railroad at the western edge of Kearny Marsh West and the Keegan Landfill (Kearny). Arguably the most interesting intermittent pools were those on the service road of the Paterson Lateral Gas Pipeline crossing the Kane Natural Area (Carlstadt and South Hackensack), destroyed in 2010 by construction of the mitigation bank. There were perhaps 40 pools on 1 km of road, in the range of 2–30 m^2 in area and with a maximum depth when full about 15 cm. The pools contained water most of the time, and the water was turbid and slightly brackish (Schmidt and Kiviat 2007, Orridge et al. 2009). These pools supported a large population of the globally rare Mattox's clam shrimp (*Cyzicus gynecius*; see Chapter 10). We were unable to find this invertebrate in intermittent pools on the former clay flats east of the pipeline, or in the other intermittent pool locations mentioned above. Subsequently we discovered Mattox's clam shrimp and another, apparently rare, species, of either the genus *Limnadia* or *Eulimnadia*, in rain pools on two dirt roads in Lyndhurst. One road was recently re-surfaced, eliminating clam shrimp habitat. Long-lasting rain pools are evidently critical habitats for clam shrimps in the Meadowlands and elsewhere in North America and deserve study and conservation. Other biota of the Paterson Lateral Pipeline road pools included muskrat, songbirds, snapping turtle, dragonflies and damselflies, ostracods, wetland seed plants, and algae. We never saw mosquito larvae or pupae there.

STREAMS

As is the case in many coastal cities, most of the smaller streams of the Meadowlands have been altered by channelization, culverting, tidegating, salinization, and pollution. Losen Slote is also spelled Losen Slofe (USGS 1967). The lower reaches are in Losen Slote Park, the BCUA property, and the east end of the Kane Natural Area where they are ponded on the flat topography by a tide gate just upstream of the Hackensack River. This stream is broad, muddy, and in many places bordered by marsh plants such as common reed and swamp rose mallow or by woody vegetation. In 2007, dike and tide gate failure reestablished estuarine conditions in lower Losen Slote as well as portions of the adjoining marshes.

The middle reach of Wolf Creek, on both sides of Slocum Avenue in Ridgefield, is a small perennial stream with predominantly rocky substrate. A mostly narrow, wooded riparian habitat bordered the stream, and had some large hardwood trees and an interesting mixture of native plants (such as skunk-cabbage [*Symplocarpus foetidus*] and may-apple [*Podophyllum peltatum*]) and nonnative plants (knotweed [*Polygonum cuspidatum*] and lesser-celandine [*Ranunculus ficaria*]). However, the stream itself appeared species-poor. During visual inspection on two visits, we saw only one fish (American eel [*Anguilla rostrata*]) and virtually no macroinvertebrates.

WETLAND BUFFERS

The *buffer zone* comprises the upland soil and vegetation bordering a wetland or stream. Buffers remove nutrients, chemical contaminants, and silt from waters entering wetlands and streams, reduce exposure of wildlife to the visual disturbance and noise from developed areas, and provide wetland-upland and riparian habitat combinations for many animals, as well as support seed plants and cryptogams that require these transition zones. Different buffer zone widths provide variable protections; whereas 10 m of upland soil and vegetation can filter runoff waters significantly, 100 m are commonly not enough for the upland activities of some wetland animals such as terrestrially nesting turtles and non-breeding amphibians. The pervasive urban-industrial development in the Meadowlands in many places approaches closely to the wetlands, even to the very wetland boundary. In the Meadowlands District, it was estimated that less than 25% of wetland edges have a 100 m buffer (Tiner et al. 2005).

CONCLUSIONS

All surface water habitats in the Meadowlands are affected by excessive, polluted runoff from impervious surfaces, and many of these habitats are also intruded by brackish waters from the estuary. This is a challenging combination for organisms. Yet some habitats support rare species such as the clam shrimps and the Atlantic Coast leopard frog. Thus, all surface waters, as well as the tidal marshes and estuarine channels that attract the majority of discussion, deserve detailed study and appropriate management.

The wetlands and surface waters of the Greater Meadowlands are of crucial importance to biodiversity in the region and far beyond. In context, during

2002–2012, 1544 ha of wetlands were destroyed in the New Jersey portion of the New York–New Jersey Harbor and Estuary watershed, including 79 ha of "estuarine emergent" wetland and 119 ha of "freshwater emergent" wetland (Stinette et al. 2018). Most of this wetland loss was a result of urbanization. We do not know how much of this loss occurred in the Greater Meadowlands, but the less wetland there is in the region or in the whole watershed, the more the remaining wetlands matter for their provision of ecosystem services, including biodiversity support.

Chapter 3

Uplands and Forested Wetlands

In addition to estuarine and freshwater marshes and waterbodies, there are other major types of habitat that contribute to the value of the Meadowlands as an urban biodiversity hotspot. These include many seminatural and developed upland habitat types as well as forested wetlands. Nineteen of the 37 mammal species that have been documented in the Core Meadowlands require upland or forested habitat but may also venture into adjacent open wetland habitats to forage or use tidal wetlands during drier periods; only 4 mammals, excluding seals, whales, and dolphins, are exclusively wetland or aquatic species. Of the 149 species of passerine birds recorded in the Meadowlands, 101 are mainly associated with upland or forest habitats. Seventeen of the 25 species of raptors that occur in the Meadowlands are associated with uplands or forests. Of the shorebirds, the killdeer and American woodcock use both wetlands and uplands, and certain other species roost on uplands. Among the wading birds black-crowned night-heron and yellow-crowned night-heron use trees in uplands as roosts and nest sites. Least terns use bare dry ground for nesting. All of the 4 turtle species and 3 snake species that are common in the Meadowlands require upland habitat; of these species, the turtles only require it for nesting. Most butterfly and moth species in the Meadowlands rely on plants in meadows, gardens, and forests as caterpillar hosts or as nectar and other food resources for adults, and most overwinter in upland habitats. Very large numbers of both native and nonnative plants are upland or swamp inhabitants. The many plant and animal species that rely upon uplands and forests make these habitats important to the overall biodiversity of the Meadowlands. Moreover, because they are relatively accessible to people, uplands are the habitat type most frequented by urban residents.

Uplands are defined here as *mesic* (medium moisture) to *xeric* (dry) habitats, which unlike wetlands do not usually have standing water or saturated surface soils (i.e., all habitats not classified as wetlands or open water). Mesic uplands are flatlands, hills, and slopes that are not excessively drained (Collins and Anderson 1994). Xeric habitats are found on ridges and slopes of higher elevation or on excessively drained flatlands where rapid runoff, exposure to sun and wind, and soils result in drier conditions than on mesic sites (Collins and Anderson 1994). Sipple (1972) described 14 plant communities in the Meadowlands. These include various types of non-vegetated areas, forests, meadows, and shrubland that might be classified as upland vegetation habitats. About 60% (4,670 ha; 47 km^2) of the Meadowlands District comprise upland habitats (NJDEP 2015). We discuss forests in a separate section later in this chapter; we combined upland forest and forested wetlands (i.e., wooded swamp, shrub wetland and riparian forest) into one category of forests given the limited extent of forests in the Meadowlands, their commonalities in habitat structure, and the fact that many forested wetlands in the Meadowlands were ditched and partially drained long ago. In addition to these spontaneous plant communities, we also consider the habitat value of plantings and built structures within human environments as part of the category of upland habitats.

Many of the most valuable, and most threatened, upland habitats are located on vacant lots, landfills, and *brownfields* (abandoned, often contaminated post-industrial sites; Kattwinkel et al. [2011], Venn and Niemelä [2004]). Some are the result of artificial filling of wetlands to elevate the land above the water table or above normal tidal influence as part of building, road, and railway construction or as a result of waste disposal. Vegetated upland habitats have formed or been created on inactive landfills, road and rail beds, pipeline corridors, abandoned industrial sites, backyards, parks, and higher-relief landforms such as Laurel Hill and Little Snake Hill. Corridors of habitat associated with land uses such as highways and pipelines, as well as the margins of parking areas, building sites, and highways, have not been described in the formal literature on the Meadowlands yet have a high value for some animals, plants, and other organisms (Kattwinkel et al. 2011, Venn and Niemelä 2004). Notwithstanding, these habitats are often undervalued because of lingering biases in the minds of scientists and decision-makers as well as the public (Harrison and Davies 2002, Callus 2006). The persistence of habitats in urban brownfields and other abandoned areas is threatened by redevelopment projects, which provide opportunities for urban renewal but often overlook the value of the habitats that have regenerated in these places. In addition, the importance of maintaining upland habitats adjacent to wetlands is often overlooked. Forested wetlands are to some extent protected from development under state and federal wetland laws, but they are threatened by encroaching development in some places and alteration of hydrological conditions such as stream flow, groundwater levels, and soil moisture, and some have been filled illegally.

The aim of this chapter is to provide an overview of the types of upland habitats and forests in the Meadowlands. We provide a general description of common plant and animal species associated with each habitat. In addition, important opportunities for enhancing biodiversity in the Meadowlands by protecting key upland areas and forests of all types are discussed. Finally, recommendations are outlined for restoration and management of present and future upland and forest habitats.

TYPES OF UPLAND HABITATS

Non-vegetated Areas

Non-vegetated areas in the Meadowlands include portions of railroad embankments, road and pipeline berms, trails, and dikes, as well as newly disturbed or eroded surfaces of landfills, construction sites, soil piles, active dumps, equipment parking areas, recently abandoned facilities such as parking lots, recently disturbed ORV riding areas, bedrock ledges, and dredge spoil disposal areas and other recent fill that have not been colonized by vascular plants. The Meadowlands District included about 665 ha of this land cover type, noted as barren land in the NJDEP (2015) Land Use–Land Cover Data Set. In these areas, soils are often coarse-textured and droughty at the surface, although compacted, fine-textured materials may be present and may even hold rainwater in temporary pools.

Non-vegetated areas provide exposed soil or fill that may be important nesting habitat for several species of turtles and birds, as well as many bees, wasps, ants, and other invertebrates in the Meadowlands. Species known to use non-vegetated areas in the Meadowlands are the diamond-backed terrapin, least tern, spotted sandpiper, killdeer, horned lark, and other birds (see appropriate chapters and Appendix 2 for scientific names). Turtles typically nest in well-drained but not extremely dry, *friable* (loose) soils. Diamond-backed terrapins, an important species in the Meadowlands, spend most of their time in estuarine channels and wetlands but nest in adjacent sandy habitats and along road and rail beds with sparse to moderate vegetation cover (Day et al. 1999).

Least tern, a species endangered in New Jersey, has nested on dredged material deposits and commercial building roofs in the Meadowlands (Day et al. 1999). North of the Vince Lombardi Service Area, a large sandy area historically provided nesting habitat for least tern (NJTA 1986). Also, non-vegetated areas provide resting, basking, dustbathing, and foraging habitat for upland species that also use wetland habitats (e.g., ring-necked pheasant, wild turkey, snakes, and small mammals). Waterfowl have nested on artificial islands consisting of vegetated and non-vegetated areas at the Western Brackish Marsh (Mill Creek) mitigation site (Kane and Githens 1997). Bare, friable soils are important for ground-nesting native bees including bumblebees, as well as some wasps and other insects. However, non-vegetated habitats are often short-lived, either because of spontaneous vegetation or land development.

Xeric, Poorly Developed Soils and Associated Vegetation

A community of plants adapted to extreme temperatures, high winds, poor and often shallow soils, and xeric conditions has developed on the rocky crests and quarry faces within and along the edges of the Meadowlands. This type of community is found at Laurel Hill, Little Snake Hill, Granton Quarry, and the North Arlington and Lyndhurst cliffs, among other locations. Laurel Hill was extensively mined from the late 1800s until 1982. It now supports areas of closed-and open-canopy forest, meadows, and nearly bare rock. The Torrey Botanical Society reported 115 species of plants and the New Jersey Meadowlands Commission reported 145 species of plants from Laurel Hill alone (Quinn 1997). Trees included chestnut oak (*Quercus prinus*), red oak (*Quercus rubra*), hackberry (*Celtis occidentalis*), and black cherry (*Prunus serotina*; Quinn 1997). Tree-of-heaven (*Ailanthus altissima*), white mulberry (*Morus alba*), white ash (*Fraxinus americana*), princess tree (*Paulownia tomentosa*), and eastern red cedar (*Juniperus virginiana*) were also present. A small area of chestnut oak woodland with dense low grasses on the north-eastern knoll of Laurel Hill appeared to represent natural vegetation on a non-quarried area (Kiviat and MacDonald 2006). Also, there was a stand of mature sugar maple (*Acer saccharum*) on the steep south side of Laurel Hill between the ridge top and the Turnpike. A dry meadow on fill or mine tailings at the southwest foot of Laurel Hill bore a community of eastern cottonwood (*Populus deltoides*), staghorn sumac (*Rhus typhina*), white mulberry (*Morus alba*), common mullein (*Verbascum thapsus*), mugwort (*Artemisia vulgaris*), switchgrass (*Panicum virgatum*), and other herbs.

Meadows and Shrublands

Meadow and shrub habitats were the most common types of naturally occurring upland habitat in the Meadowlands. There were about 202 ha of old field habitats and 305 ha of shrubland (included in the broad category of forest) in the Core Meadowlands (NJDEP 2015). These habitats are located principally along roadsides, on abandoned lots, on dredge spoil and other fill, in abandoned properties in residential, commercial, and industrial areas and on the slopes of landfills. Much of the 648 ha of inactive (or capped) garbage landfill has developed upland meadow and shrubland and in many places these two habitat types were intermingled (Kane and Githens 1997). Plant associations in meadows and shrublands vary locally and range from dense stands of mugwort or stunted common reed to diverse assemblages of forbs and shrubs. Young woodland (e.g., quaking aspen [*Populus tremuloides*] groves) may also be present (Figure 3.1). Brownfields, which are sites with an industrial history, commonly contaminated, that are now abandoned and inactive, often have a developing plant community of meadows and shrubland that support many plant and animal species (Kattwinkel et al. 2011, Venn and Niemelä 2004). Brownfields are common in the Meadowlands.

During the first year or so after fill is deposited, meadows are often dominated by exotic warm-season grasses with other herbaceous plants and vegetation is always changing as conditions change. Although this first stage of *vegetation development* (often called *succession*) has not been described specifically for the Meadowlands, typical pioneer species of similar habitats in New Jersey include common ragweed (*Ambrosia artemisiifolia*) and wintercress (*Barbarea vulgaris*; Collins and Anderson 1994). Mowed roadside habitats may contain species such as common cinquefoil (*Potentilla simplex*), dandelion (*Taraxacum officinale*), plantains (*Plantago* spp.), chicory (*Cichorium intybus*), common ragweed, and several grasses (Collins and Anderson 1994). After one or a few years, perennial species typically begin to dominate the meadow vegetation. In the Meadowlands common meadow forbs include goldenrods (*Solidago* spp.), common mullein (*Verbascum thapsus*), yarrow (*Achillea millefolium*), common milkweed (*Asclepias syriaca*),

Figure 3.1 Young hardwoods on landfill sideslope in North Arlington. Photo 2019 by Erik Kiviat.

Queen Anne's lace (*Daucus carota*), black-eyed Susan (*Rudbeckia hirta*), sunflower (*Helianthus*), mugwort, sweet Annie (*Artemisia annua*), and poke-weed (*Phytolacca americana*; Kane and Githens 1997; E. Kiviat, pers. obs.). In addition, we would expect to find asters, oxeye daisy (*Leucanthemum vulgare*), and spreading thistle (*Cirsium arvense*; Collins and Anderson 1994). Commonly there are the native grasses switchgrass and little bluestem (*Schizachyrium scoparium*). Patches of dense, short, slender common reed also occur. Meadows, especially those dominated by grasses, are often called grasslands. Grasslands are most extensive on inactive and capped landfills in the Meadowlands.

Eventually shrubs and trees take hold in these young habitats and over time they become the dominant plant form (this can take many years on poor fill soils). Shrubland communities can be locally diverse in the Meadowlands ranging from meadows with sparse low shrubs to scrubby habitats containing dense patches of small trees. Common shrub species included gray dogwood (*Cornus racemosa*), smooth sumac (*Rhus glabra*), staghorn sumac (*Rhus typhina*), winged sumac (*Rhus copallinum*), steeplebush (*Spiraea tomentosa*), groundselbush (*Baccharis halimifolia*), common elderberry (*Sambucus nigra*), blackberries (*Rubus* spp.), and autumn-olive (*Elaeagnus umbellata*; NJTA 1986, Kane and Githens 1997; E. Kiviat, pers. obs.). The trees that colonize meadows and shrublands included tree-of-heaven, quaking aspen, eastern cottonwood, princess tree, black cherry, white mulberry, and pin oak (*Quercus palustris*; NJTA 1986, Kane and Githens 1997; E. Kiviat, pers. obs.; see Figure 3.1).

Examples of upland meadow and shrubland on dredge spoil were present in the northwestern part of Oritani Marsh along Berry's Creek Canal. Meadow vegetation was also present at numerous locations associated with dirt roads, pipeline corridors, railroad verges, construction sites, vacant lots, and soil storage areas, as well as abandoned and capped garbage landfills. The greatest concentration of large, abandoned landfills extends in the lowlands from Kearny northward through North Arlington, Lyndhurst, Rutherford, and East Rutherford, west of the Hackensack River (Figure 1.2). Near Kingsland Marsh, the Erie Landfill was active until recently and there were several overgrown closed landfills (NJTA 1986, Kane and Githens 1997). The eastern slopes of the Kingsland Landfill in DeKorte Park were planted with flowers, shrubs, and trees to create habitats much frequented by songbirds, flower-visiting insects, and people. At Cromakill Creek, low-lying wetland fill was grown with quaking aspen, tree-of-heaven, and princess tree. South of the Meadowlands Exposition Center there was a large meadow on fill dominated by groundselbush, mugwort, dense low grasses, patches of stunted common reed, and groves of eastern cottonwood. In addition, there were grassland and open woodland communities on the rock outcrops and shallow soils of Laurel Hill and Little Snake Hill. Overpeck Park contained open meadow and meadow that was overgrown by gray birch (*Betula populifolia*; Kane and Githens 1997). Approximately 30-year-old dredge spoil wildlife islands in the Western Brackish Marsh (Hartz Mountain) wetland mitigation site were variously dominated by gray birch, tree-of-heaven, and other woody plants (E. Kiviat, pers. obs.). Upland vegetation on various kinds of fill occupied many other areas of the Meadowlands.

There was a large meadow at Overpeck Park notable for its size and its intensive use by fall migrant birds (Kane and Githens 1997); we are not sure of the status of the site today. An area of dry meadow and (now mostly) shrubland and young woodland on natural sandy soil, unusual in the Meadowlands, was in Hackensack River County Park in Hackensack. Some of the species in this area were pin oak, gray birch, black locust (*Robinia pseudoacacia*; some of which was stunted and shrub-like), bayberry (*Morella*), yarrow (*Achillea millefolium*), blue curls (*Trichostema dichotomum*), sandbur (*Cenchrus longispinus*), and a flatsedge (*Cyperus lupulinus*; E. Kiviat and other participants in Torrey Botanical Society field trip 2 September 2001). Losen Slote Creek Park contained a large shrubby meadow community with interspersed patches of eastern cottonwood and gray birch. At other sites, common reedbeds in upland areas containing a low density of woody plants, such as tree-of-heaven, princess tree, and common elderberry, occurred frequently.

Shrublands and meadows are particularly important habitats for birds in all seasons (Askins 2000). In the Meadowlands, shrubs and trees found along ditches and canals, and on landfills and other uplands, combined with meadows, provide breeding and foraging habitats for several species including indigo bunting, brown thrasher, barn swallow, tree swallow and possibly eastern bluebird (NJTA 1986, Wargo 1989, Kane and Githens 1997). Fruits of sumacs are important food for birds in winter and spring (e.g., at DeKorte Park), and we expect that fruits of common elderberry and gray dogwood, as well as other fleshy tree and shrub fruits and grass and forb seeds, are also important fall, winter, and spring foods for birds and mammals. In a study of early-development habitats in central New Jersey, Suthers et al. (2000) found that resident and autumn migrant birds used open shrubland habitat more than young woodlands or shrubland with colonizing trees. This was attributed to a lower abundance of fleshy-fruited shrubs and vines in young forests as compared to open shrubland. Furthermore, shrubland dominated by gray dogwood was favored over shrubland dominated by red cedar and multiflora rose (*Rosa multiflora*). The vine species most favored by birds in the study were gold-and-silver honeysuckle (*Lonicera japonica*), Virginia creeper (*Parthenocissus*), poison-ivy (*Toxicodendron radicans*), and grape

(*Vitis*). Bricklin et al. (2016) found that nonnative shrubs (winged euonymus, *Euonymus alatus*) and vines (porcelainberry, *Ampelopsis brevipedunculata*) were used by migrant birds, especially when native plant fruits were unavailable.

Not only do birds and other animals benefit from the rich food resources found in these habitats, but they also facilitate the arrival and establishment of fleshy-fruited shrubs and vines. Robinson et al. (2002) demonstrated that a large percentage of early-colonizing woody plants on a landfill in the Meadowlands were due to birds' importing seeds from natural source habitats.

Early-development habitats such as meadows, shrublands, and young woodlands are also needed to maintain populations of small mammals that are important prey for raptors and mesomammals (Litvaitis 2001). Upland meadows such as those found on abandoned landfills in the Meadowlands are important foraging habitats for raptors preying on the meadow vole (*Microtus pennsylvanicus*), short-tailed shrew (*Blarina brevicauda*), or other abundant small animals. In the Meadowlands, wintering long-eared owls (*Asio otus*) and short-eared owls (*Asio flammeus*) appeared to diverge from the diet of their conspecifics elsewhere which consisted mainly of voles. In the Meadowlands the diet mainly consisted of house mouse (*Mus musculus*), brown rat (*Rattus norvegicus*), and birds (Bosakowski 1982). Decreases in winter communal roosts of northern harrier and short-eared owl have been attributed to the closing of most landfills and a subsequent reduction in rodent densities (H. Carola, pers. comm.). However, red-tailed hawks (*Buteo jamaicensis*) were abundant in winter 2003–2004 (Hugh Carola, pers. comm.; E. Kiviat, pers. obs.) and often since. This indicates that an adequate prey base of small mammals exists for some species of raptors.

There is evidence of a rich insect fauna in meadows and shrublands of the Meadowlands. A study on the Kearny Landfill documented a diverse, and almost exclusively native, bee assemblage (Yurlina 1998; see Chapter 10). This is likely due to the variety of nectar plants and nest microhabitats in eroding soil and hollow plant stems. An area of the Meadowlands (East Rutherford) was surveyed for lady beetles (*Coccinella septempunctata*) and their most important prey, aphids (Aphididae; Angalet et al. 1979). The mealy plum aphid (*Hyalopterus pruni*), which alternates between common reed and woody plants, was found to be very abundant in the Meadowlands, as were several aphids found on mugwort. These aphids were the principal prey of the lady beetles. A native and an introduced lady beetle were found overwintering in association with redtop (a tussock-forming grass, *Agrostis gigantea*), common mullein, and planted pines (*Pinus sylvestris*, *Pinus resinosa*) in the Meadowlands (Angalet et al. 1979). The 15 species of aphid host plants reported in this study were mostly introduced, weedy species of *ruderal* (roadside-type) habitats or upland meadows.

The diversity and abundance of plants and animals of meadow and shrubland habitats are an integral component of biodiversity in the Meadowlands. Early developmental upland habitats are becoming rarer in the Meadowlands and throughout the northeastern United States because of intensifying land use as well as spontaneous transition to forest.

Built Habitats

Built areas such as city centers, residential neighborhoods, commercial-industrial areas, public parks, and cemeteries are human-created, upland habitats that support many species of plants and animals; these comprise about 45% of the Meadowlands District (NJDEP 2015). Human-created habitats exist on a gradient of vegetation composition and structure, and while there are some characteristics that typify certain land use types, there can be great heterogeneity at landscape scales with a mosaic of areas entirely devoid of vegetation (e.g., parking lots) to those with a rich mixture of plant communities (e.g., suburban residential areas).

Built land use types typically have considerably less vegetation cover than unbuilt areas (greenspaces) and are dominated by buildings, paved surfaces, ornamental vegetation, and highly discontinuous vegetation strata and cover as compared to natural forests (Beissinger and Osborne 1982). Many species have been planted and maintained by people and there are also many ruderal plants that spontaneously colonize abandoned lots, borders of many kinds, and untended gardens and lawns. Decreased abundance of native woody plants is associated with sharp declines in numbers of lepidopteran (moth and butterfly) larvae, the effects of which cascade up the food chain (Burghardt et al. 2010). However, just because human-created habitats are highly altered by people does not mean they do not contribute to biodiversity. Urban residential areas often have a higher diversity of plant species than would be found in nearby seminatural habitats because of the myriad of planted native and exotic species (Turner et al. 2005). On the other hand, most animal species are not intentionally brought into human-created habitats by people; exceptions are pets and animals attracted by intentional or unintentional feeding. Animal species are present because they are able to exploit resources found in urban environments and tolerate urban stressors. Most of the research relating the structure of human-created habitats to animal communities has focused on birds. Studies of human-created habitats have not been conducted in the Meadowlands with the exception of ongoing studies of bird use of landfill habitats (e.g., Seewagen and Newhouse 2018). Therefore, we provide a few examples of studies conducted in urban settings elsewhere, which should be applicable to developed portions of the Meadowlands.

Birds vary greatly in their ability to use human-built habitats and several bird species are highly *synanthropic* (associated with human environments). These include the house sparrow, rock pigeon, and, to a lesser extent, European starling. Not surprisingly, these species dominate city centers in both Europe and North America (Rosenberg et al. 1987, Clergeau et al. 1998). In more suburban areas of the Northeast, American robin is the most characteristic, if not most abundant, suburban bird species. Cavity-nesting species may be absent or in low abundance in heavily urbanized areas because of a shortage of older trees with cavities (Rottenborn 1999); however, moderately large (e.g., 40+ cm dbh) planted and spontaneous trees are fairly common in the Meadowlands. There is a high interspecific variability as to which habitat resources are most important. Different animal species respond differently to aspects of the incredibly variable human environment. For example, Savard and Falls (2001) demonstrated that birds in urban environments may be categorized based on their dependence on built structures versus natural features. European starling, rock pigeon, and chimney swift were positively associated with building density and negatively with lawn area. Several other species including mourning dove, American robin, common grackle, northern cardinal, blue jay, and song sparrow were positively associated with shrub and lawn area and negatively associated with building density.

Urban parks and cemeteries with high densities of tall trees have been shown to support nesting great horned owls in other areas of New Jersey. This species may also use such areas for breeding in the Meadowlands (Smith et al. 1999) and has been reported from Borg's Woods and other Greater Meadowlands forests. Long-eared owl and short-eared owl roosted in low coniferous shrubs and trees planted in developed areas in the Meadowlands when snow cover prevented them from using their ground roosts in the marshes (Bosakowski 1986).

Several characteristics of built habitats that are associated with increased diversity of birds are summarized here:

1. Forest present within the suburb or in close proximity (DeGraaf and Wentworth 1986, MacDonald-Beyers 2008). This should also increase diversity of mammals, amphibians, reptiles, invertebrates, plants, and fungi. Other types of natural habitats in close proximity to urban neighborhoods (e.g., meadow, wetland) presumably influence the diversity of animal species in human habitats as well.
2. Residential neighborhoods with greater cover of trees and shrubs (Guthrie 1974, MacDonald-Beyers 2008).

3. The presence of large, mature street trees and increased volume of woody vegetation (Tzilkowski et al. 1986, Goldstein et al. 1986, MacDonald-Beyers 2008). Bats may also use mature trees as roost and nursery sites.

4. Lower number of pedestrians (Fernández-Juricic 2000a). Avoidance of areas with high pedestrian traffic has also been shown to influence insect and mammal densities and behavior.

5. Higher density and larger size of conifers and shrubs (Savard and Falls 2001, MacDonald-Beyers 2008).

6. Planting of native tree, shrub, and herbaceous plant species as opposed to exotic species in suburban areas (Rosenberg et al.1987).

7. Less vehicular traffic (Tzilkowski et al. 1986). This should also affect amphibian, reptile, and mammal diversity.

8. Intermediate levels of urbanization have highest diversity because of increases in edge habitat, structural diversity including buildings and vegetation, and primary productivity due to fertilizers in these environments (Blair 1999).

9. Presence of water in suburban areas because of irrigation of lawns during periods of water scarcity (Guthrie 1974, Rosenberg et al. 1987). Presence of natural ponds, streams, and stormwater structures in urban neighborhoods should also affect insect and amphibian diversity.

10. Lower density of houses and smaller housing area (Sewell and Catterall 1998).

The myriad buildings, bridges, communications towers, highways, railroads, piers, and other artificial structures in the Meadowlands, both active and abandoned, provide a variety of habitat functions for certain taxa, in particular birds. Some species directly benefit from built structures and urban habitats. Chimney swifts nest in chimneys and other vertical structures with low light. Peregrine falcons benefit from nesting sites on large buildings and bridges (Wander and Wander 1995) as well as increased abundances of prey such as rock pigeons and European starlings. The least tern, common nighthawk, and killdeer potentially nest on flat (usually gravel) roofs, and there has been evidence of roof-nesting killdeer and least tern in the Meadowlands (see Fisk 1978 about roof-nesting in general). The double-crested cormorant, several raptors, and swallows, among other species, use towers and wires for perching and roosting. The northern rough-winged swallow nests in artificial structures such as drain pipes or masonry cracks in bridges, and presumably does so in the Meadowlands. The barn owl, barn swallow, eastern phoebe, and bats may nest or roost on and in bridges and abandoned or occupied buildings. Introduced monk parakeets have nested on a variety of structures including a street overpass, streetlights, and powerline installations in the Meadowlands.

The abundance and distribution of animals in built habitats vary by species. Bird abundance in built habitats is usually higher than in many natural habitats because some synanthropic species thrive in urban areas. The high densities of birds are believed to result in intense predation on insects, which may decrease the diversity of insects in built habitats; though evidence of this is anecdotal. Also, subsidized predators such as house cats take large numbers of birds, small mammals, reptiles, and amphibians (Churcher and Lawton 1987, Baker et al. 2005). This might affect diversity patterns in the Meadowlands where feral cats are common even though direct studies have not been conducted there (see van Heezik et al. 2010).

Upland Forests and Forested Wetlands

We combine upland forests and forested wetlands in this discussion of forest habitat and this land use type includes forest of variable crown cover. Upland shrubland is discussed with meadows, and shrub wetland is included here with forested wetland. In the Meadowlands District alone, there are about 150 ha of upland forest and 99 ha of forested wetland (including shrub wetlands; NJDEP 2015).

Mature, upland forests are very restricted in the Meadowlands but may increase over time as forests develop on abandoned landfills, and inactive industrial areas, similar to other urbanized areas in New Jersey (MacDonald and Rudel 2005). There were areas of upland oak forest and mature beech forest at Old Tappan, just north of the Meadowlands (Kane and Githens 1997), as well as around portions of the Oradell Reservoir. In the Greater Meadowlands, mature upland forests, mostly small patches of ca. 1.5 to 14 ha, existed at Borg's Woods, Ridgefield Nature Center, Matthew Feldman Preserve, Teaneck Creek Conservancy, Overpeck Creek, Laurel Hill, and Little Snake Hill. Within the Meadowlands, the woodlands on shallow soils at Laurel and Little Snake hills represent regionally rare forest communities (see section on xeric habitats). Among the tree species growing on these rocky crests were chestnut oak, red oak, hackberry, black cherry, eastern red cedar, tree-of-heaven, white mulberry, and white ash (Kane and Githens 1997). Remnant mesic hardwood forests occurred in the Ridgefield Nature Center (Ridgefield) and Borg's Woods (Hackensack), just outside the core Meadowlands. Borg's Woods (in 2021) had many large hardwood trees. Finally, areas categorized as forested wetlands (see section below) actually contained upland forests at their fringes and as small pockets on higher elevations within the wetland portions. Younger upland forests were more common throughout the Meadowlands; some of the stands had smaller trees that may be slower-growing due to stressful soil conditions on dredge spoil, landfill cover, or shallow soils. White-tailed deer densities are increasing in the Meadowlands (H. Carola, pers. comm.; E. Kiviat, pers. obs.). In the future, deer may affect forest structure and composition by reducing tree regeneration through browsing on seeds, seedlings, and saplings and by overbrowsing shrubs and consuming most of the native forbs resulting in reduced understory and increasing dominance of less preferred, nonnative plants as seen in other areas of New Jersey (e.g., Kelly 2019).

Most of the forest within the Core Meadowlands is forested wetland. This includes riparian or floodplain forests and swamp forests (Figure 3.2) with soils that are saturated for at least a portion of the growing season. Riparian forests are important in that they provide regionally-rare habitat and are hotspots for migrating and breeding songbirds and other wildlife (Naiman et al. 1993, Sabo et al. 2005, Pennington et al. 2008, Bennett et al. 2014, McClure et al. 2015). In addition, they may act as habitat corridors for movement of other organisms along the river. East-west and south-north corridors along streams and other natural and artificial linear features may allow animal and plant populations to adjust spatially to changing land use, sea level, seasons, and climate. Riparian forests provide other important functions including stabilizing banks, filtering water, slowing and absorbing flood waters, shading and cooling water, providing food resources for stream organisms

Figure 3.2 Old hardwood swamp remnant in Moonachie where trees on lower elevations died after Hurricane Sandy in 2012. Photo 27 September 2019 by Erik Kiviat.

via leaf litter, and providing habitat among their roots and downed logs for birds, fish, mammals, reptiles, amphibians, invertebrates, plants, and fungi.

Riparian or floodplain forests were located mainly along the Hackensack River and some lengths of its tributaries north and south of the Meadowlands District. These forests contained sycamore (*Platanus occidentalis*), silver maple (*Acer saccharinum*), basswood (*Tilia americana*), red maple (*Acer rubrum*), river birch (*Betula nigra*) and pin oak and in some places willows (*Salix*), black walnut (*Juglans nigra*), eastern cottonwood, box-elder (*Acer negundo*), white mulberry, and shadbush (*Amelanchier*; Kane and Githens 1997). Riparian forests farther from the river may be flooded episodically in high-rainfall years. The most notable remnants of riparian forest were in several riparian corridors along the Hackensack River at the Oradell Reservoir, Van Buskirk Island, New Milford, River Edge, Teaneck, Hackensack, and Overpeck Creek as well as Wolf Creek (Ridgefield).

Swamp forests (forested wetlands) are generally flooded for longer periods (more than two weeks during the growing season) and the soil remains saturated close to or at the surface for most of the year (Tiner 1998). Soil, hydrology, and plant community characteristics are highly variable among forested wetlands in urban areas, especially those that have been altered by ditching and draining (Ehrenfeld et al. 2003) as in the Meadowlands. Red maple, pin oak, and sweetgum (*Liquidambar styraciflua*) were typically the dominant tree species but, in general, tree species richness is high in these forests. Other typical tree species were gray birch, eastern cottonwood, princess tree, American elm (*Ulmus americana*), slippery elm (*Ulmus rubra*), black cherry, black willow (*Salix nigra*), white oak (*Quercus alba*), red oak, ashes (*Fraxinus* spp.) and tupelo (*Nyssa sylvatica*; Kane and Githens 1997). Shrubs and woodland herbs in these forests were probably more diverse in comparison to the suburban areas to the west and south due to low levels of white-tailed deer activity in the Meadowlands, yet deer activity has increased during the course of our work in the Meadowlands (1999-present). In places, American hornbeam (*Carpinus caroliniana*) was common in the understory. Shrubs included arrowwood (*Viburnum dentatum* s.l.), common elderberry, spicebush (*Lindera benzoin*), silky dogwood (*Cornus amomum*), highbush blueberry (*Vaccinium corymbosum*), and sweet pepperbush (*Clethra alnifolia*; Kane and Githens 1997). Vines were abundant, including poison-ivy (*Toxicodendron radicans*), riverbank grape (*Vitis riparia*), Virginia creeper (*Parthenocissus*), and common greenbriar (*Smilax rotundifolia*). Herbs on the forest floor included netted chain fern (*Woodwardia areolata*), cinnamon fern (*Osmunda cinnamomea*), Canada mayflower (*Maianthemum canadense*), trout-lily (*Erythronium americanum*), violet (*Viola*), bellwort (*Uvularia*), swamp lily (*Lilium superbum*), may-apple (*Podophyllum peltatum)*, skunk-cabbage (*Symplocarpus foetidus*), sedges (*Carex* spp.), and a variety of other woodland plants (Kane and Githens 1997; Kiviat pers. obs.). The most significant forested wetlands in the Meadowlands were at Teterboro Airport, Losen Slote, and Schmidt's Woods. There were also swamp forests along the upper Hackensack River estuary (e.g., New Milford).

Teterboro Airport Woods included large remnants of lowland forest totaling almost 100 ha in five patches, some of which was wet shrubland and marsh. The canopy of these forests was dominated by pin oak and gray birch with a variety of other species. Historically, the forests were extensively ditched, resulting in drier soils than once existed there. Teterboro Woods supported a variety of mosses, ferns, woodland wildflowers, and woody plants, but in recent surveys no amphibians or reptiles were detected except the Atlantic Coast leopard frog (E. Kiviat, pers. obs.). East and northeast of the commercial building at 115 Moonachie Avenue, in the southeastern outpost of Teterboro Woods, a 1.6 ha mature swamp forest was dominated by large pin oaks most of which died after Hurricane Sandy (2012), probably due to brackish water collecting in the shallow depression containing the forest. This swamp, during the several years following the storm, developed a dense shrub layer of sweet pepperbush, sweetgum seedlings, red maple saplings, and festoons of the nonnative herbaceous vine mile-a-minute (*Polygonum perfoliatum*). When observations were made between 2016 and 2021, there was a well-developed flora, diverse macrofungi, and moderate songbird activity as well as American woodcock in winter (E. Kiviat, pers. obs.).

Losen Slote Creek Park and the contiguous Bergen County Utilities Authority–Mehrhof Pond supported ca. 12 ha of mature lowland forest with peripheral shrubland and meadow. The forest had a variety of tree species and was dominated by pin oak, sweetgum, and red maple. The well-developed shrub layer was at one time dominated by sweet pepperbush, highbush blueberry, and sapling-size sassafras. However, highbush blueberry was not common when the site was visited more recently (E. Kiviat and K. MacDonald pers. obs.). At least six species of fern were present including a tiny patch of Virginia chain fern (*Woodwardia virginica*), a species typically associated with acidic bogs. Stiltgrass (*Microstegium vimineum*) and knotweed (*Polygonum cuspidatum*), two highly invasive exotic species, were locally prevalent and will likely spread.

Schmidt's Woods Park in Secaucus contained a ca. 5.5 ha lowland forest of pin oak, sweetgum, and red maple and a rich variety of other tree, shrub, and herb species as well as diverse macrofungi. The shrub layer was dominated by spicebush, arrowwood, silky dogwood, common elderberry, and blackberry (*Rubus*). Herbs including jack-in-the-pulpit (*Arisaema triphyllum*) and jewelweed (*Impatiens*, presumably *capensis*) were present. Thunberg barberry shrubs (*Berberis thunbergii*), an invasive nonnative species, was also present. The northern half of the park has been developed for passive recreation beneath the canopy of large trees, and the forest shrubs and herbs in that area have been mostly removed.

Large areas of the Core Meadowlands supported Atlantic white cedar (*Chamaecyparis thyoides*) swamps in the 1800s with smaller stands surviving into the early 1900s (Vermeule 1897, Heusser 1963, Zimmerman and Mylecraine 2003). Only subfossil stumps and logs remain from these forests. The cedar swamps would have been freshwater and nontidal, with organic soils. Cedar swamps presumably supported many species of organisms no longer present in the Meadowlands, possibly including some of the orchids collected a century ago (see Chapter 4).

Forested wetlands, like other forests in the Meadowlands, provide breeding habitat for many resident and Neotropical migrant species of birds. Schmidt's Woods Park had nesting yellow-crowned night-herons in recent years despite the proximity of human visitors to the park and adjoining recreation facilities. Observations at Teterboro Airport included varied resident and migrant birds that were likely breeding there including cavity-nesting species (K. MacDonald, pers. obs.; see Chapter 7). Dead, standing trees and other sources of nest cavities in forest habitat are probably somewhat rare in the Meadowlands landscape and the remaining forests may be critical for populations of the five cavity-nesting birds observed as well as others that may have gone undetected. One migrant breeding at Teterboro, the wood thrush (*Hylocichla mustelina*), is a forest-nesting, Neotropical migrant, which would not be breeding in the Meadowlands without mature forest habitat (K. MacDonald, pers. obs.). In addition, the red-tailed hawk (*Buteo jamaicensis*) and great horned owl are likely to be nesting in these forests. Small urban forests also provide important stopover sites for migrating forest songbirds (MacDonald-Beyers 2008). Kane and Githens (1997) counted 55 species of migratory forest birds in April-May 1995 in the forest at Losen Slote. We would expect additional birds, mammals, invertebrates, and plants that do not occur elsewhere in the Meadowlands (e.g., the forest herb ghost pipe [*Monotropa uniflora*], Virginia chain fern [*Woodwardia virginica*]). For example, red-backed salamander (*Plethodon cinereus*) and spring peeper (*Pseudacris crucifer*) have been shown to persist in forests surrounded by urban development in southern Connecticut (Gibbs 1998); the salamander occurred in the forest at Borg's Woods and the peeper at Long Swamp (Palisades Park), outliers in the Greater Meadowlands. A rich diversity of moths has been documented in other New Jersey forests (Moulding and Madenjian 1979) and would be expected in the Meadowlands. Remnant mesic and swamp forests in the Greater Meadowlands have diverse macrofungi.

TARGETS FOR CONSERVATION

Non-vegetated Areas

Because non-vegetated areas (bare soil) provide a critical resource for nesting turtles, certain birds, and a variety of insects in the Meadowlands, as well as small seed plants and mosses, these habitats should be assessed and maintained. Sites adjacent to wetlands are of high value because they are easily accessible to nesting aquatic turtles and waterbirds. These sites are also most at risk of being lost due to sea level rise.

Preservation of the two prominent geological features Laurel Hill and Little Snake Hill, which support rare communities and species, should be a priority. Laurel Hill and Little Snake Hill support species of woody and herbaceous plants, mosses, and lichens found nowhere else in the Meadowlands, as well as nesting ravens. The New Jersey Turnpike Authority has removed rock from the eastern slopes of Laurel Hill to reduce hazards of falling rock to the Turnpike below. An equipment road and parking area, as well as the rock removal activities, caused loss of vegetation and may have affected rare species. Expansion of recreational facilities has also affected the seminatural habitats on the abandoned mine floor on the north side of Laurel Hill, and habitat has been lost to recent construction of a school.

Meadows and Shrublands

The loss of meadow and shrub habitats in the eastern United States is recognized as a major threat to biodiversity. Fifteen of the 19 grassland and meadow specialist bird species found east of the Mississippi River have declined since 1966 (Askins 1993, Peterjohn and Sauer 1994). Similarly, 15 of the 16 shrubland species have declined (Askins 2000). In the Meadowlands, non-forested upland habitats support many species of flowering plants and specialized breeding birds, as well as provide food resources to seed and fruit-eating birds during migration and winter. Meadow voles, grasshoppers, and other small animals in these habitats are prey for raptors, snakes, and mammals. Also, grasses and forbs in meadows and shrublands are substantially responsible for the diversity of butterflies, bees, and many other arthropod taxa in the Meadowlands.

Because the vegetation of meadow and shrubland habitats changes rapidly, it may require active management such as mowing to inhibit development to woodland. Mowing of grasslands and meadows should be planned outside the breeding season for birds, and alternating sections of a meadow should be mowed at, for example, 2–3 year intervals. Restoration of meadow habitat on inactive or capped landfills, other abandoned areas, and highway verges should be a priority. Because of poor soils in many areas, restoration may require that the area be seeded with native species (such as little bluestem) tolerant of soils that are droughty, low in organic matter, and lacking natural structure. Planning for development of any kind on landfills, notwithstanding the risks of instability and subsidence (see Chapter 1), should include a significant component of habitat preservation and habitat management, including planting natives. If nonnative weeds such as common reed, barberry, knotweed, mile-a-minute, and mugwort, among many others, are to be managed, non-chemical techniques and a region-wide integrated weed management approach need to be implemented to avoid herbicide toxicity to nontarget flora and fauna. In some cases, highly targeted use of herbicide in cut-stem or basal-bark treatments, may be warranted after assessing neighboring organisms. Finally, there is potential to expand grasslands as a prominent habitat type through management of abandoned and remediated landfills.

Forests

Forests provide important habitat for many species in the Meadowlands including forest-breeding and migrating songbirds and raptors. Forests support many woody and herbaceous plant species and specialized animal species not found in other habitat types. Because of the scarcity of mature forest in and around the Meadowlands, all existing forest stands should be preserved. Increasing the amount of forest through afforestation should be a high priority where it does not conflict with other biodiversity management priorities. Forest health should be managed by monitoring and controlling pests and diseases, invasive plants, and deer densities to a practical extent. Forest patch area, which is related to increasing diversity of birds and other animals, should be increased where possible. It should also be evaluated whether creating or maintaining connections between forests and other upland habitats via corridors as well as between uplands and wetlands is desired. Restoration of contiguous riparian forest corridors along the Hackensack River and its tributaries would provide habitat for many species including plants, birds, reptiles, amphibians, and mammals and would also contribute to habitat connectivity and enhance movement of species among patches while helping to maintain water quality and slow floodwaters. Bird diversity in urban parks is higher when the parks are connected by tree-lined urban streets (Fernández-Juricic 2000b). Forest restoration should be planned in conjunction with other upland habitat management to ensure that the various upland habitat types (meadow, shrubland, and forest) are adequately represented in the landscape. This will require large-scale, inter-municipal planning, an approach NJSEA is equipped to initiate.

Hydrologic function in forested wetlands needs to be restored. Most sites were ditched or water was diverted long ago; some extant ditches are as deep as 1 m. Long-term changes in plant community composition will result if this is not corrected; however, increasing swamp hydroperiod (duration of flooding) will be adverse to some species as well as good for others. Management practices such as maintaining dead standing trees (i.e., snags) for cavity nesting birds, foraging birds, fungi, and other organisms, and coarse woody debris for small mammals, reptiles, salamanders, invertebrates, bryoids, and fungi should be implemented where dead wood is not a hazard to people.

Restoration of Atlantic white cedar has been suggested repeatedly and reportedly tried a few times although documentation seems to be lacking. Unfortunately, this iconic tree is sensitive to salinity, high nutrient levels, fire, logging, and deer browsing, and reestablishment with long-term survival would be extremely challenging if not impossible.

There are several nonnative plant species that are of concern in forests in the region. Colonization of forest understories and other upland habitats in the Meadowlands by species such as harlequin maple (*Acer platanoides*), Thunberg barberry, garlic-mustard (*Alliaria petiolata*), stiltgrass, and knotweed need to be assessed objectively to determine if important elements of biological diversity are threatened. For example, knotweed stands often have nesting songbirds, hence any management treatments should be performed after the bird breeding season or stands inspected for nesting birds prior to treatment.

Tree diseases and pest outbreaks are also a threat to trees in New Jersey (New Jersey Forest Service 2019). Asian long-horned beetle (*Anoplophora glabripennis*) infestation has resulted in the removal of large numbers of trees in the New York metropolitan region including Jersey City. Emerald ash borer (*Agrilus planipennis*) is killing the three native species of ash in the region and will result in large-scale die-offs of ash trees in forests and built areas of the Meadowlands. Extensive ash mortality will affect the composition of eastern forests on a scale similar to chestnut blight in the early 1900s. Spotted lanternfly (*Lycorma delicatula*) has spread into western and central New Jersey from Pennsylvania where it was first detected in 2014 and will likely colonize the Meadowlands. This sap-feeding planthopper prefers tree-of-heaven, an abundant, nonnative species in the Meadowlands, as host and will use many other woody species. The lanternfly is known to decrease survival in a variety of plant species including many native trees and agricultural crops. Trees in forests as well as street trees in the Meadowlands should be monitored for signs of these pests, as well as other tree pests and diseases, so that appropriate measures can be taken.

The Urban Matrix

The value of human-built habitats (including residential, industrial, commercial, transportation, and parkland) to maintaining biodiversity in urban areas is often underestimated. Many native species of plants, birds, mammals, invertebrates, and other organisms are able to survive and reproduce in human habitats. Furthermore, built environments are made much more valuable habitat for birds and other organisms by simply increasing the density of large native trees and shrubs. Neotropical migrant songbirds use large trees in neighborhoods and parks as stopover sites where they rest and feed, which makes management of shade- or street trees very important. Supplemental feeding of birds may aid in winter survival and productivity of songbirds (though this activity may have negative impacts on birds and also increase the density of rodents and other pests). Plantings of large pines and shrubs in landscaping provide refuges for birds and mammals in winter. Gardens provide flowering plants that are used as nectar sources by butterflies, bees, other pollinating insects, and hummingbirds. It is important to encourage planting of native plant species because they are adapted to local conditions so require less water and fertilizers, and in many cases native wildlife, especially some herbivorous insects, relies upon native plants. Managing stormwater on residential and business properties through green infrastructure (e.g., rain gardens) will greatly reduce polluted stormwater runoff from entering streams and rivers while also providing habitat for wildlife. Road culverts and bridges may provide passage for aquatic and terrestrial species if proper connections are maintained from one side of the structure to the other; this may require retrofitting of outdated culvert and bridge designs in critical wildlife corridors such as along Hackensack River tributaries. Many programs exist to guide home- and business-owners on ways to manage property to benefit wildlife and water quality (e.g., https://www.nwf.org/garden-for-wildlife/create).

In conclusion, there are various targets for improving natural and human-built upland and forest habitats in the Greater Meadowlands in order to protect and restore this important reserve for urban biodiversity and the ecological conditions that support it. Natural uplands and forested habitats including wetland and upland are relatively rare in the Meadowlands and should be priorities in conservation planning. In addition, human-built habitats and brownfields have great potential to provide habitat features for a variety of native species if a few best management practices are initiated by individual landowners.

Chapter 4

Seed Plants

Seed plants (the spermatophytes) are those plants that produce flowers and seeds for sexual reproduction. They do not produce spores as do the cryptogams (Chapter 5). Seeds are durable, protected, reproductive structures that in many species enable wide dispersal or persistence in the *seed bank* through adverse periods, in some species for many years. A few species do not actually produce, or rarely produce, flowers or seeds and instead reproduce asexually. Some spread extensively by underground *rhizomes* or aboveground runners (*stolons*) or by dispersal of fragments or specialized vegetative structures such as *bulblets*. The tiny, free-floating duckweeds (Lemnaceae) usually reproduce by budding off daughter *thalli* (plant bodies). Some species, such as common reed, spread both by seeds and by *vegetative propagules* (asexual structures for dispersal, such as rhizome fragments).

Seed plants include *woody plants*, which have parts that overwinter above ground and produce increments of tissue year after year, as well as winter buds that give rise to new twigs, leaves, and flowers the following growing season. Woody plants are the trees, shrubs, and woody vines. Seed plants also include *nonwoody* (*herbaceous*) plants that either die at the end of the growing season or overwinter beneath the ground or on its surface, and do not produce woody tissues nor have winter buds more than a few centimeters above the ground. Herbaceous plants include annuals that start from seeds each year, and biennials or perennials that live for more than one year. A few seed plants are intermediate between woody and herbaceous plants, being woody only at the stem bases and roots, or weakly woody higher up, and behaving in an ecologically intermediate fashion. Swamp rose mallow (*Hibiscus moscheutos*), knotweed (*Polygonum cuspidatum s.l.*), and purple loosestrife (*Lythrum salicaria*) are examples of semi-woody species that are prominent and important ecologically in the Meadowlands.

The *flora* is the array or list of plants of an area, and biologists sometimes also refer to the fungal flora, lichen flora, or bacterial flora, although those organisms are not plants. The flora is different from the *vegetation*, which

85

is the total plant cover of an area, often described by measures of plant frequency or abundance. Each U.S. state has a Natural Heritage Program (in
some states under a different name) that ranks the rarity of each native plant
species in the state based on the number of recently confirmed occurrences
of each species (see NJDEP 2019). The state ranks are S1 through S5, from
the most to the least rare statewide.

Our surveys were often relatively short, reconnaissance-type surveys in
which we observed a wide variety of groups of organisms; an exception was
Hudsonia's intensive two-month flora and vegetation study of the 445-ha
Berry's Creek area (Kiviat and Graham 2016). Therefore, we did not necessarily detect a plant at every locality where it occurred, nor did we identify
or document every plant species in the Meadowlands during the past two
decades. We endeavor to use common names of plants that are in regional
use in the New York metropolitan area (exceptions below). In Appendix 1,
we use scientific names from the nationwide USDA Plants Database (USDA
2021). Orthography of common names is not standardized, and we use
narrow-leaved rather than narrowleaf (and likewise for similar constructions).
Where part of a compound name refers to an unrelated taxon we hyphenate
(e.g., wild-rice instead of wild rice, because wild-rice is *Zizania* and rice is
the cultivated rice and its relatives, *Oryza*). We avoid combining words in
most cases, thus wild-rice rather than wildrice, and floating marsh pennywort
not floating marshpennywort. We also eschew geographic names for nonnative weeds that imply only part of the native range and suggest ethnic bias
(e.g., we use harlequin maple in place of Norway maple, box-leaved holly
instead of Japanese holly). Some of these common names are in common
usage, whereas a few are infrequent or modified for our purposes. We also
avoid a few names of native plants that contain ethnic references (e.g., wood
grass instead of Indian grass).

In this chapter we draw on specimen data from the New York Metropolitan
Flora Project (Brooklyn Botanic Garden 2010), observations from Torrey
Botanical Society field trips (NY-NJ-CT 2021), and our own collections
and observations. This chapter focuses on native species; details of the nonnative component of the flora will be discussed in a future book. Voucher
specimens from our surveys are in the Hudsonia—Bard College Field Station
Herbarium.

THE MEADOWLANDS FLORA

The documented flora of the Meadowlands constitutes 1065 taxa (species
or genera) of seed plants, including 287 historically reported taxa not documented in a century or more (Appendix 1). The year of the most recent record

known to us, in Appendix 1, demonstrates which species have not been documented in many years and are probably extirpated from the Meadowlands. The number of species grows whenever experienced botanists go in the field, and especially when they survey the less extensive or less typical habitat types such as the bedrock ridges (e.g., Laurel Hill) or outlying forested preserves. Some of these apparently missing species may yet be rediscovered. Since our studies began in 1999, 706 taxa have been found by Hudsonia and other botanists. Hough (1983) reported more than 2,600 species of vascular plants from New Jersey (including pteridophytes [ferns and fern allies], which are cryptogams and therefore excluded from Appendix 1; see Chapter 5).

Botanists sometimes analyze the numbers of species and proportions of the flora in certain plant families, especially those species-rich families such as the aster family, the grass family, and the sedge family. In Appendix 1, we count (including both native and nonnative species, historically or recently documented) 123 composites (Asteraceae, members of the aster family), 116 grasses (Poaceae), and 96 members of the sedge family (Cyperaceae). These three families typically contain the largest numbers of species in northeastern floras. The Meadowlands seed plant flora, irrespective of historic or current status, contains 118 trees, 101 shrubs, 39 woody vines, 37 herbaceous vines, and 770 other herbs. Likewise, the flora contains 735 native, 318 nonnative, and 12 taxa of uncertain or dual origin.

Although some species are missing and some groups of plants are poorly represented in the Meadowlands due to urban stresses and historic alterations including wetland fill and drainage, the urban environment may present advantages for certain species. Possibly the as-yet relatively low level of herbivory by white-tailed deer releases some species from grazing or browsing pressure. It may well be that the fragments of forested habitat have deer-sensitive species that would not thrive outside the urban area. In the Bronx River Parkway Reservation in Westchester County, New York, and in Van Cortlandt Park in New York City, there are forest wildflowers that appear to be thriving at low deer population levels. (The Westchester site is a fragment of habitat bordered by parkway, railroad, and river, the combination of which seems to discourage deer activity.)

The urban heat island effect, along with climate warming, although adverse for plants of northern affinities and cool microclimates, may be favorable for certain species of southern affinities, among them floating marsh pennywort (*Hydrocotyle ranunculoides*) for which the Meadowlands are a northeastern range outlier. Calcium in concrete, mortar, and gypsum potentially encourages *calcicolous* (calcium-associated) species. Thus, the urban environment has advantages as well as disadvantages for plants (and other organisms) that must be considered in analyzing the Meadowlands biota.

We are fortunate to have several historical surveys of the Meadowlands flora (Sipple 1972). Granted, different botanists or botanists at different time periods may focus on different taxa; ferns and orchids have always appealed to botanists, but interest in the sedge flora has only increased recently as other, easier-to-identify taxa became better understood. No one survey is complete. Yet a comparison of species collected recently (e.g., during the past 20 years) vs. species collected historically is intriguing. There are various factors that contributed to species loss: hydrological alterations, salinization, loss of the cedar swamps and other nontidal peatlands, destruction and fragmentation of remaining habitats, loss of natural upland soils, contaminants, and climate change. Since the times of Torrey (1819) and even Heusser (1949), many species have apparently disappeared with the loss of the cedar swamps, fens, and freshwater tidal wetlands, the destruction of much of Laurel Hill, and the general intensification of land use. Many of the species reported by Torrey (1819) that have not been found in recent years are species of undisturbed habitats some of which were cool sites (such as wet forests, cedar swamps, and bogs). Many apparently new species have been found in the past two decades; some of these are recently arrived nonnative species, whereas others are native species that may have taken advantage of the changing climate, dispersal pathways, and newly created habitats. Of course, many species such as sedges and dodders, and small plants in general, may have been overlooked historically or recently, because they are localized and hard to find and identify.

Many species with affinities for northern regions or cool microclimates were recorded historically but not recently; for example, wild calla (*Calla palustris*), bristly sarsaparilla (*Aralia hispida*), yellow birch (*Betula alleghaniensis*), twinflower (*Linnaea borealis*), goldthread (*Coptis trifolia*), red-berried elder (*Sambucus racemosa*), bunchberry (*Cornus canadensis*), small cranberry (*Vaccinium oxycoccos*), and several orchids. Somewhat surprisingly, some species of southern or coastal affinities have also disappeared, for example, goldenclub (*Orontium aquaticum*), Virginia snakeroot (*Aristolochia serpentaria*), strawberry bush (*Euonymus americana*), man-of-the-earth (*Ipomoea pandurata*), Collins' sedge (*Carex collinsii*), marsh fimbry (*Fimbristylis castanea*), and butterfly pea (*Clitoria mariana*). Goldenclub presumably disappeared with most of the nontidal peatlands and fresh-tidal marshes, the habitats in which it occurs in the Hudson Valley and southern New Jersey. Virginia snakeroot probably occurred on Laurel Hill, which has some similarity to localities where this species has been found in New York. Collins' sedge is present in cedar swamps in southern New Jersey (William Standaert, Salisbury University, pers. comm.).

Some of the species recorded historically but not recently are associated with freshwater organic soils or acidic freshwater habitats, such as Virginia chain fern (*Woodwardia virginica*; a single plant was still present a few years ago), poison-sumac (*Toxicodendron vernix*), bog goldenrod (*Solidago uliginosa*), round-leaved sundew (*Drosera rotundifolia*), and large cranberry (*Vaccinium macrocarpon*). Not only are these habitats gone, but the extirpated plants were probably sensitive to salinity and high nutrient levels as well. However, the reasons for many deletions and additions are unclear. A remnant, freshwater, peatland habitat dominated by common reed and peat mosses, discovered in 2015 in Upper Penhorn Marsh, may be found to support some of the missing species of fens or bogs.

Which taxa are currently underrepresented by native species? The following are some of the families or genera that appear to be *depauperate* (poor in species and abundances) compared to a hypothetical, rural, northeastern region of similar size: the sedge family, bur-reeds (*Sparganium*; no species reported), bedstraws (*Galium*), willow-herbs (*Epilobium*), willows (*Salix*), violets (*Viola*), bur-marigolds (*Bidens*), orchids (Orchidaceae), and the conifers (several genera). It is among these and similarly depauperate taxa that we should look for indicators of urbanization stress. On the other hand, taxa such as the nettle family (Urticaceae) with 5 species, the parsley family (Apiaceae) with 19 species, the cashew family (Anacardiaceae) with 8 species, the grass family with 116 species, and the aster family with 123 species, are well represented. The numbers of species, of course, do not tell the whole story. Many native plants are scarce in the Meadowlands despite being common elsewhere in surrounding regions. These scarce plants either occur at very few localities or form very small stands. We found sassafras (*Sassafras albidum*), fragrant sumac (*Rhus aromatica*), and New York ironweed (*Vernonia noveboracensis*) in very few places, and there were few stems at each occurrence.

Not all weeds are nonnative species. There are native plants that are weedy in the sense of exploiting human-created or altered habitats, reproducing and growing rapidly, and sometimes displacing less competitive or less common native plants. Some of these weedy (or *ruderal*) natives are common evening primrose (*Oenothera biennis*), switchgrass (*Panicum virgatum*), saltmarsh-fleabane (*Pluchea odorata*), common milkweed (*Asclepias syriaca*), hemp dogbane (*Apocynum cannabinum*), some of the smartweeds (*Polygonum s.l.*), pilewort (*Erechtites hieraciifolius*), and goldenrods such as rough goldenrod (*Solidago rugosa*). Many of these species are large herbs (1–2 m tall), and some grow in dense patches or abundantly mixed with other plants.

SEDGE FAMILY (CYPERACEAE)

Sedges do not attract the attention of amateur botanists and naturalists the way that ferns, orchids, and trees do. Many of the *true sedges* (*Carex*) are difficult to identify, even for experts, and they must be collected when the fruits are ripe or nearly ripe but before the fruit clusters have shattered. Nonetheless, compared to many regional floras, the sedge family and especially the true sedges are underrepresented in the Meadowlands flora. Torrey (1819) reported 4 true sedges, Britton (1889) reported 11 species, Harshberger and Burns (1919) reported none, Sipple (1972) reported none from his own field surveys, and our 2002 Metropolitan Flora list contained 4 species (Kiviat and MacDonald 2002). Our current plant list (Appendix 1) contains at least 48 true sedges, including historical records and those reported through 2021. We were unable to search every likely habitat for sedges or return to every site to collect them at the right stage of maturity for identification, so we think there are undocumented species in wet woods, swamps, wet meadows, dry rocky crests (in places like Laurel Hill), Mehrhof Pond, remnants of the meadows west and southwest of Harrier Meadow, and Teterboro Airport Woods. Hough (1983) reported 135 species of true sedges for New Jersey, 68 of them from Bergen and Hudson counties (but many of the Bergen-Hudson sedge records were based only on old collections). Buegler and Parisio (1982) reported about 54 true sedges for nearby Staten Island including historical records, although only 11 were known extant in 1981. (The Staten Island numbers seem unreasonably small and may be due to an incomplete compilation.)

Many true sedges are specialized to *oligotrophic* (low nutrient) wet or dry habitats. Also, many of the true sedges seem to have limited seed dispersal. Numerous forest sedges (as well as other forest herbs) have ant-dispersed seeds (see Gaddy 1986) and may not readily travel long distances. Most wetland sedges are probably dispersed by water or ice. That very few true sedges are considered nonnative species is probably testament to the limited dispersal and competitive abilities of plants in this genus. Tussock sedge (*Carex stricta*), one of the most frequent and abundant species in the Hudson Valley and other rural areas, is so rare in the Meadowlands that we have only seen a few plants at Hackensack River County Park (Hackensack) and Losen Slote Park (Little Ferry). This is all the more interesting because tussock sedge is resistant to fire, livestock grazing, and fluctuating water levels (Costello 1936). We do not know what limits it in the Meadowlands. Most of the extant true sedges are present in very limited numbers of localities and individuals.

Beyond the true sedges, the bulrushes (*Scirpus, Bolboschoenus,* and *Schoenoplectus*) are also poorly represented in the Meadowlands (Figure 4.1). Although numerous species were reported historically (Appendix 1),

Figure 4.1 Wet meadow and temporary pool with threesquare (*Schoenoplectus*), **Losen Slote Creek Park. Photo 2021 by Erik Kiviat.**

we found most of those species to be quite rare. Other Cyperaceae include one species, coast flatsedge, that may be *Cyperus polystachyos* var. *texensis*, which is ranked S1 Endangered in New Jersey. Coast flatsedge was collected at the Kane Natural Area before mitigation; we do not know if it persists there. The genus *Cyperus* includes many species that are weedy, urban-tolerant, or nonnative as well as some rare species; we found 16 species of *Cyperus* overall. Lastly, 12 spike-rushes (*Eleocharis*) have been recorded in the Meadowlands. The tiny dwarf spikerush (*Eleocharis parvula*) volunteers abundantly in the tidal wetland mitigation projects.

THE GRASSES (POACEAE)

The grasses are generally weedier and better adapted to habitat disturbance and nutrient enrichment than the true sedges. Grasses are moderately diverse in the Meadowlands, with 116 native and nonnative species.

Bluejoint (*Calamagrostis canadensis*) is a native grass of somewhat northern affinities. Half a century ago, a bluejoint-dominated plant community occupied scattered wet meadows in the northern Meadowlands (Sipple 1972). We found bluejoint only in the Kane Natural Area where small patches of bluejoint meadow about 5–10 m in diameter were common in 10 ha of diked marsh within 100 m of the west side of the Paterson Lateral Gas Pipeline. In 2007, when the berms separating this wetland from the Hackensack River were breached by a storm, brackish water flooded the bluejoint meadows and may have eventually killed all the bluejoint. (This area was subsequently treated with herbicide to kill common reed and modified in additional ways to create cordgrass tidal marsh.) Bluejoint may have been declining even earlier due to continued salinization and competition with common reed. Bluejoint is fairly fire-tolerant (Tesky 1992); perhaps fires started by ORVs or their riders, or by vandals, helped this species persist amid the common reed matrix near a road with much ORV use. Bluejoint might be a suitable species for use in wetland management projects if it is not at high risk of being marginalized by climate warming, salinization, or fire suppression.

Wild-rice (*Zizania*) was apparently once abundant and widespread in the Meadowlands (Chapman 1900, Brooks 1957); it was found as recently as 1949 by Heusser (1949). The Metropolitan Flora reported two species of wild-rice, and listed a 1948 specimen (Appendix 1). A specimen of wild-rice was collected in 2019 in Carlstadt; it awaits specific identification. Wild-rice is sensitive to salinity, sulfur salts, turbidity, animal consumption of seeds and shoots, some herbicides, and possibly high temperatures (Vennum 1988, Fort et al. 2014, Myrbo et al. 2017). It is a very important wildlife food species, especially for southward-migrating ducks, rails, and blackbirds in late summer and fall. On the Hudson River, wild-rice is a characteristic middle-intertidal zone plant of freshwater tidal and slightly brackish tidal marshes (Kiviat et al. 2006). It was common in the Hudson River in the 1930s and 1940s, had declined substantially by the early 1970s, and after herbicide (2,4-D) spraying of water-chestnut (*Trapa natans*) ceased in 1976, wild-rice increased. Probably this taxon was abundant in the freshwater tidal marshes that would have been common in the Meadowlands before the Oradell Dam increased salinity. The nearest major potential sources of wild-rice propagules (seeds or live plant bases) are probably the Hudson River at Manitou Marsh near Garrison, and the freshwater tidal marshes of southern New Jersey, but many kilometers of waters that are too saline for this species lie between those potential sources and the Meadowlands. Seeds might arrive accidentally with plant materials used in mitigation projects; however, seeds must be stored in water to maintain germinability (Muenscher 1936). Not only are current conditions of salinity and water quality largely unsuitable for wild-rice, but its germination requirements and attractiveness to herbivores make it a poor candidate for restoration.

The cordgrasses are an important genus in the Meadowlands. Only 60 years ago there were four species: smooth cordgrass (*Spartina alterniflora*), saltmeadow cordgrass (*Spartina patens*), big cordgrass (*Spartina cynosuroides*), and freshwater cordgrass (*Spartina pectinata*). Smooth cordgrass is still common and is the dominant plant in the upper intertidal zone in the more brackish marshes (often referred to as salt marshes). Saltmeadow cordgrass was the dominant plant in the supratidal or high marsh in the more brackish marshes, and presumably was the valuable salt hay species historically. Now saltmeadow cordgrass is declining and is mostly limited to small patches at the more saline sites (and a few places where it has been planted). Big cordgrass was probably common in the Meadowlands in the early 1900s and was an important feature of the area around the Atlantic white cedar swamp that Heusser (1949) studied in Secaucus. Big cordgrass resembles common reed at a distance and might be hard to spot among the nearly ubiquitous reedbeds of the Meadowlands. In 2015, big cordgrass was found in small numbers in the Berry's Creek area (Kiviat and Graham 2016). Freshwater cordgrass was also present in Heusser's (1949) study area but is probably gone from the Meadowlands. These two species are also quite rare on the Hudson River estuary and may be sensitive to high nitrogen levels or another urban factor.

Switchgrass (*Panicum virgatum*) on the East Coast is commonly found in a border at the upper edge of the salt marsh. Stuckey and Gould (2000) stated that it occurred landward of permanently saline habitats. Switchgrass is tall and can form dense clumps. We found this species common in the Meadowlands on wetland fill and other habitats, but not in shady or frequently flooded areas. Recently, switchgrass has been touted as a bioenergy crop that can be grown on poor soils; this might be an option in the Meadowlands on capped landfills. Many other native and nonnative grasses also occur in the Meadowlands.

CATTAILS (TYPHACEAE)

The cattails are important marsh plants of freshwater and brackish wetlands (up to about one-third seawater concentration) in our region. We have seen little cattail in the Meadowlands. Most of the cattails we examined in the field appeared to be hybrid cattail (*Typha ×glauca*), the hybrid of broad-leaved cattail (*Typha latifolia*) and narrow-leaved cattail (*Typha angustifolia*), based on naked-eye morphology. This hybrid can resemble either of the parent species or may have an intermediate appearance. The cattails are a *hybrid swarm* displaying a spectrum of morphology and genetics from one parent species to the other (Witztum and Wayne 2015).

Narrow-leaved and hybrid cattails dominate the upper intertidal zone of many fresh and brackish Hudson River marshes, with dominance by hybrid cattail typical of the more altered marshes (Mihocko et al. 2003). Kerry Barringer (pers. comm.) has observed that narrow-leaved cattail is more frequent in disturbed or artificial habitats, whereas broad-leaved cattail is more frequent in the more stable habitats. McCormick (1970) found narrow-leaved cattail more common in tidal habitats and broad-leaved cattail more in diked habitats. A few decades ago, cattail was the dominant marsh plant in large areas of the brackish Croton Point and Croton River tidal marshes on the Hudson in Westchester County. Cattail died off and was replaced by common reed, possibly due to leachate from the 40 ha garbage landfill on Croton Point. It is unclear how abundant or dominant cattails were in Meadowlands marshes historically. Perhaps Meadowlands cattails have declined due to contamination from landfill leachate or other sources, as well as competition from common reed. Heusser (1949) reported both narrow-leaved and broad-leaved cattails at his Secaucus study area, although narrow-leaved cattail was much more common. Cattails may have difficulty establishing in some areas (as more than very small patches) because of feeding by muskrats, which generally prefer hybrid over narrow-leaved cattail.

Both broad-leaved cattail and narrow-leaved cattail have usually been regarded as native to the United States. Some ecologists have asserted that narrow-leaved cattail was introduced from Eurasia (e.g., Galatowitsch et al. 1999). The species has been identified in pre-Columbian pollen records at a number of northeastern U.S. sites and was believed present in low abundance prior to European colonization (Shih and Finkelstein 2008). There is still uncertainty about the origin of northeastern narrow-leaved cattail (Bansal et al. 2019) and even its taxonomy (Ciotir and Freeland 2016). The pollen of narrow-leaved cattail and broad-leaved cattail differs in that the former species has single pollen grains (*monads*) whereas the latter has pollen grains joined together in fours (*tetrads*). Narrow-leaved cattail pollen resembles bur-reed pollen; thus, additional work on the history of narrow-leaved cattail in North America would be desirable. Given the mobility of the wind-dispersed seeds and pollen of all cattails, it seems unlikely that narrow-leaved cattail would have stayed put in Europe or the northeastern states before European settlement of the U.S. Selbo and Snow (2004), working in Ohio, did not diagnose hybrids on the basis of DNA where both parental species occurred in close proximity, and opined that hybridization events may not be common.

THE COMPOSITES (ASTERACEAE)

Many composites are weedy and urban-tolerant. Many species also have showy flower heads and provide nectar and pollen to flower-visiting insects. Ten species of goldenrods have been reported in the Meadowlands. Seaside goldenrod (*Solidago sempervirens*) is tolerant of salinity, fire, coarse and medium-textured soils, and fluctuating soil moisture, does not require high soil fertility, and is intolerant of shade (Snell 2010). This species is also urban-tolerant; for example, it is common on the estuarine shorelines of New York City. The large, slightly succulent goldenrod was common in the Meadowlands but apparently nowhere abundant. On Staten Island, Cheplick and Aliotta (2009) found that seaside goldenrod appeared to inhibit density of two native grasses, and size of one of them, within 2 m; they suspected *allelopathy* (release of chemicals toxic to other plants). In the Meadowlands, other goldenrods were common on dry wetland fill and the drier portions of formerly tidal wetlands; rough goldenrod (*Solidago rugosa*) may be the most important species.

Tall sunflowers (*Helianthus annuus, Helianthus giganteus,* and probably other species or hybrids) occurred at the edges of garbage landfills, near stormwater ponds, and on dry wetland fill. Several large and conspicuous white- and purple-flowered species of thoroughworts, bonesets, and Joe Pye weeds (*Eupatorium* and *Eutrochium*) also occurred. All of these plants provide resources for flower-visiting insects. New York ironweed, superficially similar to Joe Pye weeds, is a tall (up to 2 m) forb of wet meadows that was rare in the Meadowlands and generally uncommon in the surrounding region.

Saltmarsh-fleabane was common to abundant in brackish tidal marshes (including recently managed tidal marshes such as Skeetkill Marsh in 2006), a few shallow stormwater ponds (e.g., at the northern foot of the 1-E Landfill), and on floating peat masses in slightly brackish, nontidal Kearny Marsh West where patches of common reed had died. The flowers of this plant, like those of many composites, are attractive to flower-visiting insects and are probably an important resource for pollinators in the Meadowlands.

We found giant ragweed (*Ambrosia trifida*) at a single location (the Paterson Lateral Gas Pipeline at the Kane Natural Area before mitigation). Kerry Barringer (pers. comm.) found it along the abandoned railroad right-of-way west of Kearny Marsh West, and farther south along Bergen Avenue. This plant can exceed 2 m in height. McCormick (1970) found it locally abundant in Tinicum Marsh (now John Heinz National Wildlife Refuge), an altered and disturbed, fresh-tidal, urban site in Philadelphia on the Delaware River estuary. At Tinicum, giant ragweed flourished in the uppermost intertidal zone and in the supratidal zone, and on acidic soils of old mosquito ditch spoil or old garbage fill. This species was rare in the freshwater tidal wetlands at Tivoli North Bay on the Hudson River (1970s–2010s), although it occurred in occasional large patches in the nontidal, drained peatland of the Black Dirt wetlands in Orange County, New York, in the 2010s. We were surprised not to see this species more often in the Meadowlands. Possibly it is limited by bird predation of the seeds (see Kiviat 1982a).

All 10 species of the Joe Pye weeds, bonesets, and thoroughworts recorded in the Meadowlands have been found during the past two decades. Contrast this with the 6 species of bur-marigolds and beggarticks (genus *Bidens*), only

3 of which have been found since 1990. It would be interesting to know if this is an artifact of survey effort, or if the non-detection of these species is repeated in New York City and other urban areas (also, some *Bidens* species are difficult to identify). In 2017–2019, excepting the ambiguously native Spanish needles (*Bidens bipinnata*), *Bidens* spp. were rare at two estuarine shoreline sites Hudsonia studied in New York City. Oddly, Spanish needles, as far as we know, has not been collected in the Meadowlands since 1895 (Appendix 1). We discuss a composite vine, climbing hempweed, and two composite shrubs, groundsel-bush and high-tide bush, later in this chapter.

ORCHIDS (ORCHIDACEAE)

We have seen no native orchids in the Meadowlands, and the nonnative helleborine (*Epipactis helleborine*) very rarely. Torrey (1819) reported 9 species of native orchids, Britton (1889) reported 10 species, Harshberger and Burns (1919) reported 3 species, and Sipple (1972) found none. Overall, 16 native orchids were found historically in the Meadowlands. Interestingly, Brown (1997) stated that Atlantic white cedar swamps in New England, due to their acidic soils, supported few orchids; thus, many of the species historically present in the Meadowlands may have been associated with other habitats, such as fens and hardwood swamps. Of the 16 orchids known historically from the Meadowlands, 5 were associated with upland forests in New England, and 10 were wetland species (Brown 1997). New York City had at least 30 orchids historically, but in recent years only 8 have been documented (DeCandido et al. 2004).

Most orchids have tiny, wind-dispersed seeds (Dressler 1981). Orchids should be able to disperse into the Meadowlands from rural areas and wildlands within several kilometers. However, few northeastern orchids are adapted to a *ruderal* (roadside or disturbed-area) life, orchids have a critical need for a fungal association immediately following germination, most are sensitive to fire and logging, they do better in less-fertile soils, and may require particular species of pollinators absent from fragmented habitats (Dressler 1981, Gilbert 1989, Sheviak and Young 2010). All these characteristics evidently add up to pronounced urban-sensitivity.

In spite of the presumed urban-intolerance, orchids are sometimes found in altered or degraded habitats. Soil mines, roadside ditches, powerline rights-of-way, oldfields, and pastures are habitats for several orchid species (Brown 1997, Sheviak and Young 2010). Brown also mentioned the Lynnfield Marshes near Boston as a productive locality for orchids; these wetlands were degraded by a golf course, hydrological alterations, and other urban-fringe influences when studied in the late 1980s (Tashiro et al. 1991).

Brown (1997) implied that the dense human population on Long Island, New York, was a negative influence on orchids, although the less-developed eastern end of the island had high orchid diversity.

OTHER WOODLAND, RIPARIAN, AND WETLAND FORBS

Apart from woodlands developing gradually on garbage landfills and wetland fill, most of the forests of the Meadowlands are partly-drained swamp forests or low-lying riparian forests. All of these wet forests occur in small fragments, with the exception of the Teterboro Airport Woods. The largest patch there covered ca. 60 ha in 2018 but is threatened by airport expansion. Forest fragments are more affected by external influences such as drying winds and colonization by weeds compared to extensive, unbroken forests. This fragmentation may have eliminated, or prevented colonization by, specialized pollinating insects and some of the plants themselves. Some forests, including the Teterboro Woods, were probably farmland until a century ago, and post-agricultural colonization by the herbaceous plants may be lagging behind that of trees. Woodland herbs are generally slow to colonize (Gilbert 1989). As with the woodland sedges, many forest forbs are adapted for ant dispersal (Gaddy 1986) and may have difficulty traveling long distances. Even corridors, such as fencerows, may not facilitate dispersal well, because some forest interior herbs disperse very slowly (e.g., less than 5 m/year; Jolls [2003]). Fragmentation of forest habitats may increase predation of seeds by rodents and competition from shade-intolerant native or nonnative plants (Jolls 2003). Many forest herbs can recover following fire, depending on fire intensity, the survival of vegetative parts or seeds below ground, and other factors (Roberts and Gilliam 2003). Yet shade-tolerant forest herbs may take many years to regain diversity after logging or other severe disturbance (Gilbert 1989, 81; Roberts and Gilliam 2003). Following agriculture, forest vegetation may develop relatively rapidly, but recruitment of many herbs and other plants may lag behind the dominant trees (Flinn and Vellend 2005, De Frenne et al. 2011).

In spite of these problems, several forbs typical of northeastern upland forest or swamp forest interiors occurred locally in the northern Meadowlands. These included skunk-cabbage (*Symplocarpus foetidus*) at several sites. Teterboro Woods had wild sarsaparilla (*Aralia nudicaulis*) and Canada mayflower (*Maianthemum canadense*). Losen Slote Park had may-apple (*Podophyllum peltatum*) and abundant trout-lily (*Erythronium americanum*). In addition to skunk-cabbage, Wolf Creek had false-hellebore (*Veratrum viride*), may-apple, trout-lily, and false Solomon's-seal (*Maianthemum racemosum*). Lizard's-tail (*Saururus cernuus*) and great blue lobelia (*Lobelia siphilitica*), not forest herbs strictly speaking, occurred at a wooded spring in Ridgefield Park but nowhere else; the locality appeared to be isolated from the estuary. Species present in outlying Borg's Woods (Hackensack) that were not otherwise in the Meadowlands included dwarf ginseng (*Panax trifolius*), spring beauty (*Claytonia caroliniana*), and springcress (*Cardamine bulbosa*). Many of these forbs seem to do well in habitats referred to as rich, apparently meaning rich in both cationic nutrients (especially calcium) and anionic nutrients (nitrogen, etc.). If certain woodland wildflowers are absent from the Meadowlands due to dispersal limitations, it might be possible to introduce them to suitable forest interior habitats. The current (so far) low density of deer populations could be favorable to these herbs. However, if pollution, trampling, herbivory, excess nitrogen, or lack of pollinators is the limiting factor, restoration of these species could be difficult or impossible.

Wet meadows, of limited occurrence in the Meadowlands, supported a few noteworthy forbs including Canadian burnet (*Sanguisorba canadensis*), swamp lily (*Lilium superbum*), and swamp candles (*Lysimachia terrestris*). (The first two, at least, appear to be regionally-rare and were in Little Ferry; the last was in Moonachie.)

Swamp rose mallow (*Hibiscus moscheutos*) is a robust, semi-woody plant with clustered stems and gigantic pink or white flowers. It was fairly common in nontidal fresh and mildly brackish tidal marshes of the Meadowlands. Like purple loosestrife and tussock sedge, swamp rose mallow is one of the few semi-woody or herbaceous plants that builds large, elevated bases in habitats with fluctuating water levels. These *hummocks* or *tussocks* are root crowns or masses of roots and short rhizomes enhanced by the accumulation of dead stems and leaves and may support other plants (and animals) that require slightly elevated substrates. In turn, swamp rose mallow is one of the plants that has established on old stumps of Atlantic white cedar (such as in Upper Penhorn Marsh).

Tidewater-hemp (*Amaranthus cannabinus*) is an amaranth that occurred in natural marshes and has volunteered in mitigation projects. It tends to grow along tidal creek banks or in areas of disturbance such as around muskrat lodges.

We did not find sea-lavender (*Limonium*) in the Meadowlands, and it was not reported historically (but see unpublished report, Chapter 2). There are two species in the general region, *Limonium carolinianum* and *Limonium nashii* (considered a single species, *L. carolinianum,* by USDA [2021]). Buegler and Parisio (1982) considered *L. nashii* common on Staten Island historically (1879, 1930) but rare in 1981. This species is a minor component of the high salt marsh assemblage along the northeastern coast and may be an indicator of the biotic integrity of that community (as may the native plant species diversity overall in the high salt marsh). We do not know if sea-lavender is sensitive to pollution, if it was overharvested for dried flower arrangements (as *statice*), or if the fragmentation of the high salt marsh community by hydrological changes and common reed dominance resulted in habitat units that were too small for this plant, which was perhaps always uncommon. In Nova Scotia, sea-lavender was overharvested near roads, greatly reducing reproduction of this species whose seeds do not disperse far (Baltzer et al. 2002).

Orache (*Atriplex prostrata*) is another plant of brackish and sometimes freshwater marshes. It has a broad habitat niche (Stuckey and Gould 2000) and often occurs in Meadowlands reedbeds (where it is one of the few regularly associated plants) as well as in disturbed tidal marsh edges.

Floating marsh pennywort is a rare species (S3 Threatened) in New Jersey but is locally abundant in the Meadowlands in at least four known localities. It is at the northeastern limit of its geographic range. An unidentified insect mines the leaves, making complex patterns of irregular lines. Floating marsh pennywort has been introduced widely in Europe and is considered invasive or potentially invasive there; it benefits from high nutrient levels and warm temperatures (EFSA 2007). Taylor (1915) reported this species only as far north as southeastern Pennsylvania and not at all in New Jersey; it may have extended its range northward since then or simply been unrecorded previously.

In 2019, we found another state-listed plant in Elizabeth near the western shore of Newark Bay on probable old fill. Torrey's rush (*Juncus torreyi*) is ranked as New Jersey S1 Endangered.

WILDFLOWERS OF ROCKY CRESTS

The mostly shallow, dry, probably nutrient-poor soils on the predominantly diabase bedrock of disturbed rocky crest habitats at Laurel Hill, Little Snake Hill, and Granton Quarry constitute a very different habitat complex than the wetlands and wetland fill soils. Some of the interesting and perhaps regionally-rare plants there were pale corydalis (*Corydalis sempervirens*), wild-coffee (*Triosteum perfoliatum*), starry campion (*Silene stellata*), and blue toadflax (*Nuttallanthus canadensis*). Wild-coffee was scattered around the ridgetop at Laurel Hill, but the largest number of plants was on the abandoned quarry floor where much of the habitat was lost to expanding park facilities during 2006–2012. The diabase uplands probably support other species of conservation concern.

SUBMERGENT PLANTS

Submergent plants, often called *submerged aquatic vegetation*, are species that are underwater most of the time, although they may survive stranded on mudflats or shores at low tide or during droughts, and many species have emersed flowers that are pollinated in the air. Submergent plants are considered important components of northeastern estuaries. In the Connecticut River estuary, larger and less urbanized than the Hackensack River estuary, the presence of submergent plant beds was considered a positive indicator of water quality (Rosza et al. 2001). In the Hudson River estuary, submergent plant beds of wild-celery (*Vallisneria americana*, not known in the Meadowlands) provided a benefit to fish by increasing the dissolved oxygen content of the surrounding water (Caraco et al. 2006). Many submergent species are important foods for water birds (Martin et al. 1951), and are also eaten by turtles, muskrat, and other animals. Submergent plant beds provide crucial habitat for fishes and aquatic invertebrates.

The Meadowlands supported few species of submergent plants, and the beds seemed very restricted in extent. The only native submergents we found were horned-pondweed (*Zannichellia palustris*), leafy pondweed (*Potamogeton foliosus*), common coontail (*Ceratophyllum demersum*), Nuttall's waterweed (*Elodea nuttallii*), and wigeon-grass (*Ruppia maritima)*. We found all of these plants except wigeon-grass in the slightly brackish waters of Overpeck Creek. Leafy pondweed occurs in all 50 U.S. states and all but 2 Canadian provinces (USDA 2021). It was the only pondweed found in a study of 53 ponds on southwestern Staten Island (Hendricks and Behm 1976). Only one other pondweed, curly pondweed (*Potamogeton crispus*, nonnative), has been reported from the Meadowlands, although rural areas have many additional species in this large genus. Horned-pondweed also occurred in brackish waters at Mill Creek and Kingsland Impoundment; at the latter site it varied greatly in abundance year-to-year (Kyle Spendiff, formerly New Jersey Meadowlands Commission, pers. comm.). This is a widespread and common plant in both fresh and brackish waters and is a good waterfowl food (Martin et al. 1951).

In our limited experience, Nuttall's waterweed was uncommon in Overpeck Creek (and we have not seen common waterweed [*Elodea canadensis*] in the Meadowlands). In the freshwater tidal Hudson River, both are present but common waterweed is much more common than Nuttall's waterweed (Kiviat, pers. obs.). Oddly, Beal (1977), based on North Carolina data, attributed narrow affinities for water quality and soil organic matter to common waterweed and rather broad affinities to Nuttall's waterweed (perhaps due to a smaller sample of localities for common waterweed or the more southern location).

Submergent plants are adversely affected by high turbidity and appear to respond to sedimentation (erosion and deposition) patterns; turbidity may prevent development of submergent plant beds in the main channel of the Hackensack River.

The nonnative species hydrilla (*Hydrilla verticillata*) was reported in 2019 for the first time in the Meadowlands, forming dense beds in Overpeck Creek (Fallon 2019a). Identification was confirmed by Emily Mayer (SOLitude Lake Management). This plant is a pest by virtue of interfering physically with boating and fishing (and, in other water bodies, swimming and drinking water intakes). Prior to the establishment of hydrilla, we found submergent plants scarce in Overpeck Creek.

TREES AND SHRUBS

The trees, shrubs, and woody vines of the Meadowlands constituted a diverse assemblage of native and nonnative species adapted to a variety of wet and dry habitats and including many highly urban-tolerant species as well as species that are only moderately urban-tolerant. Our plant list (Appendix 1) includes 258 woody species, comprising 118 trees, 101 shrubs, and 39 woody vines. Of the 258 species, 170 are native and 86 nonnative with 2 indeterminate. This woody flora comprises 67 native and 50 nonnative trees, 80 native and 20 nonnative shrubs, and 23 native and 16 nonnative woody vines. (See below on the ambiguity of the shrub-woody vine categorization.) A few of these species are based on historic records and no longer occur in the Meadowlands (e.g., Atlantic white cedar, black spruce), but the majority of woody species that has been found in the Meadowlands still occurs there. Overall, the Meadowlands had, and still has, a large diversity of woody plants.

From a floristic viewpoint, based solely on the numbers of species recorded, the woody flora of the Meadowlands is predominantly native (about 2:1 native to nonnative species); however, several of the nonnative woody plants are very abundant. Yet the diversity of native species, many of which occur in small woodland or scrub patches or in borders along highways or water edges, is a valuable resource. Woody patches such as these provide seeds for spontaneous revegetation of altered habitats (Robinson and Handel 1993, Handel 2013) and resources for specialist invertebrates. Here we discuss a few of the noteworthy species.

We found native, non-planted conifers virtually absent from the Meadowlands. Even eastern red cedar (*Juniperus virginiana*) was rare except where planted. White pine (*Pinus strobus*) occurred at a very few places. Both species are abundant in the Hudson Valley of New York. Unlike white pine, red cedar is fairly salt-tolerant and is common just inland of the salt marshes on rural northeastern coasts. Pitch pine (*Pinus rigida*), a common species of acidic rocky crests in the Hudson Valley and dominant in the Pine Barrens of southern New Jersey, is absent from Laurel Hill and Little Snake Hill—perhaps because the soil pH is too high or the air quality insufficient, or because there was no seed source to repopulate after institutional use and quarrying ceased. Nonnative pines have been planted widely, such as along the Turnpike. A single American yew (*Taxus canadensis*) was found at the Sports Complex where it may have escaped from cultivation (Kiviat and Graham 2016).

Hackberry (*Celtis occidentalis*) is an urban-tolerant tree that is usually small (less than about 20 cm *diameter-at-breast-height* [dbh]). Hackberry in the Hudson Valley is typically associated with calcareous soils and this species may benefit from the calcium derived from clay, concrete, mortar, and brick in the Meadowlands. This is a tree of uplands and wetland edges and was uncommon but widespread in the Meadowlands. Its fruits, which often stay on the twigs into winter, have a human-edible pulp and are potentially important food for wild mammals and birds. Hackberry leaves are the sole diet of the larvae of three rare butterflies, all found in the Meadowlands (see Chapter 10).

Black walnut (*Juglans nigra*) is fire-resistant, wind-firm, and shade-intolerant, and grows in a variety of non-wetland habitats but does best on deep, well-drained, circumneutral, moist, fertile soils (Fowells 1965). This tree is strongly allelopathic and many other plants are harmed by contact with its roots (*ibid.*). Black walnut is undergoing a population increase in the Hudson Valley that may also occur in the Meadowlands in the near future. We have seen black walnut in at least Oradell, Hackensack, Teaneck, and Carlstadt. Black walnut was characterized as one of the most valuable eastern timber trees (Brown 1975). Thefts of this tree have occurred in public parks in the United States and should be guarded against in the Meadowlands. Butternut (*Juglans cinerea*; New Jersey S3, NJDEP 2019) is closely related but much less common in rural areas and likely less urban-tolerant. We have seen it only in Carlstadt and Teaneck. Butternut is vulnerable to mortality caused by a canker fungus (Ostry et al. 1996) and may be declining in the northeastern states. Fruits of both butternut and black walnut are edible to humans and rodents that can chew through the thick shells. Butternut hybridizes with the nonnative *Juglans ailantifolia* (Zhao and Woeste 2011), and Meadowlands occurrences of *Juglans* should be checked for hybrids.

The oaks are important in the Meadowlands, and most oaks can reach large size there (about 60+ cm dbh). We found pin oak (*Quercus palustris*), and to a lesser extent, swamp white oak (*Quercus bicolor*), common in the wet woods and swamp forests. Chestnut oak (*Quercus prinus*) was prominent on parts of Laurel Hill especially in areas that appear not to have been quarried. Red oak and white oak were common trees on upland soils in general; black oak was less common. Scarlet oak (*Quercus coccinea*), sometimes difficult to identify in the field, seemed to be rare. There was a small population of post oak (*Quercus stellata*) on Little Snake Hill; this southern species is uncommon in the New York City region.

Pignut hickory (*Carya glabra*) occurred at Little Snake Hill. We found shagbark hickory (*Carya ovata*) rare on Laurel Hill. Both mockernut (*Carya tomentosa*) and bitternut (*Carya cordiformis*) were documented in the 1990s by the New York Metropolitan Flora. Pignut and mockernut were confirmed recently on Laurel Hill (Gerry Moore *fide* Kerry Barringer).

Several maples were also important in the Meadowlands. Red maple (*Acer rubrum*) was very widespread and often common in swamp forests, wet woods, and other wet habitats and also occurred in upland habitats, as is typical of this species. Silver maple (*Acer saccharinum*) is usually a riparian and floodplain tree, and it was uncommon in the Meadowlands. There was at least one large individual at Teaneck Creek Conservancy. Sugar maple (*Acer saccharum*), a species with northern affinities and often not particularly urban-tolerant, dominated a forest stand on the steep southern side of Laurel Hill. We consider the abundance of sugar maple at that location unusual; the shade of the ridge on the north side of the stand and of the elevated Turnpike on the south side may keep the habitat a little cooler and moister than its surroundings, and thus more favorable to this species. Box-elder (*Acer negundo*), an atypical maple with compound leaves, is a weedy, urban-tolerant, native species that occurs at water edges and in disturbed areas. We found it widespread on fill and on more-or-less natural soils. Box-elder is sometimes considered nonnative this far east, although its natural range appears to be very extensive in North America (Maeglin and Ohmann 1973).

Eastern cottonwood (*Populus deltoides*) was abundant in the Meadowlands. Many trees were of good size (ca. 30–60 cm dbh). This is a fast-growing species that often attains large size in areas of natural disturbance such as riverbanks and in urban or otherwise damaged areas. Eastern cottonwood bark is a favorable substrate for mosses and lichens. Swamp cottonwood (*Populus heterophylla*) has not been documented in the Meadowlands since 1884, although it still occurs rarely in deep-flooding nontidal wetlands of the Hudson Valley. Quaking aspen (*Populus tremuloides*) was common, often at small diameter, and formed groves (presumably colonies connected by underground stems) on the sideslopes of inactive landfills as well as on low-lying wetland fill and other habitats. Quaking aspen groves may be used by American woodcock and songbirds. Quaking aspen is fairly common on sandy and gravelly soils, in disturbed areas, and sometimes in old fields in the

Hudson Valley. Big-toothed aspen (*Populus grandidentata*) occurred rarely in at least three Meadowlands municipalities.

Gray birch (*Betula populifolia*) was very common and widespread in the Meadowlands. This was a typical small tree on wetland fill, wetland edges, rocky crests, and other stressful habitats. Gray birch, for example, had formed dense stands on some of the fill islands created as part of the Western Brackish Marsh (Hartz Mountain) wetland mitigation project in Secaucus. Gray birch tolerates heavy metals well (Gallagher et al. 2010), which may help it thrive in the mitigation project where some of the pre-mitigation sediments had very high levels of toxic metals. It is a common tree in old fields, disturbed soils, and on hummocks in swamps in the Hudson Valley. We found black birch (*Betula lenta*) only at Little Snake Hill and Granton Quarry, although this species commonly thrives in young forests and forest edges outside the Meadowlands. Yellow birch (*Betula alleghaniensis*), a tree of cool moist environments, was last documented in the Meadowlands in 1894 (Appendix 1). We found naturally occurring river birch (*Betula nigra*) only on Van Buskirk Island in Oradell. (It was planted elsewhere, including at the Mill Creek Trail in Secaucus.) This species is a common wild tree along New Jersey rivers such as the upper Passaic River.

Two other native trees are of particular interest due to their abundance. We found black cherry (*Prunus serotina*) rather common in a variety of upland habitats. In the Hudson Valley this tree does well on sandy and gravelly soils as well as in fencerows in livestock-grazed areas. Sweetgum (*Liquidambar styraciflua*) was widespread and locally common in swamps and wet woods, as is typical on the Coastal Plain of New Jersey.

A single wafer-ash (*Ptelea trifoliata*) had been reported previously from Laurel Hill (NY-NJ-CT 2021), and we found a small group of fruiting stems on the ridge top in 2008. Because of the historic development of public facilities on Laurel Hill and the likelihood that ornamental trees and shrubs had been planted there, botanists have debated whether wafer-ash is native at this site. Rhoads and Block (2000) considered wafer-ash native in Pennsylvania where it was "rare in old fields, stream banks, and alluvial thickets." These habitats are all subject to natural or human disturbance; however, in other respects they are quite unlike the pocket of deeper soil on a rocky crest where the species occurred on Laurel Hill. In New York, wafer-ash was considered native to the western half of the state where it occurred on "rocky, upland slopes" (Brown 1975). Brown also stated that the species was occasionally planted as an ornamental in the eastern states. Hough (1983) reported wafer-ash scattered in six New Jersey counties (including Hudson County which may refer to Laurel Hill) in both alluvial and rocky upland habitats, and considered the species both native and escaped. Given the unusual geochemical character of Laurel Hill, that there were at least several fruiting stems in 2008, and that Laurel Hill was not overrun with ornamentals not found commonly elsewhere in the Meadowlands, we believe wafer-ash should be considered native unless proven otherwise. Wafer-ash is listed as S1 Endangered in New Jersey. The species can form colonies, thus the stems on Laurel Hill may belong to a single individual.

Additional species were uncommon or rare in the Meadowlands. We found American hornbeam (*Carpinus caroliniana*) only at the upper Hackensack River estuary in New Milford. The most recent Metropolitan Flora record for this species is from 1970 (Appendix 1). We found flowering dogwood (*Cornus florida*) only on Van Buskirk Island. This is a common tree in the Hudson Valley, although its abundance has been reduced by disease in recent decades. Basswood (*Tilia americana*) occurred at Van Buskirk Island, Andreas Pond (Teaneck), Mehrhof Pond (Little Ferry), and Little Snake Hill. In the Hudson Valley, basswood is an uncommon tree associated with fertile calcareous soils. We found sassafras uncommon at the upper Hackensack River estuary, Teterboro Airport Woods, Granton Quarry, Mehrhof Pond and Losen Slote Park, Ridgefield Park Nature Preserve, Little Snake Hill, and Kearny Marsh West. Sassafras is uncommon to locally common species in the Hudson Valley, on both fertile low-elevation soils and rocky ridgetops. We found tuliptree (*Liriodendron tulipifera*) at Van Buskirk Island, the upper estuary in New Milford, Teterboro Airport Woods, Granton Quarry, and Wolf Creek (Ridgefield). There was a large specimen, for example, at Wolf Creek. In the Hudson Valley, tuliptree is principally a species of the inner valley at lower elevations (i.e., near the Hudson River estuary), and does best on deep fertile soils.

Several native hardwood trees seemed conspicuously absent from the Meadowlands. These included alternate-leaved dogwood (*Cornus alternifolia*), hop hornbeam (*Ostrya virginiana*), and the *arborescent* (tree-size) shadbushes (*Amelanchier* spp.). There is no record of common shadbush (*Amelanchier arborea*) in the Meadowlands, and Canadian shadbush (*Amelanchier canadensis*) was last documented in 1948. Common nonnative trees included princess tree (*Paulownia tomentosa*), catalpa (*Catalpa*), and especially tree-of-heaven (*Ailanthus altissima*).

Native shrubs were diverse in the Meadowlands. Arrowwood (*Viburnum dentatum s.l.*) was fairly widespread in swamps and lowland woods. We found blackhaw (*Viburnum prunifolium*) at only three sites: the upper estuary in New Milford, Teterboro Airport Woods, and Little Snake Hill. This is a small viburnum of lowlands and wetland edges. The Metropolitan Flora has 5 records, a relatively large number. Spicebush (*Lindera benzoin*) occurred at 3 sites: Teaneck Creek Conservancy, a site in Ridgefield Park, and Schmidt's Woods (the last in Secaucus), although the Metropolitan Flora does not have a recent record. Surprisingly, we only found sweet pepperbush (*Clethra alnifolia*) at 3 sites, all near each other: Teterboro Airport Woods, Losen Slote Park, and Mehrhof Pond. However, this species was abundant in the outlying portion of Teterboro Airport Woods where canopy hardwoods were dead following Hurricane Sandy. Silky dogwood (*Cornus amomum*), principally a wetland species, was widespread in wetland edges. Staghorn sumac (*Rhus typhina*), smooth sumac (*Rhus glabra*), and winged sumac (*Rhus copallinum*) were all widespread and fairly common on upland soils including various kinds of fill.

Several native shrubs were missing from the Meadowlands flora, among them bladdernut (*Staphylea trifolia*; recorded in 1889), New Jersey tea (*Ceanothus americanus*; 1894), downy arrowwood (*Viburnum rafinesquianum*; 1919), and speckled alder (*Alnus incana*; no record). Other common rural species were very rare in the Meadowlands, for example, smooth alder (*Alnus serrulata*). Also conspicuously missing was mountain-laurel (*Kalmia latifolia*; we do not know whether Laurel Hill was named after this plant). Although a protected species in New Jersey, mountain-laurel is fairly common and even grows along heavily traveled sections of the Garden State Parkway farther south; Laurel Hill and Little Snake Hill seem eminently suitable habitat. Mountain laurel grows on rocky or sandy soils in pastures, forests, and roadsides, but it requires a mycorrhizal fungus, as do most other members of the heath family (Ebinger 1997). Some mycorrhizal fungi may be susceptible to urban conditions such as contaminated soils.

Eastern prickly-pear (*Opuntia humifusa*), which is semi-woody, has been documented on a number of warm, sunny rock outcrops near the Hudson River as far north as Columbia and Greene counties, New York, and including at least one New York Palisades locality on diabase bedrock. Given the Hudson River occurrences on bedrock and the similarity of the Palisades to Laurel Hill, it is surprising that this species has never been reported from Laurel Hill, Little Snake Hill, or Granton Quarry. Perhaps early quarrying activities and other human impacts eliminated prickly-pear from the Meadowlands, or it was dug up for transplantation to gardens.

VINES

Our Meadowlands list (Appendix 1) contains 39 species of woody vines; 23 are native and 16 nonnative. Our list also contains 37 herbaceous vines of which 24 are native, 10 nonnative, and 3 both native and nonnative. (Some wide-ranging species have recognizable forms that evolved in Eurasia and the United States. Thus, in a given locality, both a native form and an introduced, nonnative form can occur.)

Where vines are prominent, they are often considered indicators of disturbance. Some vines can take advantage of nutrient-rich soils at wetland edges and disturbed forests where the vines are able to climb up to the light and at least in some situations compete with trees for nutrients (see Dillenburg et al. 1993). In a sense, large areas of the Meadowlands are ideal vine habitats, and many species are indeed present. There are both woody and herbaceous vines, and vines can also be categorized on the basis of how they attach to their supports. Climbing vines have tendrils (e.g., grapes) or aerial rootlets (poison-ivy) that attach to the host. Twining vines (e.g., round-leaved bittersweet, bindweed) spiral around the host. Clambering vines (e.g., brambles, vetches) simply lean on the host, although multiflora rose can ascend several meters up trees by means of shoot extension and backward-pointing prickles. Most vines are just using their hosts for support and many vine species are as likely to use a fence or building as another plant. Species of a single genus in our region, the non-woody dodders (*Cuscuta*), are fully parasitic and depend on the host plant for nutrition.

Herbs are not typically hosts for woody vines, as most herb species are small, and any robust aboveground structures generally do not stand through the winter. A few robust herbaceous (or semi-woody) herbs, however, make good vine hosts. These are plants such as purple loosestrife, common reed, and knotweed, with sturdy aboveground structures 2-3+ meters tall that stand through the winter and well into, or through, the following growing season. This generational overlap of robust aboveground material in the hosts allows vines such as grape to grow from support on last year's host stems to the current year's stems. Robust common reed at the wetland-upland edge often supports lush growths of vines, and overall, we have seen perhaps 30 vine species using reed for support in the Meadowlands. Many of these vines add substantial structure and biomass to their host plants, and some bear fruits that are eaten by birds and mammals or leaves that are eaten by insects; thus vine-carrying marsh edges, shrubland, or forest may provide more complex and attractive breeding, refuge, and foraging habitats for many small animals.

The parasitic dodders are unusual in the flora of our region. These small vines look much like slender orange, yellow, or purple spaghetti twined around their host plants or hanging in festoons. Seven native and one nonnative dodders have been found in the Meadowlands. Swamp dodder (*Cuscuta gronovii*), although by far the most common species in the Hudson Valley, did not stand out as especially more frequent than the others in the Meadowlands. We found the dodders using a variety of hosts, although common reed and knotweed were not often parasitized. The dodders of the northeastern states rarely do much visible damage to their hosts, although in the Hudson Valley swamp dodder is occasionally lush enough to prevent purple loosestrife from flowering.

GENERAL DISCUSSION OF THE FLORA

The sketches of various taxa in this chapter illustrate the importance of certain sites for plant diversity, such as Little Snake Hill, Laurel Hill, Teterboro Airport Woods, Mehrhof Pond—Losen Slote Park, Ridgefield Nature Center—Wolf Creek, Schmidt's Woods, and Teaneck Creek Conservancy. It is interesting that these are not the tidal and formerly tidal marsh sites that so much political and management attention has focused on. Marshes are important to biodiversity in other ways, but they commonly do not support the high diversity of seed plants that the unusual, remnant, outlying, and northern areas do. Salinity intolerance among many species is the most likely explanation; contamination, management impacts, and habitat fragmentation probably are also factors.

The loss of urban-intolerant native species and the gain of urban-tolerant native or nonnative species during the past two centuries are noteworthy, and this is especially striking for the orchids (see above). The numbers of native orchids declined from perhaps 16 or more species collected in the 1800s to none found in the past 40 years. Probably the loss of sheltered, infertile freshwater wetlands resulted in environments that did not have the mycorrhizal fungi, low conductivity (low cation and anion levels), or other requirements for this often *stenotopic* (having narrow habitat affinities) group of plants. Some species of mycorrhizal fungi may be at a disadvantage in the Meadowlands and other urban areas with cut-and-fill soils, often contaminated, low in organic matter, and lacking the physical and chemical characteristics that require centuries or millennia for development in natural soils (Figure 4.2).

The seed plants illustrate some important features of the Meadowlands and likely other urban-industrial regions. Diversity depends on the presence of a variety of habitats and microhabitats. Urban conditions eliminate many species and apparently foster others. Many species are uncommon or rare and may be found at very few locations and, where they occur, in very small numbers. Occurrence of species within the Meadowlands may be less predictable (e.g., on the basis of habitat or soil) than in non-urban regions. Populations of rare or uncommon species, or species found very locally within the region, are at the mercy of land use and habitat management, including park management and wetland mitigation. And much is still unknown about the Meadowlands flora despite two centuries of biological exploration. The most important conservation actions for seed plants are the thorough, expert survey of the flora wherever changes in land use, direct human activity, or habitat management are anticipated, and the preservation of many small and large habitat units.

Figure 4.2 Plant habitats on cut-and-fill soils in North Arlington: looking from the old mine face that constitutes the western scarp of the Meadowlands, across a sparsely vegetated area of poor mineral soil to a spontaneously vegetating landfill. Photo 2021 by Erik Kiviat.

Chapter 5

Cryptogams

The cryptogams are the pteridophytes (ferns and fern allies, i.e., the horsetails, clubmosses, spikemosses, quillworts), bryophytes (mosses, liverworts, hornworts), algae, and (no longer considered plants) the fungi, lichens, and blue-green "algae" (*Cyanobacteria*). Bryophytes and lichens may be referred to collectively as *bryoids*. Most cryptogams discussed here reproduce by means of spores or by fragments. The lichens also reproduce by means of specialized vegetative structures containing both fungal (*mycobiont*) and algal or cyanobacterial (*photobiont*) partners. Spores are sexual (products of meiosis), or asexual (products of mitosis), microscopic propagules many of which can survive adverse conditions and allow a species to move from one place to another. Spores can be blown long distances, and in some instances resist extremes of temperature and dryness better than the seeds of seed plants. Some of the asexual propagules of lichens containing both mycobiont and photobiont can do the same. In theory, many cryptogams should be able to readily disperse into urban habitats, and some species establish and thrive there. There has been no dedicated survey or research on cryptogams in the Meadowlands, and very little for other North American cities.

Compared to higher plants, relatively few species of cryptogams are urban-tolerant. Lichens are especially sensitive to sulfur oxides, fluorine, strongly oxidizing substances such as ozone, certain metals, and excess nitrogen (Wetmore 1988, Brodo et al. 2001, Gutiérrez-Larruga et al. 2020). Many mosses are also sensitive to sulfur dioxide and dryness (Kimmerer 2003), and the effects of these two are synergistic in suppressing photosynthesis (Winner 1988, Winner et al. 1988). Heavy metals, especially mercury, inhibit germination of moss spores (Kaur et al. 2010). Liverworts require a stable moist environment and most apparently need good water quality; as a group they are not urban-tolerant. Few ferns and fern allies are urban-tolerant. Urban parks should be given high priority for cryptogam surveys because those habitats are highly susceptible to pollution, overuse, and conversion to other land uses (Brodo 2017) while often supporting important cryptogam diversity.

Some habitats that are scarce in wildlands abound in cities, such as buildings and statues that are cliff-like or rock-like (Kimmerer 2003). Many cryptogams are *saxicolous* (rock-inhabiting) in the wild. Thus, some species that
can tolerate dryness and pollution do well in cities on cliff-like structures as
well as natural rock surfaces. Concrete, mortar, and marble objects in cities
are substrates for *calcicolous* (calcium-associated) cryptogams. Trees outside
of their natural ranges tend to support fewer *corticolous* (bark-dwelling)
cryptogams than native trees, although some native trees, such as the aspens,
are also poor hosts (Barkman 1958), and large specimens of tree-of-heaven
(*Ailanthus altissima*) can have extensive bryoid coverage (E. Kiviat, pers.
obs.). Birches, white pine, and eastern red cedar are other native trees that
are poor hosts for corticolous bryoids, probably because of exfoliating bark
(red cedar in the Hudson Valley of New York sometimes supports lush lichen
growth on small branches). Higher pH and greater water-holding capacity
of the bark of certain species are generally favorable for bryophyte diversity
(Studlar 1982); large trees are often important bryoid hosts. Corticolous bryophytes and lichens may be inhibited more by prolonged low sulfur dioxide
concentrations than brief high concentrations (LeBlanc and De Sloover 1970).

Surveys rarely find all the species of an area. Unlike many animals and
many herbaceous seed plants, the lichens, bryophytes, and woody plants
(trees and shrubs) are detectable year-round; they do not overwinter out
of sight, die back to the ground, or leave the area seasonally (although the
soil-dwelling species may be obscured by fallen leaves, snow, or ice). This
advantage in surveying bryoids is balanced by the small size and cryptic
appearance of many species, restriction of many to specific habitats or
microhabitats, and the inherent difficulties of identifying species. The birds
and fishes of the Meadowlands have been surveyed fairly thoroughly and if
species have gone undetected it is because they are new to the area, transient,
very rare, or restricted to difficult-to-survey habitats. On the other hand, the
cryptogams (with the possible exception of ferns) are poorly known and
mostly had not been surveyed at all prior to our studies. Cryptogams other
than ferns are notoriously under-surveyed in general, and thus we often do
not know enough to manage rare species or other noteworthy forms (e.g.,
Löhmus and Löhmus 2009).

One might ask why cryptogams, especially small forms such as mosses and
lichens, matter enough to be surveyed and considered in management planning. Andrus (1990) stated eloquently that rare bryophytes should be studied
for their own sake. Bryophytes occur in some habitats where seed plants are
absent, and bryophytes provide information and enjoyment differently than
seed plants (*ibid.*). Lichens are even more different from seed plants than are
bryophytes.

Cryptogams have many existing and potential uses. Lichens produce many unique organic chemicals called *lichen acids* or *lichen substances*. Atranorin, for example, is under study as a COX-1 inhibitor with pharmaceutical potential. Basnet et al. (2018) reviewed numerous lichen chemicals with antimicrobial or anti-cancer properties. Most bryophytes produce antibacterial and antifungal compounds, and anticancer compounds have also been found (Hallingbäck and Hodgetts 2000). Liverworts contain essential oils that seem to inhibit feeding by animals; many of these compounds have pharmaceutical potential (Stotler and Crandall-Stotler 2006, Asakawa 2012). Many, if not most, woody seed plants, orchids, heath family species, and other seed plants benefit from, or require, mycorrhizal relationships with fungi that help plants obtain nutrients from the soil. Ferns are important garden plants. Mosses are planted on green roofs in Europe (Studlar and Peck 2009). Cryptogams, especially lichens and mosses, are sensitive indicators of pollution; a neighborhood with lush growths of bryoids is likely relatively clean (Kimmerer 2003). Moss cover reduces soil temperatures in hot weather (Chen et al. 2019) and mosses provide microhabitats for small invertebrates such as tardigrades (water-bears).

All the ferns, fern allies, bryophytes, and lichens discussed in this chapter are believed native to our region (possibly excepting red thread moss, see Eckel and Shaw [1991]). This may result from the common long-distance dispersal of cryptogams by windblown propagules, allowing many species to have ranges encompassing large areas of North America and, in many cases, multiple continents (see Frahm 2008 on mosses). The lack of nonnative species may also be because species of cryptogams that are introduced for aquaria or gardens cannot survive the harsh conditions of the Meadowlands. A few introduced cryptogams could potentially become invasive in the Meadowlands, for example, water-clover (*Marsilea quadrifolia*), a fern that is established in a few ponds in Westchester and Orange counties, New York, and Monmouth County, New Jersey (E. Kiviat, pers. obs.).

Most literature references use only scientific names for cryptogams other than the pteridophytes. Cryptogams have generally not received enough attention from non-specialists to have standardized common names. In keeping with our practice elsewhere in this book, however, we use common names in this chapter as much as possible. Most of our names for lichens are from Brodo et al. (2001), and for mosses from McKnight et al. (2013) or iNaturalist (inaturalist.org). Many of the common names are invented names rather than vernacular names. Voucher specimens from our surveys are in the Hudsonia—Bard College Field Station Herbarium.

FERNS AND FERN ALLIES

Pteridophytes are *tracheophytes* (higher vascular plants) with well-developed conducting tissues, but they accomplish long-distance dispersal by spores rather than seeds. The large visible forms of most ferns and fern allies are the diploid *sporophytes* or spore-producing stages. The haploid *gametophytes*, or sexual stages, are small and inconspicuous. Fern spores may be carried long distances by winds; however, establishment in a new locality depends on the correct conditions for germination of the spores, development of the environmentally sensitive gametophytes, and subsequent development and survival of the sporophytes (Lellinger 1985). Ferns and fern allies are collectively known as pteridophytes although they are now believed to have arisen separately in evolution (lack a recent common ancestor).

No quillworts, spikemosses, or clubmosses are currently known in the Meadowlands, although a spikemoss and a clubmoss were reported historically (see below). Two species of quillworts, 2 spikemosses, and 8 clubmosses have been found in Bergen and Hudson counties overall (Montgomery and Fairbrothers 1992); one additional quillwort was reported by USDA (2021). Besides the pteridophytes discussed below, many other species once occurred in or close to the Meadowlands according to the range maps in Chrysler and Edwards (1947). The small scale of those maps, and the commonly vague locality data on the labels of old herbarium specimens, may make it hard to determine precisely where specimens were collected (examining specimens was mostly outside the scope of this book except those which we collected ourselves). Undoubtedly the Meadowlands had a rich flora of pteridophytes before documentation began.

A single clubmoss has been reported from the Meadowlands historically. Most clubmosses can colonize open areas as well as intact forests; some species of the genus *Lycopodiella* even occur in abandoned borrow pits and powerline rights-of-way (Haines 2003). A few upland species of clubmosses occur widely in young-to-mature post-agricultural forests in the Hudson Valley but apparently are absent from the Meadowlands. Clubmosses are usually not found in saline habitats (Haines 2003). The clubmoss species that occur in northeastern New Jersey are in general associated with acidic soils (Lellinger 1985). In the Adirondack Mountains of New York, three species of clubmoss were found at median pH of 4.5 or lower (Kudish 1992). Perhaps geochemical, estuarine, and urban influences in the Meadowlands make most habitats too alkaline for clubmosses. Many clubmosses have gametophytes that require close associations with fungi to survive (Haines 2003), potentially creating an additional ecological bottleneck for establishment of the sporophytes that most botanists and naturalists study. Lellinger

(1985) considered the clubmosses the ranges of which include northeastern New Jersey to be difficult to establish in cultivation and sensitive to over-fertilization. By extrapolation, these plants may have difficulty establishing in urban areas and are subject to harm from pollutional nitrogen.

Quillworts could occur in the claypit lakes (such as Mehrhof Pond) and a few of the other low-salinity waterbodies, but quillworts are inconspicuous plants, are often hidden among other aquatic plants or in deep water, and are hard to identify to species (Cobb et al. 2005). Shore quillwort (*Isoëtes riparia*) tolerates freshwater tidal and slightly brackish water (Lellinger 1985); this species was known historically from the Passaic River above Newark (Chrysler and Edwards 1947) and the fresh-tidal Hudson River. Quillworts are sensitive to high temperatures (Montgomery and Fairbrothers 1992) and may require water that is fresher or of better quality than most of that available in the Meadowlands. Appalachian quillwort (*Isoëtes appalachiana*), a rare species in Indiana, was considered threatened by rapid water level changes, water pollution, and competition (Hauser and Haulton 2008). Lellinger (1985) considered the quillworts fairly easy to cultivate, which suggests they may establish and survive readily in the wild given suitable conditions. Habitats of Engelmann's quillwort (*Isoëtes engelmannii*) in New Jersey included ponds and ditches (Chrysler and Edwards 1947).

Rock outcrops at Laurel Hill and Little Snake Hill (see Figures 1.3–1.4 for locations) seem suitable habitat for rock spikemoss and there is a historic record from Laurel Hill. This species was considered sensitive to treading (Chrysler and Edwards 1947), and most of Laurel Hill has either been quarried, eroded by ORVs and foot traffic, or formerly built on. Air pollution may have contributed to decline of this species in northeastern New Jersey (Montgomery and Fairbrothers 1992). Although rare in the Hudson Valley, rock spikemoss grows sparingly on a sedimentary rock ledge in urban Kingston, New York, ca. 120 km north of Laurel Hill (E. Kiviat, pers. obs.).

The only horsetail we found in the Meadowlands is the common, disturbance-tolerant field horsetail. It was reported by Harshberger and Burns (1919) but not by Torrey (1819) or Britton (1889). Two horsetails, scouring-rush (*Equisetum hyemale*) and marsh horsetail (*Equisetum palustre*), now considered common and rare, respectively, in the northeastern states (Cobb et al. 2005), were found in the Meadowlands by Torrey (1819), and a third species, now considered common by Cobb et al., water horsetail (*Equisetum fluviatile*), was found in the Meadowlands by Britton (1889). Five species of horsetails have been reported from Bergen and Hudson counties (Montgomery and Fairbrothers 1992). Salinity intrusion and poor water quality possibly limit the establishment of most horsetails in the Meadowlands. It is odd, however, not to find scouring-rush, a large horsetail that occurs, albeit uncommonly, on seepy *calcareous* (calcium-rich) soil and even roadcuts in the Hudson Valley, where it is presumably exposed to high levels of motor vehicle pollution and sodium chloride used for de-icing. Scouring-rush was considered a regular inhabitant of coastal brackish habitats (Stuckey and Gould 2000). This species is fairly easy to transplant (Montgomery and Fairbrothers 1992) and might be transported by earth-moving or road maintenance. Lellinger (1985) deemed the horsetails generally weedy and stated that some disperse by means of vegetative fragments. Horsetail spores are short-lived (Whittier 1996), which likely inhibits dispersal into urban habitats. Curiously, field horsetail occurs in tree planters in the urban core of White Plains (Westchester County, New York) where the plants may have arrived with the soil rather than by means of windblown spores.

Field horsetail thrives on a variety of moist, often sandy, soils, including along roadsides, and is widespread and weedy (Cobb et al. 2005). It is abundant and widespread in New Jersey (Montgomery and Fairbrothers 1992). In the Meadowlands this horsetail is rare; we have seen it on a dredge spoil mound at Berry's Creek Canal, next to a railroad in Leonia, in the mixed vegetation at Teaneck Creek Conservancy, and at the edge of a constructed stormwater pond in Moonachie. Lellinger (1985) stated that field horsetail is associated with mildly acidic soils. Given the extent of the Meadowlands region and the types of habitats present, if the Meadowlands were in the rural Mid-Hudson Valley there would probably be at least 3 species of horsetails (field horsetail, scouring rush, and water horsetail), as well as three clubmosses (staghorn clubmoss [*Lycopodium clavatum*], ground-cedar [*Lycopodium complanatum*], and tree clubmoss [*Lycopodium obscurum s.l.*]). Brundrett (2002) stated that horsetail sporophytes are often mycorrhizal; possibly the necessary fungi are absent from the Meadowlands.

We found 14 species of ferns in the Meadowlands and 6 additional species have been reported, mostly historically (Table 5.1). Thirty-nine species of ferns were reported from Bergen and Hudson counties by Montgomery and Fairbrothers (1992). Most of the ferns we found were associated with moist to wet soil in sheltered areas that were mostly subject to occasional intrusion of low-salinity estuarine waters. Several species occurred on the remnants of talus slopes and on the old quarry faces and rubble at Laurel Hill and Little Snake Hill (Figure 5.1). The geologically similar New Jersey Palisades in Bergen County were identified as a good habitat for ferns (Slowik and Greller 2009). The Teterboro Airport Woods, a large complex of partly-drained hardwood swamps, had a moderate diversity of ferns. This area has moist to wet organic soils and little direct human disturbance due to airport security, although there is presumably pollution from the airport (see Chapter 1 regarding airport pollution).

Table 5.1 Ferns and fern allies found in the New Jersey Meadowlands.

Scientific name	Common name	Source
Ferns		
Asplenium platyneuron	Ebony spleenwort	Hudsonia
Botrychium dissectum[a]	Cut-leaved grape fern	Torrey 1819
Botrychium virginianum[a]	Rattlesnake fern	Torrey 1819
Cheilanthes lanosa[a]	Hairy lip fern	Britton 1889
Cystopteris fragilis[a]	Fragile fern	Britton 1889
Dennstaedtia punctilobula	Hay-scented fern	Hudsonia
Dryopteris carthusiana	Spinulose wood fern	Hudsonia
Dryopteris marginalis	Marginal wood fern	Britton 1889; Hudsonia
Onoclea sensibilis	Sensitive fern	Heusser 1949; Hudsonia
Ophioglossum vulgatum[a]	Southern adder's tongue	Britton 1889
Osmunda cinnamomea	Cinnamon fern	Hudsonia
Osmunda claytoniana[b]	Interrupted fern	NY Metropolitan Flora
Osmunda regalis	Royal fern	Torrey 1819; Harshberger & Burns 1919; Hudsonia
Polypodium vulgare	Rock polypody	Hudsonia
Pteridium aquilinum	Bracken	Hudsonia
Thelypteris noveboracensis	New York fern	Hudsonia
Thelypteris palustris	Marsh fern	Hudsonia
Woodsia obtusa	Blunt-lobed cliff fern	Torrey 1819; Britton 1889; Hudsonia
Woodwardia areolata	Netted chain fern	Britton 1889; Hudsonia
Woodwardia virginica	Virginia chain fern	Torrey 1819; Hudsonia
Horsetails		
Equisetum arvense	Field horsetail	Harshberger & Burns 1919; Hudsonia
Equisetum fluviatile[a]	Water horsetail	Britton 1889
Equisetum hyemale[a]	Scouring-rush	Torrey 1819
Equisetum palustre[a]	Marsh horsetail	Torrey 1819
Spikemosses		
Selaginella rupestris[a]	Rock spikemoss	Montgomery & Fairbrothers 1992
Clubmosses		
Lycopodiella alopecuroides[a]	Foxtail clubmoss	Harshberger & Burns 1919

Note: Common and scientific names follow Cobb et al. (2005).

[a] Species known only from historic records

[b] Kiviat and MacDonald (2002) included this species based on two records from the New York Metropolitan Flora, the more recent from 1991.

Figure 5.1 Natural talus at Little Snake Hill. Talus slopes support many cryptogams, plants, and animals. Photo by Erik Kiviat.

Netted chain fern, which is common in the *mesic* (medium-moisture) to wet forests of the northern Meadowlands (Teterboro Airport Woods, Losen Slote Park, and Mehrhof Pond), is a species principally of the Coastal Plain that is said to tolerate brackish water (Cobb et al. 2005). Gargiullo (2007) called this fern "intolerant of salt"; different authors may be referring to different levels of salinity. Virginia chain fern, commonly associated with acidic bogs in the northeastern states (although in both acidic and mildly alkaline soils according to Lellinger [1985]), occurred as a colony of perhaps 0.5 m^2 at one forested location where it is probably a relic of an earlier peatland flora. Cobb et al. (2005) stated this species usually occurs in standing water, but the Meadowlands site is not nearly so wet (due to historic wetland drainage). Lellinger (1985) characterized netted chain fern as associated with acidic soils, and Montgomery and Fairbrothers (1992) stated that both chain fern species were associated with acidic soils in New Jersey.

Sensitive fern occurs widely in wet and moist habitats, often disturbed, in the northeastern states and may be the most abundant fern in New Jersey (Montgomery and Fairbrothers 1992). Sensitive fern was one of the most common Meadowlands ferns. Gargiullo (2007) stated that this species was intolerant of salt, and Montgomery and Fairbrothers (1992) considered sensitive fern intolerant of pollution. This species was not noted in the

Meadowlands literature until Heusser (1949); it may have become common in the Meadowlands after considerable alteration of the landscape.

Bracken is also a very widely distributed and weedy fern occurring in a broad range of habitats (Cobb et al. 2005). In the New York City region, Gargiullo (2007) reported that bracken was often associated with infertile, sandy soils. We found bracken at three sites (Teterboro Airport Woods, Losen Slote Park, and the adjoining Bergen County Utilities Authority property) in the northern woodlands and it was rare in the Meadowlands.

Cinnamon fern and royal fern were uncommon in shrubby and wooded areas of the northern Meadowlands. Lellinger (1985) stated that cinnamon and royal ferns were associated with acidic soils, and Cobb et al. (2005) considered these species very widespread and common in the eastern states.

We found a large stand of hay-scented fern on the slope of the railroad embankment and at the edge of the freshwater nontidal Upper Penhorn Marsh, and this species also occurred in Teterboro Airport Woods. Hay-scented fern was also described as an *acidicole* (acidic habitat associated; Lellinger 1985). Cobb et al. (2005) stated that hay-scented fern is a weedy species that is associated with dry sandy soils in sun or partial shade. After several years, the Upper Penhorn Marsh stand was much diminished in extent.

We found ebony spleenwort on old mine tailings on the former quarry floor and quarry faces at Laurel Hill; this fern is associated with calcareous soil in the Hudson Valley. It also occurred on the mortar of an old brick structure next to the Hackensack River in Lyndhurst. Lellinger (1985) stated that ebony spleenwort was associated with mildly acidic soils but also masonry (which would be calcareous and not acidic, at least where mortar is present). Cobb et al. (2005) commented that ebony spleenwort was more tolerant of disturbed sites than other spleenworts. Ebony spleenwort was found on the serpentine bedrock of Hoboken, just outside the Meadowlands (Montgomery and Fairbrothers 1992); perhaps it will colonize the serpentine rock rubble dumped in the Kane Natural Area (see Chapter 1).

New York fern occurred locally in the northern woodlands of the Meadowlands, and marsh fern was widespread in the sunnier wetlands. Lellinger (1985) stated New York fern was associated with mildly acidic soils. Marsh fern tolerated "seashore" conditions and was the only fern on Sandy Hook, New Jersey (Chrysler and Edwards 1947).

Blunt-lobed cliff fern, which we found as a single tuft on the dry quarry rim at Laurel Hill, was described as occurring on a variety of soils and rock substrates although usually where moist (Lellinger 1985, Cobb et al. 2005). This species grows on mortar and concrete in the New York City region (Gargiullo 2007). Fragile fern (*Cystopteris fragilis*) was documented more than a century ago. In 2011 Eric Martindale collected a fern in Borg's Woods (Hackensack) that we identified as *Cystopteris* sp. This fern requires further study.

We have not found grape ferns (*Botrychium*) in the Meadowlands, although 2 species were reported historically (Table 5.1) and 5 species have been found in Bergen County overall (Montgomery and Fairbrothers 1992). The two species reported by Torrey (1819), cut-leaved grape fern and rattlesnake fern, are relatively common grape ferns that are associated with a wide variety of soils and habitats (Cobb et al. 2005). This group of ferns is presumably urban-sensitive although certain species occur alongside dirt roads and trails in rural areas. The grape ferns are very difficult to cultivate, probably due to complex fern-soil-fungi relationships (Lellinger 1985, Montgomery and Fairbrothers 1992). There is a historic record of southern adder's tongue fern in the Meadowlands (Chrysler and Edwards 1947). Cobb et al. (2005) noted that adder's tongue ferns and grape ferns were found in diverse habitats and often occurred in slightly disturbed soils without deep leaf litter. They commented that these ferns are often hidden among other plants and are hard to find and identify.

Generally, the fern flora of the Meadowlands shows moderate diversity and for most species low abundance, and the fern allies are extremely *depauperate* (poor in species and abundances). Some of the more sensitive species formerly occurred in the Meadowlands but appear to have been extirpated. Most of the species that occur in the Meadowlands now are common and tolerant of a wide range of conditions elsewhere; sensitive fern and field horsetail are good examples. In addition to salinity and poor water quality, the scarcity of natural upland soils and probably a scarcity of acidic soils may contribute to the absence of many species. The presence of species such as the chain ferns and the osmundas may indicate natural, organic-rich soils as differentiated from low-organic, cut-and-fill soils. Teterboro Airport Woods, which has sheltered swamp forest interior habitats and organic soils, has moderate fern diversity. That it does not have even more species of ferns, and some fern allies, may indicate pollution acting to inhibit mycorrhizal fungi or affecting another aspect of fern physiology or development. Or the agricultural use of the area a century ago may have eliminated most pteridophytes and recovery of this group of organisms (as well as woodland wildflowers, amphibians, and reptiles) is slow.

We think a qualitative comparison of Meadowlands pteridophyte diversity with the still predominantly rural Mid-Hudson Valley of New York is instructive. The commonest Meadowlands species (sensitive fern, marsh fern) are even more common in the Mid-Hudson. Many of the species that we found uncommon or rare in the Meadowlands (e.g., marginal woodfern, hay-scented fern, New York fern, bracken, cinnamon fern, royal fern, field horsetail) are more-or-less common in the Mid-Hudson. Notably, the Mid-Hudson Valley has many species, some common and some not, that we did not see in the Meadowlands. These include some of the calcicolous species (e.g., walking fern [*Asplenium rhizophyllum*], purple cliffbrake [*Pellaea atropurpurea*], maidenhair spleenwort [*Asplenium trichomanes*], water horsetail), some of the acidicoles (the upland clubmoss species mentioned above), certain other saxicolous species (rock spikemoss), and other forest species (e.g., Christmas fern [*Polystichum acrostichoides*]). Yet the Mid-Hudson is far from a fern paradise, and many species and sites are probably still recovering from the extensive agriculture of the early-to-mid-1900s, while other sites and species are newly threatened by urbanization, degradation of wetlands and forests, and perhaps the spread of nonnative plants.

Large-scale harvest for the florist industry of New York City threatened many fern species in the Berkshire Hills of western Massachusetts a century ago (Cobb et al. 2005), and fresh-cut woodferns were much-used as decorations in Vermont shops (Chrysler and Edwards 1947). Certain clubmosses were also widely harvested to produce Christmas wreaths (Chrysler and Edwards 1947, Frankel 1981). Eric Martindale (email dated 4 April 2010) reported that cinnamon fern and Christmas fern had been extirpated from Borg's Woods in Hackensack just outside the Core Meadowlands, by collecting for gardens. Although we have not found records of a historic commercial harvest of ferns or clubmosses in the Meadowlands, this use seems likely given their proximity to residential gardens and urban markets.

Many pteridophytes are associated with substrates that have a particular range of pH, sand content, organic matter, or moisture. Evidently the development of the gametophytes and the early development of the sporophytes of many species are sensitive to environmental conditions (not surprisingly, considering the small size and delicate appearance of the gametophytes, and the complexity of the process of sexual reproduction, in pteridophytes). Frankel (1981, 134), discussing fern gardening, remarked that ferns require "rich, spongy and porous soil with good drainage" and commented on the importance of organic matter supplements. The most fern-rich habitats in the Meadowlands are the remnant wet forests (e.g., Teterboro Airport Woods), albeit partly drained. Much of the rest of the Meadowlands, where not too brackish, has soils that are either low-organic-matter wetland fill, clayey or sandy deposits, or quarry cut-and-fill. Ecological restorationist Sven Hoeger (pers. comm.) also believes that abundant soil organic matter is crucial to the establishment and survival of most ferns. Ferns are generally *mycorrhizal* (associated with fungi that help the ferns acquire nutrients from the soil; Richardson and Walker 2010). Fern cover and diversity tend to be greater on higher-fertility soils (*ibid.*); this could account for greater diversity in the drained swamp forests compared to cut-and-fill soils in the Meadowlands.

MOSSES AND LIVERWORTS

Bryophytes (mosses, liverworts, and hornworts) are small, usually mat or cushion-forming plants that are commonly some shade of green (a few species are brown, gray, silver, or red). Bryophytes are fairly diverse, and nearly 400 species of mosses alone have been listed for New Jersey (Karlin no date); Barringer (2021a) listed 368 species plus another 70 that have not been collected in a century or more. Yet the bryoflora of New Jersey is poorly studied (Karlin 1990, Karlin and Andrus 1988, Karlin and Schaffroth 1992). We were interested in the bryophytes of the Meadowlands because bryophytes statewide are an important component of New Jersey's biodiversity (Karlin 1990), they are generally omitted from biodiversity studies yet are sensitive indicators of ecological conditions and environmental change and are poorly studied in American cities. Moreover, because moss spores are widely dispersed by winds and the spores of some species can survive for a century, and because mosses can also become extirpated from regions, mosses can respond to, and indicate, changes in climate and land use (Frahm 2008). Karlin (no date) found that 40% of the moss species collected at least once in the state had not been collected since 1950 and well over half of the species were known from fewer than 10 localities each in the state. It is unclear whether those species that have not been found recently have disappeared from the state or simply gone unnoticed due to lack of surveys and the scarcity of biologists who study bryophytes. An international conservation analysis of bryophytes recommended "comparing bryophyte floras of undisturbed and disturbed habitats to determine the impact of disturbance, and to identify those species unable to survive in disturbed areas" (Hallingbäck and Hodgetts 2000). American liverworts have been so poorly studied that it is difficult to assign conservation status to species (Lincoln 2008). Ecologists rarely study liverworts because they are hard to identify and culture and are not economically important (Schuster 1949).

By means of opportunistic and non-intensive collecting beginning in 2000, we found 87 species of mosses and 8 species of liverworts in the Meadowlands (Table 5.2). We collected *morphospecies* (visibly different kinds) and submitted them to specialists for identification. Bryophytes are difficult to identify, easily overlooked, and usually under-collected (Karlin 1990). Many bryophytes are similar to each other (and the same species may look quite different when dry or wet), so undoubtedly we overlooked some species in the field; inexperienced collectors tend to collect the common bryophytes and miss the rare ones (Karlin 1990, Susan A. Williams, pers. comm.). The proportion of species new to our bryophyte survey each year from 2001 to 2018 declined substantially but did not stabilize at zero.

Table 5.2 Bryophytes collected in the New Jersey Meadowlands, 1999–2018.

Scientific name	Common name	Habit	Substrate	Specimens
Mosses				
Amblystegium serpens	Delicate willow moss	P	A,I,W	5
Atrichum altecristatum	Wavy starburst moss	A	A,S	6
Atrichum angustatum	Slender starburst moss	A	I,R,S,T,W	31
Atrichum crispum	Oval starburst moss	A	S	5
Aulacomnium palustre	Ribbed bog moss	P	I,S,W	8
Barbula unguiculata	Beard moss	A	A,I,R,S	10
Brachytheciastrum velutinum (*Brachythecium velutinum*)	Velvet feather moss	P	A	1
Brachythecium campestre (*Brachythecium salebrosum*)	Golden foxtail moss	P	A,R,W	7
Brachythecium cf. *digastrum*	Two-pouched foxtail moss	P	S	2
Brachythecium laetum (*Brachythecium oxycladon*)	Pleated foxtail moss	P	R,S,W	6
Brachythecium rivulare	River foxtail moss	P	I	2
Brachythecium rutabulum	Rough foxtail moss	P	I,S,W	7
Bryhnia novae-angliae	Bonsai moss	P	A,R,S	4
Bryoandersonia illecebra	Worm moss	P	S,W	2
Bryum argenteum	Silver moss	A	A,I,R,S	16
Callicladium haldanianum	Sword moss	P	R,S,W	13
Ceratodon purpureus	Fire moss	A	A,R,S,W	34
Dicranella heteromalla	Fine hair moss	A	A,I,S	14
Dicranum montanum	Crispy broom moss	A	S,W	2
Dicranum scoparium	Windswept broom moss	A	S	1
Ditrichum lineare	Linear hair moss	A		1
Ditrichum pusillum	Dwarf hair moss	A		3
Drepanocladus aduncus	Common sickle moss	P	W	1
Drepanocladus polygamus (*Campylium polygamum*)	Stiff star moss	P	S	1
Elodium paludosum (*Helodium paludosum*)	Narrow-leaved beard moss	P	I,S,W	4
Entodon cladorrhizans	Flat glaze moss	P	T,W	8
Entodon seductrix	Cord glaze moss	P	A,T,W	18
Fissidens fontanus	Limp pocket moss	A	R	1

Scientific name	Common name	Habit	Substrate	Specimens
Fissidens taxifolius	Common pocket moss	A	S	1
Funaria hygrometrica	Water-measuring cord moss	A	S	1
Gemmabryum caespiticium (Bryum caespiticium)	Tufted thread moss	A	A	2
Grimmia sp.	Dry rock moss	A	A	1
Haplocladium microphyllum		P	W	1
Haplocladium virginianum (Bryohaplocladium virginianum)	Virginia small plume moss	P	W	1
Hedwigia ciliata	Medusa moss	P	A,R	2
Homomallium adnatum	Homomallium moss	P		1
Hygroamblystegium varium (Amblystegium varium)	Variable willow moss	P	A,S,W	27
Hygroamblystegium varium (Hygroamblystegium tenax)	Fountain feather moss	P	A,R,W	3
Hypnum imponens	Brocade moss	P	W	2
Hypnum lindbergii	Pale plait moss	P	T,W	3
Hypnum pallescens	Lesser plait moss	P	R	2
Leptobryum pyriforme	Pear moss	A	S	5
Leptodictyum riparium	Wet thread moss	A	I,S,T,W	18
Leskea gracilescens	Necklace chain moss	P	A,R,S,T	6
Leskea obscura	Obtuse chain moss	P	T	1
Leskea polycarpa	Curled chain moss	P	T	4
Leucobryum glaucum	Pincushion moss	A	S,W	2
Mnium hornum	Lipstick thyme moss	A	S	1
Mnium lycopodioides (Mnium ambiguum)	North star moss	A	S	1
Oncophorus wahlenbergii	Goiter moss	A		1
Orthotrichum anomalum	Ribbed bristle moss	A	A	1
Orthotrichum pumilum	Low bristle moss	A	T	5
Orthotrichum pusillum	Tiny bristle moss	A	R,T	5
Orthotrichum stellatum	Bald bristle moss	A	A,T	4
Oxyrrhynchium hians (Eurhynchium hians)	Spare rug moss	P	A	1
Physcomitrium pyriforme	Goblet moss	A	S	6
Plagiomnium ciliare	Saber tooth moss	P	A,T,W	4
Plagiomnium cuspidatum	Baby tooth moss	P	A,S,T,W	12
Plagiomnium ellipticum	Marsh thyme moss	P	S	1

Platygyrium repens	Oil spill moss	P	T,W	22
Pohlia nutans	Copper wire moss	A	R,S,T	6
Polytrichastrum ohioense (Polytrichum ohioense)	Ohio haircap moss	A	S	1
Polytrichastrum pallidisetum (Polytrichum pallidisetum)	Pale haircap moss	A	S	1
Polytrichum commune	Common haircap moss	A	I,R,S	5
Polytrichum piliferum	Bristly haircap moss	A	R	3
Pseudotaxiphyllum elegans	Sprouting silk moss	P	R,S	3
Ptychostomum creberrimum (Bryum lisae var. cuspidatum)	Small-mouthed thread moss	A	A,I	2
Ptychostomum pseudotriquetrum (Bryum pseudotriquetrum)	Marsh bryum	A	I	5
Pylaisia polyantha (Pylaisiella polyantha)	Stiff paintbrush moss	P	T	1
Pylaisiadelpha tenuirostris	Gentle moss	P	T	1
Rhynchostegium serrulatum (Steerecleus serrulatus)	Dark beaked moss	P	I,S,T	6
Rosulabryum capillare	Capillary thread moss	A		1
Rosulabryum rubens (Bryum rubens)	Red thread moss	A	I	1
Schistidium apocarpum	Prickly cannikin moss	A	A,R	4
Sciuro-hypnum plumosum (Brachythecium plumosum)	Rusty feather moss	P	R	1
Sematophyllum adnatum	Signal moss	P	R	2
Sematophyllum demissum	Curled scrap moss	P	R	1
Sphagnum capillifolium	Small red peat moss	S	S	1
Sphagnum fimbriatum	Fringed peat moss	S	S	7
Sphagnum palustre	Blunt-leaved peat moss	S	S	7
Syntrichia ruralis (Tortula ruralis)	Talon moss	A	A	1
Tetraphis pellucida	Four-tooth moss	A	W	1
Thuidium delicatulum	Delicate fern moss	P	S,W	4
Tortula muralis	Wall-screw moss	A	A,R	2
Ulota crispa	Crispy tuft moss	A	W	1
Weissia controversa	Pigtail moss	A	A,R	4
Zygodon viridissimus	Green yoke moss	A	T	1

Liverworts

Aneura pinguis	Common greasewort	S	1
Aneura sharpii	Sharp's greasewort	S	1
Frullania brittoniae	Britton's scalewort	T	2
Frullania eboracensis	New York scalewort	T	2
Frullania inflata	Inflated scalewort	T	1
Lophocolea heterophylla	Crested liverwort	W	4
Pellia epiphylla	Overleaf pellia	I	1
Riccia huebeneriana ssp. *sullivantii* (*Riccia sullivantii*)	Sullivan's crystalwort	S	1

Note: "cf." indicates a likely but uncertain species identification.

Scientific names follow Barringer (2021a, b), common names are from various sources.

Habit: A = acrocarpous, P = pleurocarpous, S = *Sphagnum* (see text).

Substrate (not recorded for every specimen): A = Artificial (concrete, etc.), I = on or among stem bases of purple loosestrife, common reed, cattail, or switchgrass, R = Rock, S = Soil, T = Tree bark, W = Wood (including decorticated logs, rail ties, boards).

This indicates reasonable coverage of the flora but suggests there may be a number of undiscovered bryophytes. (For example, we did not collect much at the Ridgefield Nature Center which is a promising area with mature trees and little foot traffic.) We did find that every time we visited a substantially different type of habitat we found additional species. Diversity of bryophytes is partly due to the variety of macrohabitats, both wetland and upland, in the Meadowlands, as is also true for many other taxa.

Bryophytes tend to have narrower habitat affinities than seed plants, and the habitat specificities of bryophytes are often related to their establishment requirements (Cleavitt 2005). Establishment and persistence of some species depend on microhabitats or small-scale disturbances such as chipmunk trails on logs (Kimmerer and Young 1996) or reindeer and human footprints in mud (Theakstone and Knighton 1979). The total occurrences of six types of substrate in our bryophyte collections, by species not by individual specimen, are: Artificial (solid substrates such as concrete) 25, Rock 24, Soil 41, Tree or shrub bark 22 (Figure 5.2), Wood (dead tree, lumber, rail ties, etc.) 28, and stem bases 16. Our particular interest in purple loosestrife, which often supports bryophytes on its stem bases, may have biased the last category upward.

Bryophytes are very sensitive to pollution and to decreased moisture (Hallingbäck and Hodgetts 2000, Kimmerer 2003) and they predominate more in cool, moist climates and infertile habitats (Andrus 1990), characteristics

Figure 5.2 Large willow tree (*Salix*) **at Kearny Marsh West. Large trees provide important bark microhabitats for macrofungi, lichens, and bryophytes, as well as nesting cavities and invertebrate prey for birds, bats, and other wildlife. Photo by Erik Kiviat.**

that probably underlie the urban-sensitivity of these plants. Sulfur dioxide, sewage, chemical wastes, herbicides, and fertilizers are important threats (Hallingbäck and Hodgetts 2000). At least some mosses take up nitrogen directly from the soil, probably making them susceptible to nitrogen pollution (Ayres et al. 2006). Declining species of mosses in the Netherlands are those associated with *oligotrophic* (low-nutrient) habitats (Greven 1992). High sulfur dioxide deposition rates were correlated with species-poor moss communities (Winner 1988). All of this predicts low diversity in the Meadowlands. Heavy metal and nitrogen concentrations in European mosses are declining (Harmens et al. 2015); it would be revealing to compare concentrations in modern Meadowlands mosses to specimens from a century ago.

Some of the bryophytes we found are well known for their occurrence in urban areas. Both silver moss (*Bryum argenteum*) and fire moss (*Ceratodon purpureus*) are urban-tolerant and colonize anthropogenic disturbances (Dignard 1990, Kimmerer 2003); for example, they were two of the most frequently recorded bryophytes in a survey of fourteen European towns (Fudali 2005). Li et al. (2014) suggested silver moss as a desert species was preadapted to urban life. Silver moss tolerates road salt (Ćosić et al. 2019). Fire moss is quintessentially tolerant of harsh conditions: it grows in cities, in wildland burn areas, and in one study was relatively tolerant of a herbicide (Måren et al. 2008). The habitats of fire moss in Connecticut were described as "Burnt-over woods, roadsides, waste ground, and roofs" (Evans and Nichols 1908), and in Maine fire moss was considered "A common, weedy species able to tolerate dry, open, sunny conditions. Found on soil or sandy areas often in disturbed places (roadsides, sidewalks, roofs), also on charcoal, burned over soil, rock ledges or crevices, and rotting wood" (Allen 1995). Heady (1942) considered fire moss typical on compacted soils as well as natural sandy soils at a site in the Adirondack Mountains of New York. These tolerances suggest preadaptation to the harsh conditions prevalent in the Meadowlands. There fire moss is common along the edges of dirt roads and railroads, for example, and it seems abundant on formerly quarried or otherwise damaged areas of Laurel Hill. One Meadowlands habitat of this species was supratidal. Fire moss is often associated with gravelly substrates, including roofs (Kimmerer 2003).

Silver moss is common along forest paths and in cities; it is dispersed by fragmentation as well as by spores (Kimmerer 2003, Munch 2006). Frahm (2008) considered silver moss weed-like and probably human-dispersed. This moss "grows everywhere . . . specially fond of dry compact soil in sandy fields and waste places. It grows abundantly in paths and between the bricks of sidewalks in towns and cities" (Grout 1924). Silver moss may have evolved in association with seabird guano at cliffside nesting sites (Kimmerer 2003) which are sparsely vegetated, drought-prone, and nitrogen-rich. This species is a *nitrophyte* (associated with high-nitrogen environments [Crum 1973]). Silver moss is generally tolerant of metals, and fire moss evolves metal-tolerant forms (Shaw 1994).

Variable willow moss (*Hygroamblystegium varium*), slender starburst moss (*Atrichum angustatum*), and cord glaze moss (*Entodon seductrix*) were the second, third, and fourth most frequent mosses in our collections. Variable willow moss grows in microhabitats that are moist or wet most of the time. Cord glaze moss was in supratidal habitats in some instances and clearly tolerant of irregular flooding by brackish water. In the Great Lakes area, slender starburst moss grows in dry, open woods, low quality lawns, road verges, and other disturbed habitats (Crum 1973). Slender starburst moss, fire moss, and silver moss, along with other urban-tolerant organisms, thrive in the droughty, often nutrient-rich, weather-beaten habitats so characteristic of urban environments. Variable willow moss, fire moss, and cord glaze moss are tied for the greatest diversity of substrates (5 of 6 categories, Table 5.2) in our collections.

Some other mosses on our list are common ecological generalists. Baby tooth moss (*Plagiomnium cuspidatum*) "is common in lawns and parks in moist shady corners, and is to be found abundantly in moist woods everywhere" (Grout 1924). Crum (1973) characterized several species as occurring in disturbed habitats such as roadbanks; these include dwarf hair moss (*Ditrichum pusillum*), fine hair moss (*Dicranella heteromalla*), beard moss (*Barbula unguiculata*), goblet moss (*Physcomitrium pyriforme*), small-mouthed thread moss (*Ptychostomum creberrimum*), pleated foxtail moss (*Brachythecium laetum*), golden foxtail moss (*Brachythecium campestre*), and oil spill moss (*Platygyrium repens*). Among the bryophytes of Old Québec City, all but one are distributed around the Northern Hemisphere and many are ubiquitous (Dignard 1990). Four of the moss species we found in the Meadowlands (ribbed bog moss [*Aulacomnium palustre*], fire moss, fine hair moss, and copper wire moss [*Pohlia nutans*]) were also the most frequently collected mosses in seventeen abandoned coal mines or exposed coal seams in western Pennsylvania (Davis and Atwood 2010) where they would have been exposed to toxic elements abundant in coal as well as the high acidity typical of water in coal mines.

Although most of the bryophytes in Table 5.2 are common and widespread in New Jersey, we found several species that appear to be rare. Red thread moss (*Rosulabryum rubens*) is apparently new to New Jersey, but this is a confusing taxon (Eckel and Shaw 1991). Additional species new to the state are green yoke moss (*Zygodon viridissimus*) and Sharp's greasewort (*Aneura sharpii*) (Barringer 2021a, b). Virginia small plume moss (*Haplocladium microphyllum*) was last collected in New Jersey ca. 1865 (Barringer 2021a). Not new, but rare in New Jersey are pale haircap moss (*Polytrichastrum pallidisetum*) and small red peat moss (*Sphagnum capillifolium*; Barringer 2021a). In Polish urban parks and cemeteries, in addition to common generalist urban bryophytes, Fudali (2005) found specialist species not previously recorded in densely developed urban areas. Clemants and Ketchledge (1990) remarked that, in New York, occurrence of rare mosses depended largely on specialized microhabitats.

At the other end of the urban tolerance spectrum from fire moss and silver moss are the peat mosses (*Sphagnum*). Peat mosses are stressed by increases in nutrient levels (Vitt et al. 2003), calcium (Vellak et al. 2014), desiccation, and fluctuating water levels, and in general are sensitive to human disturbance (Tousignant et al. 2010). In Swedish bogs, the growth of several species of peat mosses was reduced by experimental nitrogen fertilization (Gunnarsson and Rydin 2000). We found a single very restricted occurrence (< 0.5 m^2) of fringed peat moss (*Sphagnum fimbriatum*) in the Kane Natural Area before the berms were breached by a storm and the marsh flooded with brackish water in 2007, and then sprayed with herbicides and planted with native marsh plants in 2010; fringed peat moss is undoubtedly extirpated there. In 2015, we discovered that fringed peat moss and blunt-leaved peat moss (*Sphagnum palustre*) were abundant in a portion of Upper Penhorn Marsh. Fringed peat moss was reported by Andrus (1980) and McQueen

(1990) to be minerotrophic, for example, associated with water rich in cations (calcium, etc.), and one of the few peat mosses to occur at pH 7 (neutral). Crum (1973) considered fringed peat moss mesotrophic (occurring in moderately fertile habitats) and somewhat tolerant of drying, and Karlin and Andrus (1988) stated that both fringed peat moss and blunt-leaved peat moss do well in moderately to strongly minerotrophic habitats. Small red peat moss (*Sphagnum capillifolium*), rare in New Jersey (Barringer 2021a), was also in Upper Penhorn Marsh. Peat mosses are able to disperse long distances into new habitats, but whether this occurs by spores or fragments is unknown (McQueen 1990). Fringed peat moss produces spores more readily than other peat mosses (Andrus 1980, McQueen 1990).

Because Atlantic white cedar swamps are fresh, low nutrient, organic soil wetlands with relatively stable, humid conditions near the ground, it is likely that the historic cedar swamps of the Meadowlands had a well-developed bryoflora. In southern New Jersey, cedar swamps can support a high diversity of peat mosses (Karlin 1990), and in New England, several bryophyte species are found only in Atlantic white cedar swamps (Lincoln 2008). We wonder if any unusual bryophytes have persisted on the cedar stumps in some of the fresher, more sheltered parts of the Meadowlands such as Upper Penhorn Marsh.

Mosses (other than peat mosses) are characterized as *acrocarpous* or *acrocarps* (having more-or-less erect and unbranched shoots with sporophytes borne at tips of shoots, and often forming tufts or cushions) or *pleurocarpous* (with creeping, branched shoots, sporophytes borne on the sides of the shoots, and often forming mats). Acrocarps are more tolerant of drying and thus may be more tolerant of urban conditions (see, e.g., Alpert 1990, Kimmerer 2003, Perhans et al. 2009). Dignard (1990) found that acrocarps greatly outnumbered pleurocarps on large trees in Old Québec City, and Giordano et al. (2004) found only acrocarps in an Italian city center but an increasing proportion of pleurocarps in progressively less urban areas. However, our collections comprise equal numbers of acrocarps (43) and pleurocarps (41). Perhaps the abundance of moist-to-wet habitats in the Meadowlands fosters the pleurocarpous mosses despite the urban environment. Many specimens of the pleurocarps came from wet woods and other wet microhabitats.

Some bryophytes (and other organisms) are *calcicoles*. Urban environments are often rich in calcium due to the calcium in, and leaching from, concrete rubble, mortar, and concrete pavement. Vegetation fires leave ash rich in calcium (Crum 1973), and this would be an additional calcium source in the Meadowlands. Certain types of bedrock are also calcareous. Pigtail moss (*Weissia controversa*, from the east end of Laurel Hill), beard moss (*Barbula* cf. *unguiculata*, on concrete rubble and the North Arlington shale mine face), ribbed bristle moss (*Orthotrichum anomalum*, concrete rubble), and common sickle moss (*Drepanocladus aduncus*, riparian swamp of the upper estuary) were considered calcicoles (Crum 1973). The Laurel Hill and North Arlington collections suggest some of the Meadowlands bedrock is calcareous. However, some organic soils in freshwater wetlands, organic wetland soil exposed to air, and some rock and tree bark substrates are likely to be acidic.

Liverworts are often narrower in their affinities for microhabitats, including features of the substrate, shade, and humidity, and they are more sensitive to logging, than mosses. Most liverworts require high humidity and are sensitive to desiccation (Schuster 1949, Munch 2006, Hong 2007). Liverworts decreased even more strongly than mosses in a Swedish study of small patches of forest left after logging (Perhans et al. 2009). Human activities are generally inimical to liverworts (Schuster 1949), and liverwort species richness decreased along a gradient from outside an Italian city to the city center (Giordano et al. 2004). The scarcity of liverworts in the Meadowlands is striking. One liverwort we found was at a spring above normal tidal influence in a sheltered location in Ridgefield Park. The physically and chemically stable water that tends to issue from springs may create a safe haven for a liverwort (provided of course that the groundwater feeding the spring is not polluted). This was crested liverwort (*Lophocolea heterophylla*), a very common and widespread species (Lincoln 2008). Our second station for crested liverwort was on a wet rotting log in a small floodplain pool of the upper Hackensack River estuary in Teaneck. It is interesting that this species occurs at the edges of the region in sheltered enclaves and is a very common species outside the Meadowlands. We also found crested liverwort in Borg's Woods, an old growth forest remnant just outside the Core Meadowlands in Hackensack. Another liverwort was *Pellia*, on a floating peat mat in Kearny Marsh West. Liverworts are diverse on organic substrates (Schuster 1949). At the Kearny location, the peat may insulate liverworts from the ambient water quality in the marsh as well as hold a perennial supply of moisture. This species was the common *Pellia epiphylla*, typically found in disturbed areas (Lincoln 2008). Conspicuously absent was common liverwort (*Marchantia polymorpha*), a large thallose species described as a very common species of disturbed areas (Lincoln 2008) that we knew, for example, from the supratidal tops of old wooden pilings in Cornwall Bay, an urban and seasonally slightly brackish area of the Hudson River near Newburgh. Fudali (2005) found only 11 liverwort species or 9% of the total bryophyte species in the parks and cemeteries of fourteen Polish cities; in our survey, liverworts comprised 8.4% of total bryophyte species. Surprisingly, there were 8 liverworts and 1 hornwort along with 38 mosses in two urban parks in Queens, New York City (Morgan and Sperling 2006), thus liverworts constituted 17% of the bryophytes found. Our list of 8 liverworts has only 3 species in common with the 9 liverworts found in Hudson River fresh-tidal swamps by Leonardi and Kiviat (1990). In 352 km² non-urban Montour County, in the folded Appalachians of central Pennsylvania, only 64 mosses and 7 liverworts have been reported (Davis and Allen 2015); Montour is more than twice the area of the Greater Meadowlands. On Staten Island, with the same area as the Greater Meadowlands, Ketchledge (1980) recorded 151 mosses. Staten Island has been studied in much greater detail than either the Meadowlands or Montour County.

The 95 species of bryophytes we collected in the Meadowlands may be compared to the bryofloras documented in other urban areas. For example, Dignard (1990) found 34 mosses and 6 liverworts on stone structures and associated habitats in Old Québec City, Canada. Dignard stated that both air pollution and loss of favorable habitats limit bryophyte diversity in urban areas. Morgan and Sperling (2006) reported 77 bryophytes from specimens collected in the 1970s and 1980s in the 578 ha of two parks in a highly urbanized matrix in Queens, New York City, but recently they found only 47 species. Their study area is tiny compared to the Meadowlands. Fudali (2005) found 219 species in West Berlin, Germany. Unfortunately, comparisons such as this are imprecise because the definition of urban environment varies from one study to another as does the habitat composition of the study area, the thoroughness and time period of survey, and the regional species pools. However, we think that species richness in the Meadowlands is intermediate for an urban area. A substantial portion of the Meadowlands species list, however, comprises those mosses that we found in limited and atypical areas such as Laurel Hill and Little Snake Hill, as well as the wet forests of the northern Meadowlands. The wetland fill, developed areas, and marshes that constitute much of the Core Meadowlands support fewer species.

Little attention has been paid to bryophytes in estuarine habitats. Evans and Nichols (1908, 33) asserted that no moss was a true *halophyte* (i.e., adapted to a saline environment). However, the rare moss *Bryum marratii* in Wales occurs at the upper edges of salt marshes where salinity is close to seawater concentration (Callaghan and Farr 2018), and Ćosić et al. (2019) discussed several halophytic European mosses. In a laboratory experiment, three species of New Jersey mosses that occur in coastal dunes did not tolerate full strength (28 ppt salinity) salt spray although fire moss tolerated salt spray when also exposed at intervals to simulated rain (Boerner and Forman 1975). Limp pocket moss (*Fissidens fontanus*), which we found in Wolf Creek (Ridgefield) well above tidewater, occurs in "coastal estuaries" in North America (Pursell 2007). We did find wet thread moss (*Leptodictyum riparium*) in a brackish intertidal habitat, and the microhabitat of flat glaze moss (*Entodon cladorrhizans*) was supratidal at one site, suggesting this moss may tolerate both brackish and urban conditions. Garbary et al. (2007) reported 5 mosses from supratidal salt marshes in Nova Scotia, and commented that previously no bryophyte had been reported from salt marsh habitat in North America although bryophytes had been considered a part of salt marsh plant communities in Europe. Delicate willow moss (*Amblystegium serpens*), which we found twice in the Meadowlands, was one of the 5 mosses found in the Nova Scotia saltmarsh study (*ibid.*). Papp et al. (2016) reported 35 mosses and 5 liverworts from saline grasslands in the Balkans; among these species were 5 mosses found in the Meadowlands: beard moss, tufted thread moss (*Gemmabryum caespiticium*), fire moss, common sickle moss, and talon moss (*Syntrichia ruralis*). Hill et al. (2007) listed a number of bryophytes that tolerate some degree of salinity in the U.K. Of the 52 moss species found in two fresh-tidal swamps of the Hudson River (Leonardi and Kiviat 1990), 25 are shared with our Meadowlands list. The separation of the two floras is due at least partly to the absence of the fresh-tidal swamp habitat in the Meadowlands, and the inclusion of several upland habitats there. Bryophytes are diverse and abundant in Hudson River fresh-tidal swamps, despite high nutrient levels and circumneutral pH in Hudson River water, and tide range of 1.2 m. Mosses remove pollutants from water (Kimmerer 2003); however, the moss biomass in the Meadowlands is probably too small relative to surface water volume to have much effect except perhaps in springs and seeps.

The copper-bearing bedrock at the western edge of the Meadowlands in North Arlington could support *copper mosses* that have evolved in association with high copper levels, including copper-tolerant forms of common mosses such as fire moss and silver moss as well as more specialized and rare species (Shaw 1994) that should be looked for in the Meadowlands. In general, bryophytes accumulate heavy metals in the cell walls and some species have been used as indicators of air or water pollution (Hallingbäck and Hodgetts 2000).

Bryophytes include many opportunistic species using microhabitats or temporary habitats not favorable for other plants (Slack 1977, 62; Andrus 1990). Perhaps the difficulties that both vascular plants and bryophytes have with urban existence are responsible for the vegetation of the Meadowlands that is so often sparse in both coverage and species. This sparseness, however, may maintain space for many species that tolerate urban conditions, and may help explain the moderate levels of diversity we see in plants as well as animals. Dispersal of many bryophytes occurs by spores or small vegetative propagules, for distances of several kilometers (Miller and McDaniel 2004), across Europe (Frahm 2008), and sometimes worldwide (Kimmerer 2003); thus a large number of species should be able to reach the Meadowlands. Once there, establishment and survival must depend on the local presence of suitable macrohabitats and microhabitats as well as the quality of air, soil, and water. Enemies may limit some bryophytes, although slugs, for example, were thought to eat mosses less than seed plants (South 1992). Genetic data from fire moss and silver moss indicated a low level of large-scale population structure, suggesting that long distance dispersal occurred commonly in these species (Miller and McDaniel 2004). Likely the most prevalent urban mosses are effective long-distance dispersers and especially tolerant of urban conditions. And our species accumulation analysis suggests that the variety of potential habitats (such as moist woodlands and rocky crests) is important to the diversity of bryophytes in an urban environment. Fudali (2005) found that bryophyte species richness reflected habitat heterogeneity in Polish cities, but that some bryophytes had broader habitat affinities (e.g., substrate, moisture) within than outside cities. Fudali also found bryophyte species richness depended more on microhabitat diversity than macrohabitat diversity.

New Jersey, and the Hudson Valley, are regions of high peat moss diversity (Andrus 1980, Karlin and Andrus 1988). Although we found only 3 peat mosses in the Meadowlands, 50 species have been documented in New Jersey (Barringer 2021a). Of those, at least 10 have been documented in Bergen and Hudson counties post-1960 and 9 additional species pre-1960 (Karlin and Andrus 1988). For Bergen County, 180 species of mosses, 60 species of liverworts, and 1 hornwort are represented by specimens in the herbarium of the New York Botanical Garden (NYBG no date). The proportion of Bergen County moss species (almost half) that we have found in the Meadowlands represents reasonable diversity especially considering that the NYBG specimen records span the period 1858–1967 and some of the species represented are probably extirpated from Bergen County. However, the diversity of liverworts (8 species), hornworts (0 species), and *Sphagnum* (3 species) in the Meadowlands encompasses a small proportion of the species recorded in the county and must be considered poor. These are clearly taxa that are not doing well in the Meadowlands and probably are generally urban-sensitive.

Although we found a moderate species richness of mosses (and lichens, see below) for an urban environment, there is more to the biota than the list of species. Most of the mosses and lichens occurred sparsely and were scattered in small patches in particular habitats. Rarely did we see extensive bryoid cover on tree bark, soil, rock, or even wet rotting logs. Fudali (2005) found that bryophytes typically formed only a few small patches in the Polish parks and cemeteries studied. The mature forest remnants of the northern Meadowlands contained many trees in the 50–90 cm dbh range, and some of these large trees were surrounded by 25–50+ m of mature forest buffer. Large trees of certain species can be especially important microhabitats for bryoids, and in Sweden old beech trees with a higher bark pH and infected by fungi were most important as hosts of bryophytes and lichens of conservation concern (Fritz 2009, Fritz et al. 2009). Yet in the Meadowlands in most cases bryoids occurred on the bark only at the very base, within perhaps 20 cm of the soil. It was unusual to see a tree of any size with extensive moss or lichen cover reaching up the trunk for 2–3 m or more. Other researchers have commented that it is rare in urban areas to see corticolous mosses more than several centimeters above ground, probably due to poor air quality, while outside the city many mosses flourish on tree bark from the ground to great heights (Grout 1916, LeBlanc and De Sloover 1970). Winner (1988), studying a gradient of decreasing sulfur dioxide, found that mosses first appeared at tree bases.

FUNGI (NON-LICHENIZED FUNGI)

The fungi are a very diverse and important group of organisms. They play a crucial role in the decomposition of both woody and herbaceous plants, biogeochemical cycling, disease and death of plants and animals, diets of insects, rodents, and humans, and the establishment and survival of vascular plants. Very important pharmaceuticals have been derived from fungi.

Macrofungi, that is fungi with fruiting bodies large enough to be conspicuous to the naked eye, are almost endless in their spectra of size, shape, and color. Most macrofungi have *mycelia* (strands of fungal tissue) in the soil, wood, or other substrate, and ephemeral *fruiting bodies* (spore producing structures). The fruit bodies often appear during or after moist weather. Macrofungi inhabit forests, lawns, meadows, wetlands, wood chip piles, tide wrack and driftwood, dumps of anything organic, and waste ground. Some of the major groups of macrofungi are the agarics (gilled mushrooms), boletes, tooth fungi, polypores (including shelf fungi), coral and club fungi, stinkhorns, and the puffballs and earthstars. There has been no previous survey of macrofungi in the Meadowlands.

Slime molds, comprising three groups of organisms, are animals (the Mycetozoa), not fungi, although in some stages of their life cycle slime molds are fungus-like. Slime molds fruit on downed logs, soil, live trees and other plants, and inanimate substrates, usually after rains. Many species are small and inconspicuous, with a few large and obvious, and they seem to occur especially in the remnant swamp forests such as Schmidt's Woods (Secaucus). Slime molds are not treated here, as there is so little information from the Meadowlands.

We have collected more than 200 specimens of macrofungi in an ongoing survey of the Greater Meadowlands, with identifications in progress. Macrofungi seem moderately diverse, with many higher taxa represented. It will be interesting to know which species survive in this highly altered urban-industrial area, and whether they use altered substrates such as lawns and garden beds, landfill cover with rotting rail ties, stump and chip dumps, or natural substrates such as large living and dead trees. The conservation of mature forests (wet and dry), including the standing dead or injured trees and the downed logs and other woody debris on the forest floor, is very important for the conservation of fungal diversity. Although rare fungi have been the subject of research and conservation in Europe, very little attention has been paid to them in the United States (but see Allen and Lendemer 2015).

A tooth fungus associated solely with century-plus-old Atlantic white cedar trees, *Echinodontium ballouii*, although formerly known from the New Jersey coast has been found recently only in New Hampshire and is ranked as "Endangered" by the International Union for the Conservation of Nature (IUCN; Ainsworth 2019). This mysterious species could survive on old cedar stumps and logs in the Meadowlands.

LICHENS

The lichens are one of the most neglected macroscopic components of biodiversity in the Meadowlands, and in much of the U.S. Lichens are often scarce in urban areas (e.g., Brodo 1966, LeBlanc and Rao 1973), and many are small and cryptic in appearance anywhere. Moreover, many lichens require microscopic examination and chemical or genetic analysis to identify species. Yet these odd patches or tufts of living matter are visually beautiful and scientifically intriguing, in part because it seems that they should not be here in the city at all. Our opportunistic lichen survey discovered a surprising 58 species in the Meadowlands.

Lichens are intimate, mutualistic associations of fungi and algae (algae in the broad sense, both green algae and Cyanobacteria). Lichens are also called *lichenized fungi*. The relationship is complex and very successful ecologically, and lichens are recognizable self-perpetuating entities, with unique chemistry and distinctive physical form (Goward 2008a, b, 2009). Lichens are combinations of a particular fungus and a particular alga, although sometimes the fungus associates variably with several related algal species, and many of the algae can also be found living independently. Goward asserted that lichens should be construed simultaneously as relationships between pairs of disparate organisms, and individuated entities that are like species. Certainly for practical purposes, lichens (i.e., lichen species) can be identified, studied, used as indicators of air quality, and their ecological relationships understood.

Most lichen species take one of three forms: *crustose, foliose,* or *fruticose.* Crustose lichens occur as very thin dots or patches, sometimes extensive, usually on rocks, bark, or concrete, more rarely on soil, and they are inseparable from their substrate. (*Tree algae,* which are free-living algae without fungal associates, also form thin crusts on tree bark but are usually greener and more homogeneous than crustose lichens.) Foliose lichens often form roundish patches, have a leaf-like structure at the edges or throughout, and can be more-or-less readily separated from the substrate. Foliose species grow on rocks, bark or decaying wood, and occasionally on the soil. Both crustose and foliose lichens often occur on gravestones, depending on the sun exposure and the mineral composition, texture, and age of the stone.

Some fruticose lichens have erect projections or tufts, that may be 1–3 cm or more tall and may appear as tiny goblet, spike, or scepter shapes. Other fruticose lichens are much-branched and appear superficially mesh-like. The only fruticose lichens found in our study belong to the genus *Cladonia* (in part); we refer to them as *pixie lichens* to distinguish them from reindeer lichens, beard lichens, and other fruticose growth forms. Some fruticose lichens (especially the pixies) have basal scales resembling foliose or crustose lichens that eventually give rise to the vertical projections; these *basal squamules* may persist or disappear as the lichen develops. The fruticose lichens most often occur on soil, dead wood, or rocks, and sometimes on artificial substrates such as old outdoor wooden furniture, wooden fence rails, or weathered rail ties.

Lichens, depending on the taxon, are either *chlorolichens* which have a green algal *photobiont* (photosynthesizing partner), or *cyanolichens* which have a cyanobacterial photobiont. All the lichens we found in the Meadowlands are chlorolichens. Cyanolichens are especially sensitive to sulfur and nitrogen oxides in the air and the resulting acidity (Richardson and Cameron 2004), and also require more moisture (Ellis and Coppins 2006); they reach their greatest development in old forests (Allen and Lendemer 2015). Thus, cyanolichens are urban-sensitive.

A distinctive group of foliose lichens that is common in wildlands such as the New Jersey Highlands and Hudson Highlands, the rock tripes (*Umbilicaria*), are sensitive to air pollution (see Smiley and George 1974). Rock tripes are absent from the Meadowlands although there is apparently suitable substrate, e.g., at Laurel Hill. Also absent, the reindeer lichens (genus *Cladonia* in part), fruticose species that form highly branched masses, may be intolerant of urban conditions (*Cladonia leporina*, however, was recently documented in Brooklyn [Hoffman 2019]). They are typically associated with acidic substrates (Brodo et al. 2001) or with well-drained, coarse, mineral soils (Godeau 2019). Beard lichens (e.g., *Usnea*), complexly shaped fruticose, often pendent, species that grow on tree trunks and branches, do not occur in the Meadowlands. Both reindeer lichens and beard lichens apparently have especially high surface-to-volume ratios due to extensive and fine-textured branching, and thus may be more sensitive to pollutants that deposit on the surfaces of the *thalli* (the lichen bodies).

Because of the experience of other northeastern New Jersey forests that virtually lacked lichens ca. 1980, we expected to find species and numbers of lichen thalli very scarce in the Meadowlands and not of conservation concern. As with mosses, we opportunistically collected lichens, in total about 380 specimens, in the Meadowlands for 20 years after noticing them on tree-of-heaven (*Ailanthus altissima*) stems among common reed in the Kingsland Impoundment. Subsequently, we found lichens on the bases of trees and in places higher on trunks, especially larger tree trunks of various species, in forests such as Schmidt's Woods as well as where individual trees occurred away from the major highways and perhaps where the humidity near the ground was higher. We found many lichens on bedrock and rock fragments at Laurel Hill. We also found some of the pixie lichens on dry mineral soils and old dry decaying wood. It was intriguing to find several species on anthropogenic substrates (concrete, mine tailings, decaying rail ties in a dump, gravelly dredge spoil). However, no lichen was abundant in the Meadowlands, with coverage not extensive and individual occurrences often scattered, similar to the pattern for mosses.

Meadowlands lichens must tolerate air pollution, as well as low humidity, in habitats such as Laurel Hill and the more elevated fill areas. Sulfur dioxide, mostly originating from the combustion of fossil fuels locally or far upwind, is the air pollutant that most commonly limits the development of lichen communities (Nash and Wirth 1988). Lichens absorb nutrients principally from aerial deposition on the thalli and are not able to excrete the toxicants that are also absorbed. Thus, air pollutants, including sulfur, fluorine, and heavy metals, may injure or kill lichen thalli, and lichens are useful indicators of the levels of certain toxicants. Long-term exposure to lower levels of sulfur dioxide in air is more harmful to lichens than short-term exposure to higher levels. An annual mean sulfur dioxide level of about 0.013–0.027 ppm (parts-per-million) may be high enough to eliminate lichens (Hinds and Hinds 2007; see Chapter 1).

As air quality continues to improve, we expect lichens to become more diverse and abundant in the Meadowlands. Air quality data summarized in Chapter 1 indicate declining levels of sulfur dioxide and certain other air pollutants in recent decades. Notable lichen recovery has occurred elsewhere in the New York metropolitan area (Delendick 1994, Allen and Howe 2016) and in other regions (e.g., Loppi et al. 1997, Hultengren et al. 2004) in relation to improving air quality. Modest and decreasing levels of sulfur dioxide may help explain the moderate diversity of lichens in the Meadowlands, keeping in mind that lichens grow slowly and may take many years to recolonize and develop appreciable cover. Wirth (1988) reported that even at low sulfur dioxide levels in the Black Forest of Germany, the most sensitive lichens declined, and that "Almost all epiphytic lichen communities in Central Europe are clearly declining." Fortunately, since then lichen diversity and health have improved widely in Europe (Słaby and Lisowska 2012). Lisowska (2011) found lichen recolonization of an urban-industrial center in southern Poland during a thirty-year period of improving air quality.

In general, lichens in the northeastern U.S. are not subject to intense herbivory. Slugs, which readily feed on lichens (South 1992), might limit some species in the Meadowlands where slugs are common. Lichens are, however, sensitive to trampling because the thalli are brittle when dry (minor breakage allows dispersal of some lichens by means of fragments, but too much breakage prevents full development of thalli). Thus, lichens do not grow on surfaces where people walk or sit frequently, or where vehicles or equipment are driven. Collecting lichens for decorations such as holiday wreaths and model railroad shrubbery may be a threat to fruticose species, but they are probably too limited in cover in the Meadowlands to repay collecting effort.

Waters and Lendemer (2019) compiled a list of 479 species of lichens, lichenicolous fungi, and allied fungi, 451 of which were lichens, found in New Jersey. Hudson County is the least-collected New Jersey county (*ibid.*). Given the study area extent, uneven geographic coverage, and the opportunistic nature of our survey, we think there are many species we overlooked. Rare lichen species may be missed in surveys due to the cryptic nature of many crustose (and some other) lichens, and the lack of attention to certain habitats (Löhmus and Löhmus 2009).

A list of 58 lichens we collected in the Meadowlands is in Table 5.3; the count assumes that two tentatively identified *Cladonia* species shown with question marks were correctly named. Mealy rosette lichen, candleflame lichen, and mealy rim-lichen, respectively 2 foliose species and 1 crustose species, were the lichens we collected most frequently. The proportion of species new to our lichen survey each year from 2001 to 2018 declined substantially, as for the bryophytes, but did not stabilize at zero, again indicating an incomplete survey.

Three Meadowlands lichens may be exceptions to the pattern of common generalist species: grainy shadow-crust lichen (*Hyperphyscia adglutinata*), crowned pixie-cup (*Cladonia carneola*), and southern powdered ruffle lichen (*Parmotrema hypotropum*). All three species were considered rare in New England (Hinds and Hinds 2007) with a border only 31 km away. We found the inconspicuous grainy shadow-crust lichen in Lincoln Park (Jersey City), Mill Creek Point Park (Secaucus), Laurel Hill (Secaucus), Palmer Street (Carlstadt), Kane Natural Area (Carlstadt?), and Mt. Olivet Cemetery (Newark). The lichen was on the bark of London plane, weeping willow, tree-of-heaven, and eastern cottonwood, on both small and large trees. This lichen may be rare or just rarely collected. The Consortium of North American Lichen Herbaria (CNALH 2019) has only 3 specimens of

Table 5.3 Lichens collected in the New Jersey Meadowlands, 1999–2018.

Scientific name	Common name	Habit	Substrate	Specimens
Amandinea polyspora	Button lichen	Cr	T	8
Amandinea punctata	Tiny button lichen	Cr	R T W	13
Anisomeridium polypori	Chimney lichen	Cr	T	4
Bacidina egenula	Dotted lichen	Cr	R	1
Biatora printzenii	Printzen's dotted lichen	Cr	T	1
Candelaria concolor	Candleflame lichen	Fo	R T W	27
Candelariella efflorescens	Egg yolk lichen	Cr	W	1
Catillaria nigroclavata	Black key lichen	Cr	T W	4
Chrysothrix (Arthonia) caesia	Frosted comma lichen	Cr	T	5
Cladonia carneola	Crowned pixie-cup	Fr	S	2
Cladonia chlorophaea	Mealy pixie-cup	Fr	W	1
Cladonia conista	Cup lichen	Fr	R T	2
Cladonia cristatella	British soldiers	Fr	A S W	13
Cladonia ?incrassata	Powder-foot British soldiers	Fr	W	1
Cladonia macilenta var. bacillaris	Lipstick powderhorn	Fr	W	12
Cladonia ?parasitica	Fence-rail cladonia	Fr	W	1
Cladonia peziziformis	Turban lichen	Fr	A R W	13
Cladonia pleurota	Red-fruited pixie-cup	Fr	S	4
Cladonia pyxidata	Pebbled pixie-cup	Fr	A	1
Cladonia rei	Wand lichen	Fr	R S	7
Cladonia squamosa	Dragon funnel	Fr	R	1
Cladonia (polycarpoides) subcariosa	Peg lichen	Cr	A R S	6
Endocarpon pallidulum	Western crepes	Cr	A R	2
Endocarpon petrolepideum	Chalice lichen	Cr	R	3
Flavoparmelia baltimorensis	Rock greenshield lichen	Fo	R	1
Flavoparmelia caperata	Common greenshield lichen	Fo	T	7

Species	Common name			
Hyperphyscia adglutinata	Grainy shadow-crust lichen	Fo	T	8
Lecanora hybocarpa	Bumpy rim-lichen	Cr	T	1
Lecanora strobilina	Mealy rim-lichen	Cr	T W	18
Lecanora subpallens	Rim lichen	Cr	T	2
Lecanora symmicta	Fused rim-lichen	Cr	W	1
Lecidea (Pyrrhospora) varians	Crimson dot lichen	Cr	T W	11
Lepraria finkii (including L. lobificans)	Fluffy dust lichen	Cr	R T	4
Lepraria vouauxii	Lobed dust lichen	Cr	R	1
Myriolecis (Lecanora) albescens	Rim lichen	Cr	R	1
Myriolecis (Lecanora) dispersa	Mortar rim-lichen	Cr	R	3
Parmelia sulcata	Hammered shield lichen	Fo	W	2
Parmotrema hypotropum	Southern powdered ruffle lichen	Fo	T	1
Phaeophyscia adiastola	Powder-tipped shadow lichen	Fo	T	2
Phaeophyscia squarrosa (Phaeophyscia imbricata)	Whiskered shadow lichen	Fo	T	1
Phaeophyscia pusilloides	Pompom shadow lichen	Fo	T W	5
Phaeophyscia (Physcia) rubropulchra	Orange-cored shadow lichen	Fo	T W	15
Physcia adscendens	Hooded rosette lichen	Fo	T	1
Physcia millegrana	Mealy rosette lichen	Fo	R T W	51
Physcia stellaris	Star rosette lichen	Fo	T	13
Physciella chloantha	Cryptic rosette lichen	Fo	T	19
Placynthiella icmalea	Stiff tar-spot lichen	Cr	R T	2
Placynthiella uliginosa	Tar-spot lichen	Cr	S	1
Punctelia caseana	Powdered speckled shield lichen	Fo	W	1
Scoliciosporum umbrinum	Dot lichen	Cr	R	3
Squamulea subsoluta (Caloplaca irrubescens, Caloplaca subsoluta)	Dispersed firedot lichen	Cr	A R	3
Trapelia glebulosa	Radiate pebble lichen	Cr	R	2
Trapelia placodioides	Pebble lichen	Cr	R	3
Trapeliopsis flexuosa	Board lichen	Cr	A W	6

Scientific name	Common name	Habit	Substrate	Specimens
Verrucaria sp.	Speck lichen	Cr	A R	2
Xanthocarpia (Caloplaca) feracissima	Sidewalk firedot lichen	Cr	A	2
Xanthomendoza weberi	Bare-bottom sunburst lichen	Fo	R	1
Xanthoparmelia plittii	Plitt's rock shield	Fo	R S	5

Note: Scientific names are from Esslinger (2019); common names are from Brodo et al. (2001) or other sources if no name in Brodo. Lichen nomenclature is in flux and synonyms for scientific names are in Esslinger (2019).

Habit: Cr = Crustose, Fo = Foliose, Fr = Fruticose.

Substrate: A = Artificial (concrete, etc.), R = Rock, S = Soil, T = Tree or shrub bark, W = Wood (including logs, fences, rail ties, etc.).

Specimens = number of times collected in our survey.

grainy shadow-crust lichen from New Jersey, compared to 218 specimens of the common and conspicuous British soldiers (*Cladonia cristatella*). Grainy shadow-crust lichen has an extensive range in the eastern U.S. It is intriguing to find this little-known lichen in the Meadowlands, although in general many of the smaller lichens are under-collected and poorly studied, and this species may not actually be rare. CNALH has 132 New Jersey specimens of powdered ruffle lichen and none of crowned pixie-cup. Waters and Lendemer (2019) reported powdered ruffle lichen common in central and southern New Jersey. They did not find crowned pixie-cup and reported grainy shadow-crust without comment on abundance. Our specimens of crowned pixie-cup may be state records. We found it on soil at Laurel Hill and on the Berry's Creek Canal spoil bank in Oritani Marsh.

Candleflame lichen (*Candelaria concolor*) and grainy shadow-crust lichen were reported to be nitrophytes or usually so (Brodo et al. 2001, Hinds and Hinds 2007), and Wolseley and James (2002) considered the latter species urban-tolerant. In Italy, Frati et al. (2007) considered four species on our Meadowlands list to be nitrophytes: tiny button lichen (*Amandinea punctata*), common greenshield lichen (*Flavoparmelia caperata*), hammered shield lichen (*Parmelia sulcata*), and hooded rosette lichen (*Physcia adscendens*).

We found crowned pixie-cup in one small spot on Laurel Hill, at a trail-disturbed edge of an apparently unmined woodland. Brodo (1968) found this lichen only once in his detailed survey of the lichens of Long Island. Crowned pixie-cup is a northern species (Thomson 1967) that is approximately at its southern range margin here, and it may be sensitive to climate warming.

Surprisingly, 13 of the 58 lichen species (Table 5.3) were in the genus *Cladonia* (the pixie lichens, a group that includes British soldiers, pixie-cups, wand lichen, etc.), large lichens that appear to have high surface-to-volume ratios. None of the 13 *Cladonia* had much coverage in the Meadowlands. Thomson (1967) stated that many *Cladonia* species, including British soldiers and peg lichen, colonized formerly cultivated fields, suggesting tolerance of a harsh microclimate and high soil fertility. Brodo (1968) found British soldiers (*Cladonia cristatella*) common and widespread on diverse substrates on Long Island. *Cladonia* lichens readily colonized metal-contaminated mine tailings in Ohio (Lawrey and Rudolph 1975), although generally lichens are considered sensitive to heavy metals (e.g., Folkeson 1984). A three-day survey of several hectares on the capped Fresh Kills landfill on Staten Island yielded 17 lichen species (Allen and Howe 2016). Their list and ours share at least 12 species (possibly 2 additional shared species depending on the species-level identity of two uncertain identifications). Oddly, no *Cladonia* was reported from Fresh Kills.

In general, the order of tolerance of harsh conditions, including urbanization, among lichens is crustose more tolerant than foliose and foliose more tolerant than fruticose. Small lichens more closely attached to the substrate should be more desiccation-tolerant and less vulnerable to accumulating pollutants because their surface-to-volume ratios are smaller. Our collections (Table 5.3) contain 28 crustose, 17 foliose, and 13 fruticose species. The two most-collected species are foliose. One of them, candleflame lichen, was reported to be commonly associated with nutrient-rich and urban areas (Brodo et al. 2001). However, collection of specimens is affected by observer behavior; the larger species, especially the pixie lichens, may be easier to spot and more appealing to the non-specialist. After half of our study period elapsed, we realized we were under-collecting crustose lichens and made an effort to find them, which greatly increased their proportion in the sample. In the New Jersey lichen flora, 55% are crustose species (Waters and Lendemer 2019), compared to 48% crustose in our survey (these percentages are not statistically different).

Lichens colonize a wide variety of natural and anthropogenic surfaces. The total occurrences of five types of substrate, by species, in our collections are: Artificial (solid substrates such as concrete) 9, Rock 24, Soil 7, Tree or shrub bark 26, Wood (dead tree, lumber, ties, etc.) 18. Old, dry, partly-decayed rail ties were a frequent substrate; either the creosote treatment was not toxic to these lichens, or the toxic PAHs (polycyclic aromatic hydrocarbons) had degraded. The low proportion of species on soil may be due to heavy metal or other contamination and to the difficulty of seeing small lichens on the ground. Bryophytes (see above) were much more often found on soil, also on artificial surfaces, than lichens (47% of bryophyte species were found on soil compared to 12% of lichens). Exposed bedrock in the Meadowlands is mostly limited to the western bluffs and the Laurel Hill—Little Snake Hill area, and loose boulders and cobbles are scarce excepting those dumped or imported for landscaping purposes. Laurel Hill, with its geochemically diverse bedrock, talus, and tailings, yielded many lichens, bryophytes, and other organisms apparently absent elsewhere. Lichens do not occur on the relatively short-lived and often wet plant bases as do bryophytes (see above).

Connor (1978), in a quantitative study of bark-inhabiting macrolichens (foliose and fruticose) in northern Dutchess County, New York, about 100 km north of the Meadowlands, documented a decline in species richness (number of species) westward across the county toward the Hudson River. Connor ascribed this richness gradient to poorer air quality near the Hudson. Feuerer et al. (2003) reported 226 species of epiphytic lichens from Munich, Germany, although 102 had not been found in a century. We think the large number of lichen species in Munich may be due to a combination of a large urban area with inclusion of many parks and rural fringe environments, and to thorough surveying. Only 39 lichen species were found at a total of 434 sites in Cracow, Poland, despite a decline in sulfur dioxide pollution (Słaby and Lisowska 2012).

Conservation of lichens and non-lichenized fungi is ecologically and culturally important (e.g., Heilmann-Clausen et al. 2015). The U.S. lags behind Europe in fungal conservation (Allen and Lendemer 2015). Some of the Meadowlands lichen occurrences are susceptible to sea level rise and associated storm surge flooding, either because the lichens are on the ground or because their tree substrates will eventually be killed by brackish water flooding. Whether air quality continues to improve depends on federal and state regulation and enforcement. The biggest threat to lichens and other biota is habitat loss to infill development and redevelopment, as has been recently completed, in progress, or proposed at Laurel Hill, the area west of the I-E Landfill in North Arlington, the Lyndhurst landfills, and Teterboro Airport Woods.

STONEWORTS

The stoneworts (muskgrasses or charophytes) are a group of algae that form colonies resembling submergent aquatic seed plants such as coontail (*Ceratophyllum*). Stoneworts can be easily distinguished from coontail with a hand lens, and usually by their mildly sulfurous odor and rough feel (the latter mostly due to precipitated minerals). Recent genetic analyses have supported the hypothesis that stoneworts probably represent a link between other algae and the seed plants (Karol et al. 2001). Stoneworts are among the oldest plants in the fossil record, and the importance of stoneworts and related algae to the understanding of plant evolution makes them significant components of global biodiversity (*ibid.*).

Stoneworts grow in a variety of waters generally of low turbidity and high mineral content, particularly calcium-rich fresh waters and also brackish waters. In the U.K., they have been observed to dominate recently dredged ditches and ponds (Wade 1990). In the northeastern U.S. they are often found in calcium-rich springs or springfed ponds. Stoneworts disperse by means of oospores or fragments borne on water currents or transported by waterbirds. Stonewort oospores readily reach disturbed or temporary waters (Wade 1990, Grillas et al. 1993); however, oospores lack food storage and establishment of stoneworts may require special conditions.

Although many biologists can identify stoneworts as a group, few collect specimens or identify species. Most of the stonewort literature pertains to other countries; American stonewort floras are not well documented. We found two stonewort species, single species at each of three locations.

Opposite stonewort (*Chara contraria*) was in a single small, intermittent, rock pool on the former quarry floor at Laurel Hill Park in the Meadowlands. Ken Karol (New York Botanical Garden) identified this stonewort by morphology; there are four morphologically similar species separable only on the basis of DNA analysis. Between 2006, when we collected a specimen for identification, and 2008 when we returned to the locality, part of this pool had been filled for expansion of park facilities, and the pool seems destined for eventual destruction. Another nearby, intermittent pool that is connected to the first pool during floods, lacked the stonewort and had a muddy rather than rocky bottom. Opposite stonewort, identified morphologically, also occurred in a small temporary pool on a dirt road at the Kingsland Landfill (Lyndhurst) in 2021. Common stonewort (*Chara vulgaris*) was in a large, fresh or slightly brackish, pond at Meadowlark Marsh in Ridgefield. This pool is bordered by a street and railyard on one side and is separated from a brackish marsh by an electric transmission line service road on the other side. Crystalline, sandy, waste material of unknown origin extends to one end of the pond. No further development is planned for this site although a restoration proposal targets the brackish marsh.

OTHER ALGAE

Little has been written about algae (including Cyanobacteria) in the Meadowlands, although these small organisms are undoubtedly important components of biological communities in the waters, on the soil surface, and on the bases of higher plant stems (including tree-of-heaven and reed; E. Kiviat, pers. obs.). Mud algae (*microphytobenthos*), in particular, are likely to be important components of tidal marsh ecosystems due to the formation of large, productive mats on mudflats and creekbanks.

Suspended algae were studied in the upper Hackensack River estuary (Foote 1983). Of 232 species found, minute centric diatoms were abundant in winter and green algae were prominent in summer. Species were those tolerant of organic pollution and low concentrations of salt. Utberg and Sutherland (1982) reported that "warm summer temperatures were accompanied by a proliferation of submergent vegetation, particularly the green algae *Cladophora* sp. and *Enteromorpha* sp. in a brackish marsh in Bergen County." We have seen highly developed floating mats of algae on a shallow brackish tidal pool in North Bergen, and visible algal blooms occurred on the brackish tidal impoundment (Kingsland Impoundment) at DeKorte Park in Lyndhurst (Anonymous 2009), both indicative of high nutrient levels. Macroscopic algae were also conspicuous and abundant on the intertidal mudflats at Anderson Creek Marsh in July 2008, several months after herbicide treatment to kill common reed.

Some trees had a greenish wash of tree algae on aerial portions of the bark, but many trees (at least to the naked eye) seemed to lack algae. We have found no information on this group in the Meadowlands and it seems to be understudied in North America. Along with one of the lichen specimens, we collected a single *lichenicolous* fungus, *Lichenoconium lecanorae*. Lichenicolous fungi are fungi that are parasitic on living lichens. Undoubtedly, other algae and cyanobacteria live on soil and other substrates in the Meadowlands.

CONCLUSIONS

Cryptogams contain species, genetic, ecological, and utilitarian diversity comparable to the seed plants, and are as deserving of conservation. In general, the cryptogams are not abundant or diverse in the Meadowlands, although for a highly altered, salinized, and polluted urban environment these generally sensitive taxa are doing fairly well. If we were to guess at the number of macrofungus species in the Meadowlands (probably several hundred), species richness (numbers) of cryptogams, excluding algae, would be roughly comparable to numbers of seed plant species. Some groups are apparently absent, as yet undiscovered, or only represented by one or two species (clubmosses, spikemosses, horsetails, quillworts, grape ferns, peat mosses, stoneworts, reindeer lichens, cyanolichens); low diversity in these taxa may be generally useful as indicators of urban stress. Not only are there modest numbers of species compared to many rural areas, but the coverage or extent of patches of bryophytes, lichens, and ferns is mostly rather restricted in the Meadowlands. We have not seen large areas of forest floor, rock ledges, tree bark, or wood covered with lush mats of bryophytes, ferns, or lichens. A notable exception was a sycamore (*Platanus occidentalis*) about 1 m dbh, its lower trunk extensively grown with bryoids, at Riverview Avenue in Brett Park (2010, Teaneck). The Meadowlands feature large expanses of upland soils with sparse growth of seed plants, and if air and soil quality were better, mosses and lichens would probably exploit these spaces much more. Perhaps bryoids (and other cryptogams) will become more diverse and abundant if air, soil, and water quality continue to improve.

Certain species of cryptogams in the Meadowlands depend on small areas of suitable habitat or particular habitat features, many of which are threatened by development. For example, the mixed-hardwood forest and swamp remnants with their large living and dead trees are very important (one European study, for example, found that the abundance and diversity of bryoids on snags was favored by larger snags of certain species [Staniaszek-Kik et al. 2019]). Cryptogams are rarely considered in biological surveys or conservation planning, and we have seen no evidence of this attention in the Meadowlands. The cryptogams are, therefore, emblematic of the complexity and challenge of biodiversity conservation in the urban environment.

Laurel Hill stands out in the Meadowlands as particularly important for cryptogams because it has a relatively high diversity of mosses and lichens with one rare species of each, supports one of only two stonewort populations known in the Meadowlands, is the only locality for blunt-lobed cliff fern, and the only historic locality for a spikemoss (and historically may have yielded specimens of other ferns no longer found in the Meadowlands). Unfortunately, multiple factors threaten the greenspace at Laurel Hill, including expansion of the athletic field complex, dumping, remediation of contamination, ORVs, Turnpike maintenance, and new building construction. The cryptogams of Laurel Hill, and indeed all the Meadowlands, illustrate how biodiversity in an urban area can depend on relict scraps of habitat, and how these remnants can disappear quickly even when in a managed area such as a county park. The remnant swamp forests, such as Teterboro Airport Woods and Losen Slote Park, are also important for cryptogams but have not been surveyed thoroughly, and this is also true of outlying preserves such as Borg's Woods and Matthew Feldman Nature Preserve. Now that air quality has improved, the most important threat to cryptogam diversity is habitat loss. This is occurring via residential, commercial, industrial, recreation, and transportation development. Habitats are also being lost as abandoned garbage landfills are capped and wetlands are mitigated; conservation and reintroduction of certain cryptogam taxa could be designed into such remediation and restoration projects.

We have provided species lists of the lichens, bryophytes, and pteridophytes known in the Meadowlands, and this cryptogam biota is substantial for a highly altered, urban-industrial environment. We think many species will be added to these lists with more concerted efforts to survey areas that are less studied or less accessible, and to find inconspicuous and uncommon cryptogams in their habitats and microhabitats and at the right seasons and moisture conditions. For example, we did not survey tree crowns for bryoids although in some areas crowns are known to support bryoids not found on or near the ground (Boch et al. 2013). A survey of macrofungi is in progress and should be conducted for several years; unlike the lichens and most bryophytes, fungi are more seasonal and also rain-dependent. Expert or expert-assisted surveys for cryptogams should be conducted on large and small sites where conservation, management, or development is being planned. Research on urban cryptogams is important for the conservation of global biodiversity, illustrated by the parks and cemeteries in Polish cities that were deemed important for conservation of habitat specialist as well as generalist bryophytes (Fudali 2005). The conservation maintenance or establishment of some of the Meadowlands cryptogams would make good objectives for habitat management projects, but first we need to know more about the species that do and do not occur in the region, and what their habitat affinities and ecological tolerances are. Studying urban cryptogams is also important because derelict lands such as landfills and quarries are being redeveloped without knowledge of these elements of biodiversity and the opportunity to protect them and use them. Pending further study, protection of diverse macrohabitats and microhabitats in a wild condition, and further improvements in air quality, will do much to help conserve cryptogams.

Chapter 6

Mammals

Mammals are an important, if mostly inconspicuous, part of Meadowlands biodiversity. Though not as many species of mammals as there are fishes or birds, mammals play disproportionately large ecological roles as predators, prey, reservoirs of disease, and ecosystem engineers. Species of rodents are the prey base for top mammalian and avian predators. Mammals such as raccoons and striped skunks are major predators of turtle eggs, and white-footed mice, eastern chipmunks, and gray squirrels are important predators of bird eggs and nestlings. Bats, the only flying mammals, prey heavily upon insects. Activities of muskrats shape wetland ecosystems. Because of their secretive and often nocturnal nature, mammals seem uncommon. Several species exist at very high densities and evidently flourish in urban greenspaces, although others, like the river otter and American mink, are indeed rare finds in urban places. Mammal communities in urban areas are generally less diverse than those in more rural areas (McKinney 2008). To some extent this reflects influences that urbanization has had on distribution, persistence, and movement of individual species. However, because studies of mammal diversity are time and labor intensive, mammals are seldom surveyed. Therefore, diversity in the Meadowlands may be underestimated. In this chapter, we discuss general patterns of species richness of mammal communities relative to the regional species pool and typical densities of a species when known. We touch on ecological requirements of individual species relevant to their persistence in the urban environment. We also illustrate aspects of the urban and natural ecosystems in the Meadowlands that have shaped the mammal community and we identify targets for mammal conservation.

MAMMAL COMMUNITIES IN THE MEADOWLANDS

There have been few formal studies of mammals in the Meadowlands. Mammal surveys conducted along the New Jersey Turnpike corridor prior to Turnpike expansion in the late 1980s and early 1990s detected 16 species (NJTA 1986). Twelve species were found in two surveys conducted at Oritani Marsh (Berger Group 2001). There are also records, mostly historical and for the region just outside the Meadowlands, from the American Museum of Natural History (AMNH) mammal collections, which we obtained in 2018. Overall, about 37 species of mammals have been recently documented from the Meadowlands. These include several marine mammals reported in water-ways of Newark Bay, which we consider part of the Greater Meadowlands. Table 6.1 provides a list of mammal species documented in the Meadowlands proper and the Greater Meadowlands region as described in the Introduction. Denoted are species that are found just outside the greater Meadowlands region and those that have rare or historical sightings. Thirty-seven species is only 42% of the state's 89 mammal species (NJDEP 2004a). Given the high diversity of habitats, mammal diversity in the Meadowlands is low both in an absolute and in a relative sense (i.e., as compared to birds and fishes). This is in part due to birds and fish being more heavily studied and because these classes of vertebrates contain many more species than the class Mammalia.

Mammals are diverse in their forms and microhabitat use and require a variety of specialized techniques for surveys (Wilson et al. 1996). To our knowledge there are no quantitative or spatial data available to assess the relative abundance, population trends, and distribution for any of the mammal orders present in the Meadowlands. Although species richness is low, several species of mammals may exist at locally high densities in the Meadowlands. These are rodents and include the white-footed mouse, meadow vole, house mouse, brown rat, muskrat, and eastern gray squirrel.

Most of the reported mammals in the Meadowlands are common, urban-tolerant species of wetland or upland habitats (Table 6.1, Figure 6.1; Kiviat and MacDonald 2002, 2004), with the possible exceptions of the masked shrew, eastern mole, meadow jumping mouse, gray fox, and marine mammals. Ample studies indicate remarkable plasticity among mammals, especially carnivores, in home range size and habitat use, temporal and spa-tial activity patterns, and diet composition (George and Crooks 2006, Baker et al. 2007, Newsome et al. 2015, Šálek et al. 2015). Furthermore, mammals including the river otter, American mink, eastern coyote, white-tailed deer, muskrat and white-footed mouse have been documented using common reed (*Phragmites australis*) stands for food or refugia in North America (refer-ences in Kiviat 2019). Three mammals known to be sensitive to environmen-tal contaminants, the American mink, river otter and harbor seal, are very rare in the Meadowlands (Kiviat and MacDonald 2002). The gray wolf and mountain lion were once wide-ranging species that have been extirpated from much of their range in the eastern United States; the bobcat and black bear were extirpated from the Meadowlands but are now occasionally sighted on

Table 6.1. Mammals of the New Jersey Meadowlands.

Common name	Scientific name	Status
New World Opossums (Didelphimorphia)		
Virginia opossum	*Didelphis virginiana*	
Rodents (Rodentia)		
White-footed mouse	*Peromyscus leucopus*	
Meadow jumping mouse	*Zapus hudsonius*	SGCN
Woodland jumping mouse[a]	*Napaeozapus insignis*	SGCN
Meadow vole	*Microtus pennsylvanicus*	
Woodland vole[a]	*Microtus pinetorum*	SGCN
Black rat[b]	*Rattus rattus*	
House mouse	*Mus musculus*	
Brown rat ("Norway" rat)	*Rattus norvegicus*	
Muskrat	*Ondatra zibethicus*	*
North American beaver	*Castor canadensis*	* One carcass found away from roads
Woodchuck	*Marmota monax*	
Eastern chipmunk	*Tamias striatus*	
Eastern gray squirrel	*Sciurus carolinensis*	
Southern flying squirrel[a]	*Glaucomys volans*	Unconfirmed reports in Meadowlands
Rabbits (Lagomorpha)		
Eastern cottontail	*Sylvilagus floridanus*	
Shrews and Moles (Soricomorpha)		
Northern short-tailed shrew	*Blarina brevicauda*	
Least shrew[b]	*Cryptotis parva*	SGCN
Masked shrew[b]	*Sorex cinereus*	
Eastern mole[b]	*Scalopus aquaticus*	
Star-nosed mole[a]	*Condylura cristata*	SGCN
Bats (Chiroptera)		
Silver-haired bat[a]	*Lasionycteris noctivagans*	SGCN
Red bat[a]	*Lasiurus borealis*	SGCN
Hoary bat[a]	*Lasiurus cinereus*	SGCN
Big brown bat[b]	*Eptesicus fuscus*	SGCN
Little brown bat[b]	*Myotis lucifugus*	SGCN
Northern long-eared bat[b]	*Myotis septentrionalis*	
Tri-colored bat[a]	*Perimyotis subflavus*	SGCN
Deer (Artiodactyla)		
White-tailed deer	*Odocoileus virginianus*	*
Carnivorans (Carnivora)		
Raccoon	*Procyon lotor*	*
American mink[b]	*Mustela vison*	*
River otter[b]	*Lontra canadensis*	*
Long-tailed weasel[a]	*Mustela frenata*	*
Striped skunk	*Mephitis mephitis*	*

Eastern coyote	*Canis latrans* var. (*Canis latrans* × *lycaon*)	*
Gray fox[b]	*Urocyon cinereoargenteus*	*
Red fox	*Vulpes vulpes*	*
Domestic dog	*Canis familiaris*	
Domestic cat	*Felis catus*	
Bobcat[a]	*Lynx rufus*	E
Black bear[a]	*Ursus americanus*	*
Hooded seal[b]	*Cystophora cristata*	
Harp seal[b]	*Pagophilus groenlandicus*	
Harbor seal	*Phoca vitulina*	
Whales and Dolphins (Cetacea)		
Sei whale[b]	*Balaenoptera borealis*	E
Fin whale[b]	*Balaenoptera physalus*	E
Minke whale[b]	*Balaenoptera acutorostrata*	
Harbor porpoise[b]	*Phocoena phocoena*	
Atlantic white-sided dolphin[b]	*Lagenorhynchus acutus*	
Bottlenose dolphin[b]	*Tursiops truncatus*	

Note: E, Endangered

T, Threatened

SCGN, Species of Greatest Conservation Need (Includes Endangered and Threatened Species); NJDEP 2018a

* Species hunted or trapped in New Jersey, including game and furbearer species. Their populations are not addressed the same as other species by state agencies and many were decimated in the last hundred years in the Northeast due to overharvest and habitat loss.

[a] Species documented just outside the Meadowlands and likely to occur in the Meadowlands.

[b] Species rarely reported.

the fringes of the region due to a general population recovery. None of the terrestrial mammal species documented in the core Meadowlands is considered Endangered or Threatened. However, fin and sei whales, both of which have been found in Newark Bay in recent years, are federally endangered and are protected by the International Whaling Commission (IWC) and the Committee on Trade in Endangered Species of Wild Fauna and Flora (CITES). Also, species of furbearers, such as the river otter and beaver, reached extremely low densities in New Jersey due to over-trapping in the last century and earlier, and have subsequently increased in numbers throughout the state. Five nonnative species belong to the Meadowlands mammal fauna, namely domestic dog, domestic cat, brown rat, black rat, and house mouse.

Figure 6.1 White-tailed deer in oldfield-like vegetation on sideslope of an inactive gar-bage landfill, Lyndhurst. Photo by Erik Kiviat.

MAMMAL COMMUNITIES VARY IN SPACE AND TIME

Mammal species composition and density vary in space and time. While many species exhibit preferences for specific habitat types, many mammals in the Meadowlands are able to use a broad range of habitats. Many mammals are able to persist in heavily urbanized areas, including residential neighborhoods and industrial zones but may be present in low densities because resources such as food and nest sites are limited. Mammal community composition varies among urban greenspaces such as city parks, cemeteries, golf courses, and semi-natural areas depending on the ability of different mammals to colonize and persist in these habitats, which in turn indicates landscape permeability (connectivity) and habitat quality (Gallo et al. 2017). In greenspaces, changes in plant communities with time bring about changes in dominance of the small mammal species from the meadow vole in *old field* (post-agricultural or previously cut and filled habitat with often poor soils that are dominated by coarse herbs and shrubs) and grassy areas to the white-footed mouse in forests and shrubland. The eastern chipmunk is a forest specialist. Wide-ranging mammals such as the red fox and eastern coyote may have large home ranges and visit a variety of habitats during a single day. However, preference may be shown for areas where prey is concentrated (e.g., along habitat edges).

Mammal communities found at a particular place vary greatly with time of day and season. There are definite peaks in activity at certain times of the day for most species. Many are most active at night (e.g., bats, some rodents, opossum) while others are active during the day (e.g., gray squirrel) or are crepuscular and active mainly at dawn and dusk (e.g., eastern coyote). Most mammal species avoid humans and thus daily patterns in human disturbance may determine activity periods and spatial use (Tigas et al. 2002). Mammals may visit areas of human habitation to exploit garbage bins and gardens at times of day when there is little human activity such as night or early morning. Also, spatial behavior of mammals may vary depending on the tide (Kiviat 1989). For instance, raccoons commonly forage on exposed mudflats at low tide but are excluded from these areas at high tide. Tide also affects activity patterns of muskrats.

Seasonal variations in the mammal communities also exist. In summer and early fall there is a higher population density of most species because young of the year are included in the population. However, later in the fall and winter there is a high rate of dispersal and related mortality, which causes a decrease in population size. Habitat use varies depending on seasonal changes in food availability. With the exception of regional migrations between summer habitat and winter hibernacula of bats, there are no large-scale migration movements of terrestrial mammal species in the Meadowlands.

REVIEW OF MAMMAL FAUNA IN THE MEADOWLANDS

In this section, we briefly describe the species of mammals that have been recorded in the Meadowlands and their distribution, abundance, habitat affinities, and some relevant biological traits. Species are organized into higher taxa. As stated, little research has been done to document the mammal species in the Meadowlands and there are few specific studies of the ecology of mammals here. For these additional biological details, we frequently cite relevant research in urban areas outside the Meadowlands. In addition, we provide brief mention of basic information about the species' general habitat use, diet, behavior, and other characteristics that may be applicable to understanding their ecology in urban areas.

Virginia Opossum

The Virginia opossum, North America's only marsupial, is likely widespread in low densities in the Meadowlands; they are surprisingly difficult to find. They are mainly nocturnal and use a wide variety of habitats. Their tracks can be seen in mud along dirt roads (K. MacDonald, pers. obs.). A study of opossums in Chicago demonstrated that they were common in a variety of natural habitats and proximity to water was the biggest predictor of patch occupancy regardless of the amount of urban land use in the landscape (Fidino et al. 2016). The opossum is omnivorous, and its diet may include plants, fruits, insects, earthworms, small mammals, and amphibians (Whitaker and Hamilton 1998). They often eat carrion, and even scats of other mammals (E. Kiviat, pers. obs.). Opossums possess a prehensile tail and are able to climb trees and swim (Whitaker and Hamilton 1998). Many are killed on roads.

Rodents

Ten species of rodents are known to occur in the Meadowlands. They are found in a wide variety of habitats including along road and rail beds, pipeline corridors, and other elevated areas as well as in marshes, meadows, woodlands, and human-built habitats. Historically, high numbers of raptors at active landfills in the Meadowlands were attributed to abundant rodents there. The closing of landfills may have resulted in a decline in small mammal prey and a subsequent decline in raptor densities. Rodent species found in the Meadowlands include the muskrat, white-footed mouse, meadow jumping mouse, meadow vole, house mouse, brown rat, black rat, eastern chipmunk, eastern gray squirrel, and woodchuck (NJTA 1986) and probably southern flying squirrel. The rodents that are unknown in the Meadowlands but may be present because they have been documented near the Meadowlands include the woodland vole, woodland jumping mouse, red squirrel, and American beaver. The Allegheny woodrat (*Neotoma floridana magister*), endangered in New Jersey, has not been documented in the Meadowlands but a population of this species is in the Palisades in Bergen County—this species could once have inhabited the ledges and talus of Laurel Hill and Little Snake Hill. There is concern that the large, introduced, South American nutria (*Myocastor coypus*) will eventually expand its range into New Jersey. If the species becomes established in the state it could alter vegetation in marshes and negatively affect habitats of native species.

White-footed mouse: The white-footed mouse is likely found in greenspaces throughout the Meadowlands (NJTA 1986). The species prefers deciduous woodland habitat but is considered a habitat generalist in that it can use a variety of woodland and old field habitats independent of composition or size (Anthony et al. 1981, Matthiae and Stearns 1981, Barko et al. 2003). White-footed mice also use tidal marshes (McGlynn and Ostfeld 2000). They are semi-arboreal (Whitaker and Hamilton 1998). At the restoration site at nearby Fresh Kills on Staten Island, white-footed mice were largely restricted to a dense stand of common reed and sparse meadow areas at the top of the landfill (Sakatos 2006). They were likely excluded from lower portions of the landfill by meadow voles. The lower areas contained restored old field habitats consisting of grassy and herbaceous plant communities preferred by voles.

Urbanization is a significant barrier to movement and dispersal of the white-footed mouse (Barko et al. 2003). This species reaches very high densities in woodland patches surrounded by urban land use as compared to those surrounded by a forested landscape. The species may require corridors for movement among patches of greenspace (Wegner and Merriam 1979) but is synanthropic in its use of human dwellings and food sources. This characteristic may aid movement through urban areas. The white-footed mouse is omnivorous though a large portion of its diet is composed of tree nuts, berries, and seeds (Whitaker and Hamilton 1998). It is a reservoir for several tick-borne pathogens, including the bacterium *Borrelia burgdorferi*, which causes Lyme disease in humans.

Meadow jumping mouse: The meadow jumping mouse is locally common in the Meadowlands. Two surveys at Oritani Marsh found meadow jumping mice (Berger Group 2001, Barrett and Mcbrien 2007). The species is commonly trapped in cordgrass and common reed-dominated habitats as well as in old field habitats. We expect they occur elsewhere in the Meadowlands. In its typical old field habitats, it is often most abundant in dense vegetation along ponds, streams, and marshes and in dense herbaceous vegetation in wooded areas where the mice eat a variety of grass seeds, insects, and fungi (Whitaker 1972).

Meadow vole: The meadow vole is common and widespread in the Meadowlands. McCormick & Associates (1978) found meadow voles in common reed—saltmarsh-fleabane vegetation and what was likely meadow vole sign in a reed stand burned the previous year. Meadow voles comprised 19% of the biomass in the diet of long-eared owls in Lyndhurst in the Meadowlands (Bosakowski 1982). The species prefers grassland and meadows, especially at moister sites (Reich 1981). Meadow voles apparently exhibit some avoidance of areas with a tall, woody vegetation canopy. It is the most abundant rodent species in old field habitat where it is known to exclude the white-footed mouse (Anthony et al. 1981). At an old field restoration site at the Fresh Kills landfill, meadow vole was the most abundant small mammal (Sakatos 2006). Grassy vegetation bordering interstate highways is used as migration corridors for voles in regions of otherwise unsuitable habitat (Getz et al. 1978). The meadow vole eats mainly herbaceous plant material but will also eat woody material and animal matter (Reich 1981). The meadow vole may play the role of a *keystone species*, or a species whose presence has a disproportionate effect on other organisms in the community. In old fields, they affect vegetation development via predation on tree seedlings. Their activity can delay the establishment of trees and their selective feeding on certain tree species affects the composition and patchiness of trees (Ostfeld and Canham 1993, Sakatos 2006). This likely affects development of vegetation on land-fills in the Meadowlands.

Common muskrat: The muskrat is the flagship mammal species of Meadowlands wetlands. It occupies a wide variety of aquatic habitats, including artificial wetlands and ponds, in North America (Willner et al. 1980). In the Meadowlands, it also uses a range of wetland types including common reed marshes (NJTA 1986). The muskrat is an ecosystem engineer because its feeding and building activities have major effects on vegetation, soils, microtopography, nutrient transformations, and animal habitat (Kiviat 1978, Connors et al. 2000). Because of its important ecological role, it is a keystone species. Its lodges, built of mostly plant material provide a substrate for nesting birds, especially geese and ducks (e.g., Kiviat 1978). Rails often use muskrat-created lodges and trails in dense vegetation for movement and foraging (Conway 1995, Bannor and Kiviat 2002). Grazing by muskrats may be important in opening up common reed habitat by thinning or creating clearings. Grazing and digging may also increase rates of nitrogen mineralization in the soil, as occurs in cattail marshes (see Daiber 1982, Kiviat 1989, Connors et al. 2000). Muskrat effects on biogeochemistry may be significant in the Meadowlands because of high nutrient levels and locally or episodically dense muskrat populations. Depredation of plantings by muskrats was a problem in the Hartz Mountain mitigation project (Berger 1992), among others. Brooks (1957) opined that the "unhealthy environment" of the Meadowlands has reduced predation pressure by mammals and snakes, allowing muskrats to thrive. However, muskrats may also be affected by contaminants such as heavy metals and petroleum in their aquatic ecosystems (Willner et al. 1980 and references therein). Although tissues from muskrats in Tinicum Marsh, an urban marsh in the Philadelphia area, had high levels of various heavy metals, adverse effects on the health of individuals or the population were not detected (Everett and Anthony 1976). The muskrat has declined in Hudson River marshes since at least the 1970s (Kiviat 2010). And as noted above, there is evidence muskrat populations have declined in eastern North America and throughout the United States. Carola (2016a, b) noted that a trapper in the Meadowlands reported a substantial decline in the muskrat population that may be attributed to an unconfirmed outbreak of tularemia, a disease caused by the bacterium *Francisella tularensis holarctica*. The disease is common in other rodents and rabbits (Shimshony 2009). Muskrat populations typically fluctuate greatly as a result of weather and other factors, but the decline in the Meadowlands appears to be non-cyclic. The population of muskrats in the Meadowlands should be monitored for recovery.

House mouse, brown rat, and black rat: Three introduced rodents are found in built areas as well as greenspaces in the Meadowlands. The two species of rat are difficult to differentiate in the field and even by their skulls. In surveys of marshes in the Meadowlands, house mice and brown rats were observed (NJTA 1986). VanDruff and Rowse (1986) found that the brown rat was negatively associated with human density and was associated with greenspaces in industrial, less human-populated places. Sakatos (2006) found house mice but not brown rats at the old field restoration at nearby Fresh Kills. Analysis of pellets collected from a long-eared owl winter roost in Lyndhurst indicated that house mice comprised 47% of the diet (biomass; Bosakowski 1982). Oddly, we have not found this species in the Meadowlands but assume it is still present. Brown rats were also consumed by long-eared owls but comprised only 7% of the diet. House mice and brown rats were also important prey for the short-eared owl (Bosakowski 1982). It is believed that brown rats and possibly house mice have declined in recent years due to closing of the many garbage landfills in the Meadowlands (Quinn 1997). Black rats are thought to be distributed widely in the Meadowlands although documentation is sparse.

Eastern chipmunk: The eastern chipmunk is a woodland specialist and is likely limited to forests, thickets, and woodland strips in the Meadowlands where it seems rare. In greenspaces in Syracuse, New York, chipmunks were positively associated with areas having dead plant material and stumps and negatively associated with vegetative ground cover and water (VanDruff and Rowse 1986). They are vulnerable to predators and sensitive to sunlight, which makes it difficult for this species to successfully cross open fields and recolonize forests from which they have been extirpated (Matthiae and Stearns 1981). Therefore, chipmunks likely require corridors of semi-natural habitat such as fencerows and rights-of-way to move among patches (Wegner and Merriam 1979). It appears the abundance of the eastern chipmunk is not related to habitat area (Matthiae and Stearns 1981). This species also uses residential developments and parks where there are large numbers of tall trees and shrubs.

Gray squirrel: The highly conspicuous gray squirrel is found wherever there are trees in the Meadowlands (NJTA 1986). The gray squirrel was a visibly abundant species in wooded habitats with large trees surrounded by grass or fields in Syracuse (VanDruff and Rowse 1986). It is most numerous in built areas with artificial food sources such as bird feeders as well as natural sources from mast producing trees including oaks (VanDruff and Rowse 1986). One study demonstrated gray squirrel populations were most dense in large versus small forest islands (Matthiae and Stearns 1981). Gray squirrels are flexible feeders and may consume novel foods in urban areas, such as

Osage-orange (*Maclura pomifera*) seeds as seen in Westchester County, New York (E. Kiviat, pers. obs.).

Southern flying squirrel: The southern flying squirrel is an inconspicuous, nocturnal species found throughout the central and eastern United States in association with deciduous forest (Dolan and Carter 1977). It is very likely there are local populations of southern flying squirrel in the Meadowlands that have gone undetected or have not been documented in the recent literature. We have unconfirmed reports of southern flying squirrels in Ridgefield and Hackensack and historical records from the AMNH collections indicating their presence in the Meadowlands region. However, the species is limited in the Meadowlands by low availability of forest habitat and potential barriers of extensive wetlands. Flying squirrels cannot swim and must glide between trees rather than fly. This species uses small cavities, usually excavated by woodpeckers, as nest sites (Dolan and Carter 1977). Availability of cavities is a major limitation in urban neighborhoods where flying squirrels would likely compete with abundant eastern gray squirrel and other species for existing cavities.

Woodchuck: The woodchuck is found in the Meadowlands in a variety of open upland habitats in elevated areas such as road verges, railbeds, and berms, where burrowing above water level is possible (Quinn 1997; E. Kiviat, pers. obs.). This species is common on landfill cover, dredge spoil, and other sandy or gravelly fill soils. Woodchuck burrows can be distinguished from those of other burrowing animals by the mound of fresh soil at the main entrance (Whitaker and Hamilton 1998). The function of excavating soil may provide a critical microhabitat that would otherwise be in short supply. Woodchuck burrows serve as refugia and den sites for a variety of other wildlife including the eastern cottontail, Virginia opossum, striped skunk, raccoon, red fox, white-footed mouse, house mouse, meadow vole, short-tailed shrew, wood frog, and probably many others (Hamilton 1934, Grizzell 1955, Swihart and Picone 1995; K. MacDonald, pers. obs.). Cottontail rabbits depend on woodchuck burrows in winter (Whitaker and Hamilton 1998). Woodchucks have been observed foraging in the intertidal zone along the Hudson River (E. Kiviat, pers. obs.). They were seen climbing white mulberry (*Morus alba*) trees in the Meadowlands to eat leaves and eating mugwort (*Artemisia vulgaris*) shoots (E. Kiviat, pers. obs.). These may be novel, urban behaviors.

North American Beaver: The beaver has not been recently documented in the Meadowlands (H. Carola, pers. comm.). However, a beaver carcass was found in the Teaneck Creek Conservancy (E. Kiviat, pers. obs.) suggesting occasional dispersal into the Meadowlands. The species historically occupied all suitable habitat in the Northeast (Naiman et al. 1986) and was certainly present in the Meadowlands at one time. Trapping and possibly pollution led to their complete extirpation from the state and a moratorium on trapping. The state reintroduced beavers trapped in other areas of the country in the early 1900s and they are now more common and again trapped for fur. Beavers can persist in wetlands surrounded by urbanization in other areas (Mech 2003). Expansion of beavers into fresh-tidal and brackish tidal wetlands, as well as urban fringe areas in the Hudson Valley, suggest the Meadowlands will eventually be re-colonized by this species. As an ecosystem engineer and keystone species, beavers have significant influences on shaping ecosystems by selectively feeding on certain plant species, building underwater food caches, constructing dams, lodges, burrows, and canals, and altering water levels (Naiman et al. 1986).

Rabbits

Eastern cottontail: We refer to all Meadowlands cottontails as eastern cotton-tail because the morphologically similar Appalachian cottontail (*Sylvilagus obscurus*) and New England cottontail (*Silvilagus transitionalis*) evidently do not occur in New Jersey. The eastern cottontail is widespread and common in the Meadowlands. Although it is a habitat generalist, it requires suitable cover in the form of dense vegetation and does not use flooded areas except on the ice in winter. Without suitable cover, cottontails are present at low densities and overwinter survival declines (Litvaitis 2001). In the Meadowlands, old fields, young woods, shrublands, and reedbeds on dry as well as moist soils, and non-flooded wetlands, likely provide suitable habitat for eastern cottontails (NJTA 1986). Cottontail sign was common in an aster-ragweed meadow and in the margins of common reed marsh at the Sports Complex site (McCormick & Associates 1978). Cottontails are likely common at inactive and capped landfills. Their selective foraging on the seedlings of certain woody plants may affect development of old field vegetation (Sakatos 2006). Cottontails extensively browsed bark from sapling-size tree-of-heaven (*Ailanthus altissima*) stems at several Meadowlands sites (E. Kiviat, pers. obs.). They are important prey of the eastern coyote, red fox, and other predatory mammals and birds.

Shrews and Moles

Three shrew species, the short-tailed shrew, masked shrew, and least shrew, have been documented in the Meadowlands. The eastern mole was documented in the Meadowlands and star-nosed mole has been documented on the fringes of the region but not in the Meadowlands. Because of a high metabolic rate, these insectivores consume large quantities of invertebrates.

Shrews are prey for birds including hawks and owls, snakes, fish, and mammals; though in the case of mammals, shrews are killed and often not eaten because a musk they release makes them unpalatable (George et al. 1986). The *fossorial* (burrowing) and nocturnal habits make them difficult to study. This group warrants further research in the Meadowlands because intensive sampling in a variety of habitats, and special techniques, are needed to ensure capture of less common species.

Northern short-tailed shrew: The regionally common northern short-tailed shrew has been documented on the fringes of the Meadowlands (AMNH). Numerous skulls of this species were recovered from barn owl pellets dating from ca. 2017 and collected in the edge of the Core Meadowlands in North Arlington (E. Kiviat, pers. obs.). This species occupies a broad range of habitats with > 50% vegetation cover (George et al. 1986). VanDruff and Rowse (1986) considered short-tailed shrew a common habitat generalist of urban greenspaces. It was detected at the old field restoration at nearby Fresh Kills Landfill (Sakatos 2006).

Masked shrew: Masked shrew is a common small mammal species in North America, occupying a wide variety of moist habitats (Whitaker 2004). It was found along roads and the TransCo pipeline corridors in the Meadowlands (NJTA 1986) and there are historical records from areas just outside the Meadowlands (AMNH). The diet of this species consists of insects and other invertebrates as well as small vertebrates (Whitaker 2004).

Least shrew: The "brown shrew" (*Blarina parva*) that was listed as occurring in the Hackensack Marshes by Stone (1907) is likely what is now called least shrew (*Cryptotis parva*). Bosakowski (1982) reported a least shrew recovered from an owl pellet in Lyndhurst within the Core Meadowlands. Whitaker and Hamilton (1998) noted that the least shrew "has been trapped in the marshes about New York City, and in New Jersey and Virginia, but in the northern part of its range the species generally inhabits grassy, weedy, and brushy fields."

Eastern and star-nosed moles: The eastern mole and star-nosed mole, both widespread species, have been documented in the general region (AMNH) and eastern mole was found in the Core Meadowlands (NJTA 1986). Star-nosed mole has not been detected in the Meadowlands. Perhaps fragmentation, low prey availability, poor water quality, and lack of natural, non-compacted upland soils have limited these species. The eastern mole eats primarily earthworms and insects. Because it requires large quantities of food, it ranges over a larger area than most other fossorial mammals (Yates and Schmidly 1978). Eastern moles dig tunnels close to the surface primarily for feeding and deeper, permanent tunnels for movement and nesting. The eastern mole prefers moist, loamy or sandy soils and is scarce to absent in heavy clay, stony or gravelly soils. It is a good swimmer. The star-nosed mole is a species typical of streams, wetlands, and wetland edges in the general region.

Bats

Bats are poorly studied in the Meadowlands, and most of the species' identifications are unverified. In one study, the little brown bat was the only bat species reported in the Meadowlands where it was observed near Berry's Creek Canal, Blackman's Creek, and Cedar Creek (NJTA 1986). It preys on a variety of flying insects, especially aquatic insects such as chironomid midges (Fenton and Barclay 1980). Three or more bat species, provisionally identified as big brown bat, northern long-eared bat, and one unidentified bat were detected during a one-evening survey at Overpeck Creek in summer 2009 (Chanda Bennett, Columbia University, pers. comm.). It is likely additional species occur in the Meadowlands. Nine species of bats occur in New Jersey: little brown bat, big brown bat, northern long-eared bat, small-footed bat, tri-colored bat, Indiana bat, hoary bat, eastern red bat, and silver-haired bat (NJDEP 2004a).

A study of bats in parks and adjacent areas in the Bronx, New York City, recorded 5 species: big brown bat, eastern red bat, hoary bat, silver-haired bat, and tri-colored bat (Parkins et al. 2016). The red bat and silver-haired bat are associated with habitat adjacent to commercial-industrial land use in urban areas (Gehrt and Chelsvig 2004). Red bats typically forage along forest-field edges and near streetlights in urban areas (Mager and Nelson 2001). Many bat species, including little brown bat, use the foliage and loose bark of large, live and dead, deciduous trees and human-made structures such as buildings and bridges as roost sites in urban areas (Fenton and Barclay 1980, Mager and Nelson 2001, Duchamp et al. 2004). Large trees, suitable as bat roosting sites, are moderately common in the Meadowlands in forest fragments, residential neighborhoods, parks, and cemeteries. Bats have also been documented foraging more heavily above *green roofs* (building roofs covered in growing medium and vegetation) than traditional roofs in New York City (Parkins and Clark 2015). Bat roosting in stream culverts under roadways is also commonly reported (e.g., Meierhofer et al. 2019). Some attribute a low diversity of bats in urban areas to lack of vegetation and the resulting low abundance of insects (Kurta and Teramino 1992, Duchamp et al. 2004). Chironomid midges and mosquitoes (see Chapter 10) are abundant in the Meadowlands, but it is not known if their spatiotemporal distribution is adequate to support bat populations, and they might not constitute a suitable staple for every bat species.

Regional land use context plays a role in the effect of urbanization on bat diversity. Diversity of bats is lower in urban areas relative to adjacent forested rural areas (Kurta and Teramino 1992). Conversely, diversity is higher in urban habitat patches when adjacent rural areas are dominated by agriculture (Gehrt and Chelsvig 2004). Bats have been shown to derive little benefit from the connectivity among forests in the urban landscape provided by tree-lined streets but may instead do best within urban parks (Oprea et al. 2009).

Bat populations may also be negatively affected by chemical exposure and diseases in the Meadowlands and the surrounding region. Pesticide exposure either through direct contact or via insect food sources can cause mortality of little brown bats (Fenton and Barclay 1980). In one study, organochlorine residues (DDE and DDT) stored in fat of the Mexican free-tailed bat were released during migration and linked to high levels of these chemicals in the brain and subsequent mortality of individuals (Geluso et al. 1976).

Bats are likely affected by low insect densities in urban built areas where there is a lack of habitat containing food resources for insects and there is application of insecticides on landscaping and turf and for mosquito control. Recent, precipitous declines in flying insect abundance have taken place in many parts of the world (Hallman et al. 2017). Furthermore, outbreaks of the fungal disease *white nose syndrome* have severely affected several species of bat that use winter hibernacula in New Jersey and likely have resulted in lower summer densities of bats as has been documented elsewhere in the Northeast (Brooks 2011).

White-tailed Deer

White-tailed deer were relatively rare in the Meadowlands two decades ago, but during the past few years deer and their sign have been observed widely in the Meadowlands (Carola 2015, 2019a; E. Kiviat, pers. obs.). Deer eat a variety of woody and herbaceous plants and in high densities they greatly affect the structure of forest ecosystems. The population of deer has undergone major growth in many areas of New Jersey and New York during the last three decades due to a variety of causes including refugia in greenspaces surrounded by residential and agricultural land uses coupled with the near removal of large mammalian predators. This population growth suggests deer may become still more common in the Meadowlands. Deer browsing and grazing have adversely affected many plant species and reduced forest regeneration elsewhere in the state (Kelly 2019), and this may happen in the remnant forests and other habitats of the Meadowlands.

Carnivores

Eleven species in the order Carnivora are known from the Meadowlands. Species in this order are primarily flesh eaters but several also eat large quantities of plant material and human refuse. As discussed earlier in the chapter, the gray wolf and mountain lion have long been extirpated from the Meadowlands and throughout much of their range in the eastern United States. The bobcat and black bear are extirpated from the Meadowlands, although bears and less often bobcats are reported nearby. Most carnivores require relatively large areas in order to locate adequate prey and other food resources as well as hiding places. This makes them vulnerable to habitat loss, fragmentation, and road mortality. In addition, they have been persecuted because they are perceived as a threat to human interests due to predation upon livestock, game animals, and pets, occasional attacks on humans, and their role in transmission of diseases to humans and domestic dogs and cats. The attraction and habituation of carnivores to garbage as a food resource and intentional feeding by humans are a major source of wildlife conflict with people in urban areas. Several carnivores are trapped for their fur, which in the early 1900s resulted in some species becoming rare or locally extinct in North America (Obbard et al. 1987), and this trend was also seen in New Jersey in the case of river otter and likely other carnivores.

Raccoon: Raccoons are widespread in the Meadowlands and capable of using a broad range of habitats where woods, wetlands, and built areas provide suitable foraging and den sites (see Whitaker and Hamilton 1998). Hadidan et al. (2010) attributed the urban tolerance of raccoons to several traits including broad diet, ability to exploit a wide variety of horizontal and vertical habitats, intelligence, and high fecundity. Preference among natural habitats is for mature, deciduous woodland near a permanent source of water (Pedlar et al. 1997). While raccoons will use a variety of den sites, they often prefer hollows in trees (Oxley et al. 1974, Hadidan et al. 2010). They are highly mobile and will use not only natural areas but also resources such as garbage and storm drains in human habitats (Matthiae and Stearns 1981). Raccoons are highly opportunistic foragers and eat a wide variety of foods including fruits, vertebrates, and invertebrates, as well as refuse. Earthworms are an important food resource (Greenwood 1979). Studies have demonstrated that raccoons typically occur at higher densities with smaller home ranges in urban areas than in rural locales, but the ranges are linked to distribution and quality of food resources and den sites (Prange et al. 2003, 2004; Gehrt 2004).

Mustelids: The Mustelidae includes weasels, mink, fisher, and river otter. Of these, mink and otter are aquatic species rarely recorded in the Meadowlands. The inconspicuous long-tailed weasel is widespread in North America but has not been recorded in the Meadowlands where it likely occurs. Mustelids appear to be somewhat tolerant of urbanization. The greatest threats to this group in urban areas are road mortality and environmental pollutants in aquatic ecosystems. In more rural-suburban areas of the state, mustelids are still heavily trapped for their fur or to prevent depredation of fish and game species and small farm animals.

American mink: Mink tracks were observed in snow near the Hackensack River in River Edge in March 2006 (Carola 2006a). Mink tracks have also been observed occasionally at other locations in the Greater Meadowlands (E. Kiviat, pers. obs.). Minks are sensitive to the effects of polychlorinated biphenyls (PCBs) and mercury (Whitaker and Hamilton 1998), two pollutants prevalent in the Meadowlands. Minks use a variety of wetland and water edge habitats (Whitaker and Hamilton 1998). Their diet is broad and includes small mammals, fish, birds, and amphibians. Minks are able to use islands of natural habitat surrounded by urbanization as well as artificial aquatic habitats (Mech 2003).

River otter: Recently, snow tracks and a plunge hole of a river otter were observed at a site in the Meadowlands (Figure 6.2; E. Kiviat, pers. obs.). Prior to that the only other recent record was of a river otter found at Overpeck Creek (Quinn 1997). The species was once common throughout New Jersey until it was nearly extirpated from the state due to over-trapping. In the early 1900s, unregulated trapping, water pollution, and other degradation of aquatic and riparian habitat had caused otters to decline throughout most of their historic range (Nilsson 1980, Toweill and Tabor 1982, Melquist and Dronkert 1987). While many states instituted reintroduction programs to

Figure 6.2 Freshwater marsh habitat for Atlantic Coast leopard frog, painted turtle, mallard, river otter, and other wildlife. Photo by Erik Kiviat.

help this species recover during the 1900s, the river otter population in New Jersey is believed to be from native stock (Raisley 2001). They appear to have returned to many areas in New Jersey and are frequently reported from trapping records throughout the state. River otters use a variety of wetland and aquatic habitats (Whitaker and Hamilton 1998). Otter slides, produced when they slide down steep slopes covered in snow or mud into the water below, are sometimes the most obvious sign of their presence. Haul-outs, trails from the water's edge often littered with the remains of their prey (e.g., crayfish parts, shells, and scats containing fish scales) are another sign of their presence. Like minks, river otters are able to use fragmented remnants of natural habitat surrounded by urban land use (Mech 2003).

Striped skunk: The striped skunk is likely widespread in the Meadowlands at low density. They are not well studied in urban environments (Rosatte et al. 2010). Skunks are most abundant in grassy or shrubby areas and around agriculture and they feed primarily on invertebrates and small mammals in spring and summer and fruits in fall and winter (Whitaker and Hamilton 1998); other animals, carrion, garbage, and plant materials are also eaten (see Rosatte et al. 2010). They occupy a wide variety of urban habitats including forested parks, residential areas, dumps, industrial areas, golf courses, and vacant buildings (Rosatte et al. 1991, Rosatte and Larivière 2003). One study of striped skunks in an urban area in Arizona found they regularly crossed the interface between natural areas and built environments (Weissinger et al. 2009). Their home ranges in urban areas are smaller than in rural areas perhaps due to abundant food resources and denning areas in cities and suburbs (Rosatte et al. 1991). Striped skunks use their long front claws to forage for insects in rotten logs and under rocks and to dig earthworms and beetle and moth larvae from lawns. Their feeding on insect larvae helps reduce destruction of turf. However, their habitat of digging up lawns and occasionally spraying musk on people and pets are sources of conflict with humans (Rosatte et al. 2010). Skunks suffer heavy road mortality.

Canids: Four species of canids occur in the Meadowlands, the eastern coyote, gray fox, red fox, and domestic dog. Home range size varies between sexes and among seasons and depends on habitat quality but is large relative to most other carnivores. All four species eat a variety of foods including rodents, rabbits, snakes, birds, insects, and many plant foods. Canids are particularly susceptible to sarcoptic mange caused by the mite *Sarcoptes scabiei*, which can lead to population declines in severe outbreaks (Bornstein et al. 2001).

Gray fox: The presence of the gray fox in the Meadowlands was confirmed by a sighting in 2005 in Lyndhurst (E. Kiviat, pers. obs.). Stone (1907) noted that the earliest observations documented the gray fox as the only fox in New Jersey. This species was considered the common fox in much of the state and, unlike red fox, it was believed to avoid human habitation (Stone 1907). Despite large undeveloped areas in the Meadowlands, the gray fox is rarely seen. This may, however, be due to very low densities in greenspaces where they have escaped detection. Gray fox favors dense, woodland habitats and unlike other fox species can climb trees (Riley and White 2010). Gray foxes may be less tolerant of urbanization than other carnivores, and little is known of their ecology in urban areas. Their diet mainly consists of small mammals.

Red fox: This canid is regularly seen in the Meadowlands—often along roadways (e.g., Carola 2008a, 2016c). The red fox is successful in urban habitats because it is opportunistic and adaptable in habitat use and diet allowing exploitation of many urban habitats including city centers (Soulsbury et al. 2010, Brand 2019). It was once widely believed that red foxes in New Jersey were descendants of red foxes from Europe introduced for fox hunting (Stone 1907, Whitaker and Hamilton 1998) and possibly some that escaped from fur farms. However, a recent study comparing DNA from European and North American red foxes established that in the eastern U.S. the red fox is derived mainly from native North American populations in eastern Canada and the northeastern United States with some limited interbreeding with introduced European red foxes (Statham et al. 2012). The apparent lack of red foxes in many areas of the East Coast at the time of European colonization and their appearance during colonial times may have been due to range expansions that occurred with deforestation and the spread of agriculture or other changes that occurred at the time.

Red foxes appear to exist at highest abundance in medium-density urban development where there are ample food resources and denning sites and few free-ranging dogs (references in Soulsbury et al. 2010). Foxes are not strictly nocturnal; in their natural environments and away from human persecution, foxes are active during the day (Brand 2019). A study of red foxes in cities in the United Kingdom found that they preferred to use areas under backyard sheds for pupping dens and regularly used backyards as movement corridors (Trewhella and Harris 1990). Foxes varied in their preference for using railway embankments and rights-of-way as dens and movement corridors. Doncaster et al. (1990) studied the seasonal diet composition of red foxes in Oxford, England, and found their diet was dominated by scavenged food and earthworms. They also consumed small mammals and rabbits, fruits, birds, other invertebrates, refuse in garbage cans, and the carcasses of larger animals. Red foxes will avoid crossing secondary roads during times of peak traffic but still suffer heavy mortality on larger roads and during dispersal (Baker et al. 2007). Their populations are affected, and sometimes decimated, by outbreaks of sarcoptic mange (Gosselink et al. 2007, Soulsbury et al. 2007).

Eastern coyote: The coyote (*Canis latrans*) underwent range expansion into the eastern United States over the past century after the decline and eventual extirpation of the gray wolf (Laliberte and Ripple 2004). The eastern coyote is a coyote-wolf hybrid (Kays et al. 2010). Eastern coyotes only colonized the Meadowlands in recent decades, and they have been detected increasingly and widely during the twenty years of our studies (Carola 2006a, b, 2008a, b, 2011; E. Kiviat, pers. obs.). Coyotes have been found not only in natural areas at the fringes of cities but also in city centers including Manhattan. Home ranges in urban areas tend to be smaller than in more rural areas (Bekoff and Gese 2003) and coyotes in urban areas generally exist at higher densities than in nonurban areas, likely due to increased food resources and decreased hunting and trapping (Gehrt and Riley 2010). Territories nonetheless center on natural habitat even where there is significant overlap with the urban matrix and coyotes regularly cross between natural and built habitats (Gehrt et al. 2009). Coyotes alter their activity both temporally and spatially in response to human recreation (George and Crooks 2006) and exhibit changes in diet composition in urban areas (Newsome et al. 2015). Coyotes increase nocturnal activity and decrease diurnal activity with increasing human activity (references in Gehrt and Riley 2010). Diets are dominated by rodents and rabbits but in urban areas there is an increase in consumption of human-related foods by some individual coyotes (references in Gehrt and Riley 2010). Coyotes and foxes probably compete for prey where they overlap (Lavin et al. 2003) and in large numbers coyotes may suppress red fox populations (Harrison et al. 1989).

Domestic dog: The domestic dog is common throughout the Meadowlands as a household pet. Also, there are unsubstantiated reports of feral dogs in the Meadowlands. Domestic dogs as pets live in and around buildings but feral animals use shelters such as vegetation, dumps, or abandoned human-built structures. They play a predatory role in urban ecosystems where they may replace natural predators (Cauley and Schinner 1973, Whitaker and Hamilton 1998). Feral dogs can have a substantial impact on prey populations of rodents and rabbits. Feral dogs are usually afraid of people but may become dangerous when they form packs.

Cats: The bobcat, once likely resident in the Meadowlands, is now only reported occasionally on the fringes of the region. The domestic cat is the only species of cat known in the Meadowlands today. Cats and their tracks can be seen in greenspaces and built areas throughout the Meadowlands. Millions of domestic cats are kept as pets in the United States. Many cats are free ranging even though they are pets provided with food; in this case they act as *subsidized predators* of native wildlife. Other cats are feral and rely on their own hunting and scavenging instincts or supplemental commercial cat food supplied by well-meaning people. Both free-ranging and feral cats hunt rodents, birds, frogs, lizards, snakes, and insects, and predation by house cats in urban areas causes significant mortality of small mammals and birds (Churcher and Lawton 1987, Baker et al. 2005). In Syracuse, New York, cats were captured in urban greenspaces surrounded by high densities of people and buildings (VanDruff and Rowse 1986). Van Heezik et al. (2010) estimated that cat take of songbirds was close to city-wide population estimates of songbirds. This suggests that cities may be population sinks for some songbird species.

Seals: Four species of seals occur in New Jersey waters (NJDEP 2004a). Little is known about their population status and distribution. Most observations of seals in the Meadowlands and nearby waters are from rescues made by the New Jersey Marine Mammal Stranding Center (MMSC), National Oceanic and Atmospheric Administration (NOAA), and anecdotal reports in Hackensack Riverkeeper's newsletter *Hackensack Tidelines*. Three species have been documented in the estuaries of the Greater Meadowlands, namely harbor seal, the most common species, hooded seal and harp seal. The only other species known from the state is the gray seal, which has not been documented in the Meadowlands to our knowledge but is a rare visitor to the Hudson River.

Harbor seal: The occurrences of harbor seals have increased in the Hudson River in recent years (Kiviat and Hartwig 1994; *Hudson River Almanac* online). A harbor seal was observed in the Hackensack River by William Schultz, Raritan Riverkeeper and a staff member of the Marine Mammal Stranding Center, ca. 1 March 2001, and a harbor seal was also photographed in the late 1980s in Teaneck and reported in *The Record* (H. Carola, pers. comm.), indicating they travel far upriver. In February 2016, a harbor seal was seen in the Hackensack River in Carlstadt (Carola 2016a) and in March 2018 harbor seals were observed hauled out on docks in Secaucus and Carlstadt. The harbor seal is the only species that occurs in New Jersey waters during the breeding season (NJDEP 2006).

Hooded and harp seals: Less common in the Meadowlands and surrounding region are the hooded and harp seals. A possible hooded seal was observed in the Hackensack River in August 2006 (Carola 2006b). This species has been found in the nearby Arthur Kill municipalities of Perth Amboy and Sewaren in recent years, including a living male in June 2007 (MMSC 2008). Live harp seals were seen at two locations in Bayonne, New Jersey, in 2005 and 2006 (MMSC 2008).

Porpoises, Dolphins, and Whales

In addition to seals, marine mammals in the Greater Meadowlands include several species of cetaceans (porpoises, dolphins, and whales). Twenty-two species of cetaceans occur in coastal waters of New Jersey, with abundance widely variable by species. Seven species of cetaceans have been documented in waters in or near the Meadowlands including the harbor porpoise, bottlenose dolphin, Atlantic white-sided dolphin, fin whale, sei whale, minke whale, and humpback whale. The fin, sei, and humpback whales are listed as endangered in the State of New Jersey and federally. In addition to ship strikes, other sources of conservation concern for cetaceans include entanglement in fishing gear and the historical effects of population depletion from harvest coupled with slow recovery. In addition, dolphins and porpoises are often present in the bycatch of commercial fishing activities. Targets for conservation include determining numbers of marine mammals in bycatch, outreach to commercial fishing industry and boat operators, and regulatory incentives for changes in fishing gear and practices (NJDEP 2006). The mortality levels of fin whales and other cetaceans in Newark Bay indicate that outreach and an alert system are needed to inform pilots of the presence of whales in their vicinity, similar to the NOAA alert system for right whales. Speed restrictions on vessels are critical given the heavy shipping traffic associated with the major international ports of Newark and Elizabeth and that whales are federally listed as endangered, are protected under the Marine Mammal Protection Act, and protected through international agreements (CITES and IWC). The connection of the Hackensack River with Newark Bay and eventually New York Harbor and the Atlantic Ocean indicates that other marine mammals of coastal waters, such as the humpback whale, could at one time or another be present in Meadowlands waterways. Busy coastal ports may act as population sinks attracting marine mammals while causing high mortality rates.

Porpoises and dolphins: The harbor porpoise is the most common in the Meadowlands and surrounding estuaries. In March and April 2006, carcasses of this species were recovered at New Milford in Bergen County (French Brook, a tributary of the Hackensack River) and in Sewaren on the Arthur Kill (MMSC 2008). A harbor porpoise was sighted in February 2006 in the Passaic River off Riverbank Park in Kearny (Carola 2006c). The carcass of a harbor porpoise, which was believed to have died from exposure to oil, was found beached on Staten Island a couple of weeks later (Carola 2006a).

The bottlenose dolphin is the only cetacean that uses New Jersey waters as calving and nursery grounds. A bottlenose dolphin was found in Bayonne in 1999 (MMSC 2008). This species underwent a severe die-off in the 1980s on the East Coast of the United States (NJDEP 2006). Another die-off occurred in 2013 and is attributed to infection with cetacean morbillivirus. An Atlantic white-sided dolphin was found in nearby Perth Amboy in 2005 (MMSC 2008) and this species could also be found in the Meadowlands.

Whales: Carcasses of fin and minke whales have been found in Newark Bay, mainly near the Elizabeth Seaport (MMSC 2008). At least five fin whales were found in the waters off Port Elizabeth between 1997 and 2007. A minke whale was observed feeding in the Kill van Kull near Bayonne Bridge in May 2018 (Carola 2018a). A sei whale was hit by a ship in Upper New York Bay and brought into Newark Bay for recovery in 2005 (MMSC 2008). Humpback whales are regularly sighted in nearby New York Harbor.

CONCLUSIONS

Mammal diversity is moderate to low in the Meadowlands relative to the surrounding regions of northern New Jersey. The reasons for this phenomenon are many. There is great variation among mammal species in their responses to urbanization; as seen in the Meadowlands and in other urban regions, some mammals thrive whereas others are not able to persist, even in low numbers, relative to less developed places (Matthiae and Stearns 1981, VanDruff and Rowse 1986). In addition, there is a general lack of adequate surveys and there are no monitoring programs for mammals (not even muskrats!) in the Meadowlands, which hinders our understanding of which species are there and what their needs are. A few of the characteristics that give some species of mammals a better chance of persisting in urban-dominated landscapes include: being a generalist with a broad diet and flexible habitat use, behavioral plasticity, tolerance of human disturbance and proximity, intelligence, small home range size, low sensitivity to contaminants, and high fecundity.

As in the case of other groups of species, urbanization imposes many stresses on mammals as well as a few benefits. As discussed, the main stresses limiting mammal diversity include loss of habitat or poor quality of remaining habitats, lack of connectivity among habitat patches, industrial and stormwater pollution, low insect prey availability, subsidized predators, direct persecution through trapping and poisoning, and road mortality. Not all aspects of urbanization are bad for mammals and, in fact, urbanization provides a few unique benefits. Large wetland areas and old fields including young meadow and shrubland habitats in isolated areas are avoided by most urbanites; these isolated, urban greenspaces provide important refugia, food resources, denning areas, and other habitat functions for urban wildlife. In built areas, people provide supplemental food, water, and cover resources and this increases survival and productivity for some species (but wildlife should not be fed intentionally). Many mammal species are generalists. They are able to utilize a wide variety of habitats, including urban habitats. Also, survival of some species may be enhanced because densities of certain predators are reduced in urban areas; although domestic cats likely exert high pressure on small mammal populations.

Chapter 7

Birds

There are more species of birds in the Meadowlands than species in any other vertebrate class. Birds reach very high densities in some locations. Even the most passive observer cannot help but notice these conspicuous animals. They are popular; a large constituency of nature enthusiasts visits the Meadowlands just to see birds. An annual festival is dedicated to them, and they are usually the most sought-after sightings on the Hackensack Riverkeeper Eco-Cruises. Because they have received a lot of attention from scientists and birders, we have a relatively firm base of knowledge about when and where each species occurs. Many birds are listed as Endangered or Threatened in the state and are of regional or national priority. Furthermore, and perhaps most importantly, there have been severe declines in bird populations across North America over the past half century, and these declines are not restricted to rare or threatened species but include many that were once common and widespread (Rosenberg et al. 2019). An analysis of all 1,154 species of birds in North America (including México) demonstrated that one-third of them are of high conservation concern and without intervention are at risk of extinction due to population declines, range reductions, and loss of habitat (NABCI 2016). All these traits make birds as a group an important target for conservation planning. In fact, they are often the gauge by which the success of multimillion-dollar habitat restoration projects is measured.

In studies of urban ecosystems elsewhere, bird species richness and abundance vary when compared among gross land use types, which are usually defined as distinct points along a gradient from wildest to most urban (McDonnell and Pickett 1990, Blair 1996, Clergeau et al. 1998). Bird richness generally decreases with increasing intensity of urbanization because many urban habitat types lack the resources needed by more specialized species (review in Marzluff 2001, Devictor et al. 2007, Aronson et al. 2014). In this chapter we explore the avifauna of the Meadowlands and provide an overview of the species to address why there is such a high diversity of birds despite the urban context. We look at ways in which the urban character of the Meadowlands might impede as well as benefit birds. Finally, we identify targets for protecting and improving native bird diversity and the habitats that support it, in the Meadowlands.

BIRD COMMUNITIES IN THE MEADOWLANDS

Thanks to the work of forward-thinking conservationists, scientists, and naturalists who have studied birds in the Meadowlands for the past several decades, a good base of data exists. So far, as many as 299 species of birds are known from the Meadowlands (Appendix 2). For comparison, the state bird list for New Jersey has more than 440 species, with an average of 340 species observed per year (Walsh et al. 1999); eBird (2021) observations stand at 478 species observed in the state. That means that 63%–68% of all the bird species recorded in the state have been documented in the Meadowlands, which is less than 1% of New Jersey's land area. Even if this representation of the local species pool of birds is similar in other coastal estuaries with a comparable mixture of habitats, it is still an impressive local richness of birds in light of the intensive urban land use in and surrounding the Meadowlands. At least 88 species breed in the Meadowlands. Also, the Meadowlands provide habitat for a number of rare bird species. These include species that are listed in the state as Endangered (15 species), Threatened (13 species) and Special Concern (36 species; NJDEP 2004b, 2018a). New Jersey Species of Greatest Conservation Need (SGCN) include these 64 state-listed bird species, as well as an additional 74 species, for a total of 138 SGCN in the Meadowlands (NJDEP 2018a).

For a variety of reasons, including the number of people reporting on bird occurrences as well as potential changes in bird range distributions and accidental occurrences, we expect that the bird species reported in the Meadowlands will change from what we have reported here. Also, there is a high likelihood that bird richness and abundance will decline in the Meadowlands given the extreme declines in bird populations in North America over the past fifty years, unless the larger-scale causes are addressed (Rosenberg et al. 2019). Habitat loss, alteration, infectious diseases, and climate change are some of the likely drivers for these declines (Reif 2013).

Because of the great extent of freshwater and brackish wetlands and open water in the Meadowlands, species of birds associated with these habitats are also very widespread and diverse and occur in locally high densities. But the Meadowlands also include a wide variety of upland habitats that provide necessary breeding, foraging, and resting habitats for many species. Each of the habitat types in the Meadowlands has a somewhat distinct bird community.

BIRD COMMUNITIES VARY IN SPACE AND TIME

One will never find all 299 species of birds in a particular place in the Meadowlands or at a particular time. There is a high level of spatial variation in the species composition, species richness, and density of the bird community. In addition, bird communities vary by time of day, month, and year. Some species are year-round residents while others use the Meadowlands only as wintering or spring-summer breeding habitat. Still others are present only during spring and fall migration.

In addition to the large complex of wetland and open water habitats, the Meadowlands contain a variety of other habitats including abandoned landfills and industrial sites that are now grasslands and shrublands, a few small upland forests, riparian forests, and forested wetlands, all in the context of a variety of human-built (residential, commercial, industrial, transportation, and park) habitats.

Not all habitats are alike in the richness and abundance of species they support. In the Meadowlands, shrubland supported 122 species, open water 91 species, cordgrass (*Spartina*) marshes 87 species, and common reed marsh 81 species (Mizrahi et al. 2007). When species richness by habitat types was standardized for area, high marsh supported the most species (12 species/ha) followed by shrubland (6 species/ha); forest and low marsh only supported about 3 species/ha and open water and common reed habitats supported about 2 species/ha. In that study, the highest density of birds was in open water habitats, followed by shrubland, and then mudflats. Wetland management projects that exchange common reed marsh for tidal ponds and mudflats may be beneficial to some nonbreeding birds but detrimental to some marsh-nesting birds. In general, many species of birds use common reed stands for nesting, roosting, cover, and foraging habitats (Kiviat 2019), and this pertains to certain bird species in the Meadowlands.

A comparison of overall bird richness and diversity among habitats is interesting but may be misleading if used as the only criterion for prioritizing one habitat type over another. First, most major habitat types differ in general species composition. For example, shrub-nesting songbirds are not likely to be found in open water or mudflat habitat. In turn, sandpipers, species that typically feed on mudflats and in shallow water, are not part of the bird community in shrubby habitats. Also, some habitats have low bird diversity but support rare or highly specialized species that do not occur elsewhere. Habitat specialists occupy a narrow range of habitat types (e.g., salt marsh dominated by smooth cordgrass [*Spartina alterniflora*]). These species are, by definition, limited by the presence of their required habitat.

Within a habitat type, local variation in land use history, topography, microclimate, and other variables creates unique plant and animal communities. Thus, within broad categories of habitat there is a wide range in quality for a species and quality may be the key driver of avian diversity in cities over amount and arrangement of habitat (Dures and Cumming 2010). Freshwater marshes with a high interspersion of vegetation cover and shallow open water are preferred over habitats with low interspersion by several marsh species including the blue-winged teal, American bittern, American coot, common gallinule, and Virginia rail (Kaminski and Prince 1981, Alisauskas and Arnold 1994, Linz et al. 1997, Bannor and Kiviat 2002, Rohwer et al. 2002, Lowther et al. 2009). Even among urban neighborhoods, differences in the species richness, size, and density of large street trees are in large part responsible for differences in species composition, richness and abundance of songbirds (MacDonald-Beyers 2008). All of the habitats, in this complex, heterogeneous mosaic contribute to the overall pattern of bird diversity in the Meadowlands.

In addition to habitat type, individual habitat units and habitat complexes vary in patch characteristics such as size, shape, connectivity, heterogeneity, interspersion, and adjacency. Urbanization typically leads to loss and fragmentation of habitats (Burgess and Sharpe 1981, Wilcox and Murphy 1985). Fragmentation results in smaller patches, less connectivity among patches, and more edge habitat. Many species of birds are area sensitive (e.g., Robbins et al. 1989) in that they only occupy habitat of a substantial extent and are seldom found in small patches. The negative effect of small patch size on the richness of forest birds has been well documented in North America. There are no forests larger than about 100 hectares (the somewhat-fragmented Teterboro Airport Woods) in the Meadowlands, which should be expected to limit the use of the Meadowlands as breeding habitat for area-sensitive forest birds. Several wetland species are area-sensitive as well. Seaside sparrow and salt marsh sparrow, species that may nest in the Meadowlands, are found more frequently in large than small salt marshes (Benoit and Askins 2002). Egrets forage more frequently in larger salt marshes (Trocki and Paton 2006). Because some important habitat types in the Meadowlands are present in large patches (e.g., reed marsh, old field), many area-sensitive species are apparently supported. On the other hand, some wetland-dependent birds that breed in the Meadowlands have been shown to be area-*in*sensitive (e.g., yellow-crowned night-heron; Bentley 1994). Small fragments of urban woodland are also used by some birds for breeding and non-breeding purposes including roosting and foraging (Croci et al. 2008, MacDonald-Beyers 2008).

Tidal, daily, monthly, seasonal, and longer-term cycles greatly affect the composition of the bird community at any one time. Many birds have peak activity periods around dawn and dusk but rest during the day while others are mainly nocturnal. Many migratory birds fly at night and occupy stopover habitats only during the day. Tidal fluctuations are very important influences on daily and monthly bird distribution and density. For instance, many species forage on mudflats and shallow areas at low tide only and rest in higher-elevation habitats at high tide (e.g., egrets and sandpipers; Kiviat 1989). Birds may vary their habitat use depending on predictable patterns of human disturbance such as pedestrian and automobile traffic (Fernández-Juricic 2000a, b). There are many sources of continual, loud noise in the Meadowlands, including airports, highways, and railroads; some birds may be modifying their behavior there or avoiding those areas entirely.

There is a great seasonal flux of bird richness and community composition. Some species are present during the spring and summer breeding season in the Meadowlands but not in winter and vice versa. Shorebirds, warblers, and many hawks are mainly spring and fall migrants. Some waterfowl are only present in winter. Seasonal patterns may vary depending upon severity of weather in a given year.

Some patterns of bird spatial distribution reflect long-term temporal cycles and changes. For instance, many species undergo periodic population cycles, sometimes decades-long, which are poorly understood because most research projects are not of long enough duration to detect these patterns. Wet years and dry years differentially affect the amounts of habitat, food, and cover for certain birds. Several northward range expansions, including northern mockingbird, northern cardinal, American goldfinch, mourning dove, brown-headed cowbird, and both turkey and black vultures, have also taken place in the past century, and a few species, including common raven, peregrine falcon, and bald eagle, have recovered from long-term population declines (Aldrich and Coffin 1980, Morneau et al. 1999). Occasionally, an introduction or escape results in a new species for the region (e.g., monk parakeet, house finch). Also, some observed changes in bird populations may be due to factors far away on winter or summer ranges or migration routes.

Finally, on even longer timescales, habitat changes occur due to changing dominance of plant species and growth forms (vegetation development) at variable spatial and temporal scales. The pattern of heterogeneity in the landscape is highly dependent on the type, extent, and frequency of disturbance (e.g., periodic plant disease outbreaks). Also, changes in distribution of habitats caused by such factors as rising sea level, especially in coastal marshes and shorelines, have a great impact on species distributions. Sea level rise and the increasing frequency and severity of storm surges and more stormwater runoff promise to be critical factors influencing biodiversity in the Meadowlands.

REVIEW OF THE MEADOWLANDS BIRD FAUNA

In this section, we describe the species of birds found in the Meadowlands and their distribution, seasonality, abundance, habitat affinities, and relevant biological traits. While much has been done to document the species and their distribution and abundance in the Meadowlands, there has been little research on the behavior and ecological characteristics of bird species in the Meadowlands *per se*. For this, we draw upon research outside the Meadowlands.

Waterfowl

The term *waterfowl* refers to ducks, geese, and swans. Thirty-two species of waterfowl are known in the Meadowlands. The region is important for wintering, breeding, and migrating waterfowl. Species and densities of waterfowl vary greatly with the seasons, weather, and other factors. The most commonly encountered (i.e., widespread) species are the Canada goose and mallard (Mizrahi et al. 2007).

Many ducks that breed in northern regions of North America winter in coastal and inland wetlands of warmer, ice-free latitudes. Submergent and emergent plants in creeks, shallow bays, impoundments and mudflats of the Meadowlands provide food resources for waterfowl and protection from harsh weather and predators for wintering ducks. Tidal creeks bordered by common reed and other emergent vegetation provide important shelter from the weather for wintering waterfowl.

The Canada goose, mallard, American black duck, gadwall, northern pintail, northern shoveler, blue-winged teal, green-winged teal, canvasback, lesser scaup, and ruddy duck are the most abundant waterfowl in winter (Day et al. 1999, Mizrahi et al. 2007). In one study, daily winter counts of waterfowl in the Meadowlands averaged 2,000 birds (Day et al. 1999). The northern pintail, green-winged teal, canvasback, and ruddy duck occur in large numbers locally.

During spring and summer, breeding waterfowl are concentrated in common reed stands and other areas of dense emergent vegetation along creeks and ditches (NJTA 1986, Kane and Githens 1997). Eggs are laid in nests on or near the ground consisting of piled up plant material and feathers and are sometimes concealed beneath lodged (bent or broken) reed stems (Kiviat 2019). Studies of waterfowl in other areas have shown that during breeding season they are often found feeding in shallow open water areas where they generally rely upon animal foods, especially aquatic invertebrates such as aquatic insect larvae, snails, and crustaceans (e.g., Swanson et al. 1974,

Drilling et al. 2020, Leschack et al. 2020) in a food web based on the detritus of reed and other plants.

Species that breed in the Meadowlands in significant numbers include the Canada goose, mallard, gadwall, blue-winged teal, American black duck, and ruddy duck (NJTA 1986, Wargo 1989, Kane and Githens 1997, Day et al. 1999, Mizrahi et al. 2007); however, in recent years blue-winged teal have become very rare during the breeding season (H. Carola, pers. comm.). Studies from areas outside the Meadowlands show that blue-winged teal prefer hemi-marsh (approximately 50% emergent plant cover, 50% shallow open water) for breeding habitat (Livezey 1981, Duebbert et al. 1986, Rohwer et al. 2002). This may be true of the blue-winged teal in the Meadowlands. In 1990, a pond on top of a landfill held 144 gadwall ducklings (Day et al. 1999). Gadwalls generally nest in tall emergent vegetation near water and use shallow to deep wetlands for feeding (Leschack et al. 2020).

Regionally-rare breeding populations of the ruddy duck, a species that in New Jersey is believed to nest only in reed marshes (Kane 2001), occur at Kearny Marsh and Kingsland (Saw Mill Creek) Marsh (NJTA 1986, Day et al. 1999, Mizrahi et al. 2007). In fact, Kingsland Marsh supported one of the largest breeding populations of this species in the state (NJTA 1986) and this may still be true. For the past decade or more, the best location to observe ruddy ducks has been at Mehrhof Pond and the Hackensack River adjacent to the Bergen County Utilities Authority (BCUA) property. This species is known to feed extensively on chironomid midge larvae (Chironomidae) during the breeding season (Brua 2002; see Chapter 10); chironomids are abundant in the Meadowlands marshes but may be negatively affected by applications of insecticides for mosquito control.

Unlike other waterfowl in the Meadowlands, wood ducks nest in cavities in large trees, preferably over water, and readily use nest boxes (Hepp and Bellrose 2013). Historically, wood duck broods have been observed in Kearny Marsh West. It is likely that these wood ducks were nesting in cavities in large ornamental trees on streets or in the nearby cemetery.

The Meadowlands are visited by a number of waterfowl species during spring and fall migrations. The most numerous migrant waterfowl species include the gadwall, mallard, northern shoveler, green-winged teal, northern pintail, all three mergansers, greater and lesser scaups, American wigeon, common goldeneye, canvasback, American black duck, blue-winged teal, and Canada goose.

Wading Birds

Thirteen species of long-legged wading birds ("waders") nest, forage, and/ or roost in the Meadowlands. Common species include the great egret, snowy egret, great blue heron, green heron, black-crowned night-heron, yellow-crowned night-heron, least bittern, and glossy ibis (NJTA 1986, Kane and Githens 1997, Day et al. 1999, USACE 2000, Mizrahi et al. 2007, Tsipoura et al. 2009). The great egret and snowy egret are the most widespread and abundant species followed by great blue heron and black-crowned night-heron (Mizrahi et al. 2007, Tsipoura et al. 2009). The great egret and snowy egret were more abundant at survey sites in the Meadowlands than in other areas surveyed in 2009 throughout the NY-NJ Harbor and Raritan Bay (Tsipoura et al. 2009). The American bittern is an Endangered species in the state and black-crowned and yellow-crowned night-herons are Threatened species (NJDEP 2004b, 2018a). The least bittern and great blue heron are species of Special Concern in New Jersey. The white plumage of species such as great and snowy egrets makes them stand out against the landscape of large bodies of shallow water, mudflats, and tidal creeks; the cryptic, shy bitterns are much harder to observe and study.

Habitats that support foraging waders include the areas around tidal creeks, shallow open water, mudflats, and intertidal salt marsh (NJTA 1986, Kane and Githens 1997, Day et al. 1999). Many wading birds forage in and around shallow water and therefore their distribution and densities depend on the tides (Burger et al. 1982, Maccarone and Parsons 1994, Tsipoura et al. 2009). The diet of most herons and egrets consists mainly of fish and crustaceans such as the Atlantic silverside (*Menidia menidia*), mummichog (*Fundulus heteroclitus*), and grass shrimp (*Palaemonetes*; Maccarone and Brzorad 2002). The yellow-crowned night-heron differs from other herons in being specialized to eat crustaceans, especially crabs (e.g., fiddler crabs, *Uca* spp., and Atlantic blue crab, *Callinectes sapidus*; Watts 2011). Herons and egrets use a *sit-and-wait* or *active* hunting strategy to capture fish. The green heron is one of the few tool-using species of bird—it sometimes uses bait, such as a feather, to entice fish within striking distance (Davis and Kushlan 1994). Glossy ibises forage tactilely by probing the substrate for invertebrates such as adult and larval insects, crustaceans, and worms. Waders, especially snowy egrets and glossy ibises, tend to form mixed species feeding flocks (Erwin 1983). Wading birds may select microhabitats within particular marshes to optimize their ability to find prey. For instance, a study in the nearby Arthur Kill estuary showed that great and snowy egrets and glossy ibises preferred to forage at the mouths of small salt marsh creeks (Maccarone and Brzorad 2005), likely because small animals move back and forth there. In another study of habitat use in Rhode Island, great and snowy egrets preferred pools within marshes (Trocki and Paton 2006). They avoided ditches and stands of common reed. Interestingly, we have often seen egrets and other waders foraging within a meter or two of reedbeds in the Meadowlands, for example, in Kingsland Impoundment. There is evidence that egrets prefer larger, more extensive marshes over smaller ones (*ibid.*).

The least and American bitterns are secretive marsh species. The least bittern captures fast moving prey such as fish and dragonflies over deep water by clinging to clumps of vegetation or standing on platforms of bent reeds it builds at productive feeding sites (Weller 1961, Reid 1989). This species also actively hunts small fish at the *tideline* (water edge) in Hudson River marshes (E. Kiviat, pers. obs.). The American bittern forages along vegetated fringes and shorelines at water's edge in freshwater marshes and benefits greatly from a high interspersion of vegetation cover and open water (Gibbs et al. 1991, Lowther et al. 2009).

Wading birds typically nest in vegetated uplands or non-flooded wetlands throughout their range (Gibbs et al. 1991). Some species nest in common reed stands (Kiviat 2013, 2019). Islands often provide suitable nesting habitats that are relatively safe from predators such as raccoons. This type of habitat is very limited in the Meadowlands, thus few waders nest here and instead most of the wading birds foraging in the Meadowlands come from one of seven breeding colonies in the New York–New Jersey Harbor estuary complex (Day et al. 1999). Breeding species in the NY-NJ estuary include the great and snowy egrets, great blue heron, and glossy ibis. The black-crowned night-herons are also from rookeries outside the Meadowlands. In 2009, the Harbor Herons Rookery Complex, a series of small islands around New York City in the East River, Arthur Kill and Kill van Kull, supported 1,683 nesting pairs of herons, egrets, and ibises. The total number of island wader nests declined 16% from 2016 to 2019, from 1,420 to 1,186 pairs (Winston 2019). Waders often travel long distances from nesting to feeding areas during the breeding season because they require shallow wetlands and mudflats that are not always in close proximity to nest sites. Based on a study of flight directions of herons leaving nesting colonies for foraging grounds, the Meadowlands were identified as the likely destination for herons from South Brother Island in the East River (Nagy 2005). Also, in late summer, the Meadowlands are intensively used as a foraging area during post-breeding dispersal of adult and juvenile egrets, herons and ibises (Day et al. 1999). The Meadowlands also provide roosting habitat for these species.

Four species of wading birds are known to breed in the Meadowlands. These include the least and American bitterns, yellow-crowned night-heron, and green heron. The least bittern has been considered a fairly common breeding species in the Meadowlands (NJTA 1986, Kane and Githens 1997), although, in 2006, New Jersey Audubon detected this species only at Kearny Marsh (Mizrahi et al. 2007). The least bittern is typically found in both fresh and brackish marshes and will occasionally nest in salt marsh habitat throughout its range (Poole et al. 2009). It prefers tall emergent vegetation or clumps of woody vegetation over deep water for nest sites (Poole et al. 2009). The American bittern breeds mainly in freshwater marshes (Lowther et al. 2009). The green heron generally nests in a variety of upland and wetland habitats throughout its range, including trees and emergent vegetation with suitable foraging sites nearby (Davis and Kushlan 1994). The yellow-crowned night-heron typically nests colonially in trees (e.g., oaks and pines) and shrubs (e.g., groundsel bush and bayberry) near foraging habitat (Watts 2011). In Virginia, this species frequently nests in wooded residential areas that have open understories (Watts 1989). It has also done so in urban residential areas in the Meadowlands in Secaucus (Carola 2008c) and there is an account of a nest on an apartment building fire escape in Queens, New York (Rafter 2013). Interestingly, yellow-crowned night-herons are more commonly associated with long, narrow marshes as opposed to marshes with extensive interior habitat (Bentley 1994).

Gruiformes (Rails, Gallinules, Cranes)

Eight species of Gruiformes use the marshes in the Meadowlands as breeding and foraging habitat during spring, summer, and fall. They are the American coot, common gallinule, sora, and clapper, Virginia, and king rails, and sandhill crane (sora may be a migrant only, the clapper rail and king rail are rare, the sandhill crane is an occasional). Because many of these species are nocturnal and secretive, they are often underrepresented in bird studies. One exception is a recent study by New Jersey Audubon that utilized playback of calls to detect secretive marsh birds (i.e., rails and bitterns) at sites throughout the Meadowlands (Mizrahi et al. 2007). Rallids often use lodges and trails created by muskrats in dense vegetation stands for movement and foraging (Conway 1995, Bannor and Kiviat 2002). The king rail is a species of Special Concern in New Jersey.

American coots occupy fresh and brackish marshes outside the breeding season (Brisbin and Mowbray 2002). They are among the most abundant bird species in the Meadowlands during winter (Day et al. 1999) and also breed in the Meadowlands in freshwater marshes. Vegetation including reeds, sedges, willows, and grasses, along shorelines and at pools within the emergent vegetation, are generally preferred breeding habitat (Brisbin and Mowbray 2002). Studies outside the Meadowlands demonstrated that the highest coot densities occurred in areas with maximum interspersion of emergent vegetation and water (Bett 1983, Alisauskas and Arnold 1994). In one study, herbicide treatment of cattail to increase interspersion of vegetation and open water resulted in increased densities of coot (Linz et al. 1997). The coot diet consists of aquatic vascular plants and algae, with some terrestrial vegetation and animal matter (Brisbin and Mowbray 2002).

The common gallinule breeds in deep, fresh and slightly brackish, tidal and nontidal marshes consisting of robust emergent vegetation and open water or mudflats (Bannor and Kiviat 2002). Gallinules are often associated with cattail-dominated marsh (O'Meara et al. 1982). Gallinules in the Meadowlands have been known to breed in varied common reed habitats (Kane 1978, 2001). Gallinule diet consists mainly of small hard items, including seeds, beetles, and snails (Bannor and Kiviat 2002). Like the American coot, the common gallinule prefers habitat with a high interspersion of vegetation and open water (Brackney 1979, Bannor and Kiviat 2002). Among the breeding sites for coots and gallinules are Kearny Marsh and Kingsland Impoundment (NJTA 1986, Kane and Githens 1997). Historically, diking at Kearny Marsh decreased the salinity and increased the gallinule breeding population (Bull 1964). Also, the increased water level there fragmented the reed stands resulting in more suitable, highly interspersed habitat (Kane 1978). The common gallinule has apparently declined in the Meadowlands and along the Hudson River in recent decades (E. Kiviat, pers. obs.).

The clapper rail is a salt marsh specialist (Rush et al. 2018). The preferred nest habitat is smooth cordgrass (Mangold 1974) where few bird species nest. Nests are typically located near the edge of a tidal creek (Kozicky and Schmidt 1949). Clapper rails eat crustaceans (especially fiddler crabs) and other invertebrates by foraging along the edges of mudflats and tidal creeks during lower tide stages (Clark and Lewis 1983, Heard 1983). Two of the last reported nesting sites for clapper rails in northern New Jersey were at Kingsland Marsh (NJTA 1986) and Mill Creek (Kane and Githens 1997). More recently, they were detected at Lyndhurst Riverside, Riverbend Marsh, and Sawmill Creek Wildlife Management Area (Carola 2006d, Mizrahi et al. 2007). They are now rare in the Meadowlands.

King and Virginia rails are found in fresh and brackish marshes (Conway 1995, Pickens and Meanley 2018). Both species breed in areas of tall vegetation adjoining foraging areas. Suitable foraging areas include moist soils, shallow water and mudflats adjacent to emergent vegetation. The Virginia rail prefers areas of high interspersion of cover and foraging habitat (Kaminski and Prince 1981, Reid 1985). Like the clapper rail, the king rail is a crustacean specialist (fiddler crabs, as well as clams in marine habitats) but will also eat aquatic insects, seeds, small fish, frogs, grasshoppers, and small mammals (Pickens and Meanley 2018). The diet of Virginia rails includes a variety of small aquatic invertebrates during the breeding season (Conway 1995). Both species appear uncommon or rare in the Meadowlands.

The sora typically breeds in freshwater marshes dominated by emergent vegetation and some brackish coastal marshes (Melvin and Gibbs 2012). In salt marshes along the Atlantic coast of New York, soras breed in habitat dominated by smooth cordgrass and common reed (Greenlaw and Miller

1982). They prefer habitats with dense submerged and floating vegetation, perhaps because it increases the amount of available food (Melvin and Gibbs 2012). This species sometimes nests in very small wetlands (Crowley 1994, Robbins 1991). Soras are fairly common in the Meadowlands marshes during fall migration (Figure 7.1; Wargo 1989, USACE 2000). Chapman (1900) mentioned soras feeding on wild-rice in the Meadowlands during fall. It is unknown whether soras are breeding in the Meadowlands today. Soras ceased breeding in Tivoli North Bay on the Hudson River in the 1970s; possibly this species has withdrawn from the southern edge of its breeding range in recent years.

Figure 7.1 Sora on common reed litter in a small pool in brackish-tidal reed marsh, Walden Marsh. Pools in otherwise dense, extensive reedbeds are important for birds and other biota. Photo by Christopher Graham.

Shorebirds

Thirty-six species of shorebirds occur in the Meadowlands. Smaller shorebirds (*Calidris* spp.) can be hard to identify in the field. The Meadowlands are considered one of the most important stopovers for shorebirds migrating between South America and Arctic nesting grounds (Day et al. 1999) and are on one of eleven critical migration corridors that pass through New Jersey

(Dunne et al. 1989). Tidal mudflats and impoundments are important habitats within the Meadowlands and are used as feeding sites by many thousands of migrating shorebirds (Day et al. 1999). Shorebirds feed on a variety of aquatic and terrestrial invertebrates during migration (e.g., Macwhirter et al. 2002, Nebel and Cooper 2008). The most common fall migrants include the semipalmated sandpiper, lesser yellowlegs, short-billed dowitcher, and greater yellowlegs (Day et al. 1999, eBird 2021). Dunlin was once common but has become less so in recent years. The Wilson's phalarope is a regularly seen migrant. Some rare fall migrants include the buff-breasted sandpiper, Hudsonian godwit, and red-necked phalarope (Day et al. 1999). Among the shorebirds in the Meadowlands, the upland sandpiper is listed as state Endangered, the red knot is listed as state Threatened, and the whimbrel, sanderling, and spotted sandpiper are Special Concern species in New Jersey.

During migration, shorebirds reach high local species richness and density in the Meadowlands. For example, a recent survey counted 4,000 semipalmated sandpipers at Mill Creek Point (Tsipoura et al. 2006). Also, Kingsland Marsh and Kingsland Impoundment had 31 species of shorebirds on the tidal flats between July and November 1971–1980 (Kane 1983). Eight thousand semipalmated sandpipers were recorded in the impoundment 23 July 1994 (Kane and Githens 1997) and daily counts exceed 5,000 in most years (Day et al. 1999). Shorebirds use the mudflats exposed during drawdown of the impoundment. Other shorebirds using the impoundment include the greater and lesser yellowlegs (Kane and Githens 1997). The area of the impoundment nearest the Turnpike is used as a high-tide roost by migrating shorebirds in fall (NJTA 1986). The least sandpiper also reaches high densities locally in the Meadowlands (Tsipoura et al. 2006, Mizrahi et al. 2007).

The spotted sandpiper, killdeer, American woodcock, and Wilson's snipe are the only species of shorebird that are known to breed in the Meadowlands. Spotted sandpipers nest in a wide variety of habitats throughout their range (Reed et al. 2013). They prefer to nest in patches of dense vegetation on a shoreline of a stream or lake in semi-open habitat (Maxson and Oring 1980, Oring et al. 1983), although Kiviat (pers. obs.) found a nest 2m from the tracks of the Amtrak railroad on the Hudson River. Spotted sandpipers breed at widespread locations within the Meadowlands. Killdeer are technically shorebirds but they typically breed away from water in a wide variety of habitats (Jackson and Jackson 2000). They nest in areas of sparse or no vegetation, including sandbars, cultivated fields, parking lots, bare wetland fill, and gravel roofs (Jackson and Jackson 2000). Killdeer forage in wetlands as well as uplands.

Two anomalous sandpipers, the American woodcock and Wilson's snipe, occur in the Meadowlands. The woodcock, a species of young forests, thickets, swamps, open fields, and wet meadows, is declining in the eastern U.S.

(McAuley et al. 2013). Among other locations, we found woodcocks display-ing in a habitat dominated by small trees, shrubs, and herbs on low-lying garbage fill north of Laurel Hill Park; the habitat was subsequently destroyed for development. Other potential breeding areas include the area around Losen Slote (Kane and Githens 1997) and Schmidt's Woods (Carola 2008a). Although Wilson's snipes usually breed farther north, they were found at the Kane Natural Area, prior to mitigation in 2010, in all seasons except winter.

Gulls and Terns

Nineteen species of gulls and terns are known in the Meadowlands. Gulls and terns are fairly widespread in tidal and nontidal waters in the Meadowlands during spring, summer, and fall, and several gull species occur in winter. Ring-billed herring and great black-backed gulls are the most abundant and widespread species (Mizrahi et al. 2007). The roseate and least terns and the black skimmer are Endangered species in New Jersey; roseate tern is also listed as federally Endangered. Caspian, common and gull-billed terns are New Jersey species of Special Concern.

Most of the gulls that use the Meadowlands during spring and summer are likely from breeding colonies outside the Meadowlands, especially the islands in the New York—New Jersey Harbor Estuary. In 2007, herring and great black-backed gulls nested on islands throughout the Harbor Estuary in fairly high numbers (Bernick 2007). An exception is the ring-billed gulls that bred at one site in the Meadowlands (Kane and Githens 1997).

The least tern, a state Endangered species, has nested on dredged mate-rial deposits and may be roosting on roofs of commercial buildings in the Meadowlands (Day et al. 1999). North of the Vince Lombardi Service Area, a large sandy area historically provided nesting habitat for the least tern (NJTA 1986) but this site is no longer active. More recently, in at least one year least terns nested on the roof of a commercial building in North Bergen. Least terns that bred in the Meadowlands congregated at Kingsland Impoundment with their fledged young (NJTA 1986).

Kingsland Impoundment is an area of special note because of the large numbers of gulls and terns that use it as a foraging and resting area (Kane and Githens 1997). Also, black skimmers have been seen feeding there (NJTA 1986, E. Kiviat, pers. obs.). Landfills are known to attract many gull species and the combination of landfills and open tidal waters may have inflated the densities of gulls in the Meadowlands. The closing of landfills in the 1970s–1990s presumably reduced densities of gulls in the Meadowlands.

Black skimmers are regular pre- and post-breeding residents in the Meadowlands at Anderson Creek Marsh and the area around Mill Creek Point Park (H. Carola, pers. comm.).

Other Waterbirds

Other waterbirds in the Meadowlands include loons (2 species), cormorants (2 species), grebes (3 species), and the American white pelican. Of these, the pied-billed grebe is an Endangered species in New Jersey.

Double-crested cormorants nest in trees and on built structures on and around islands in the Harbor estuary (Bernick 2007). They often occur in flocks in the Meadowlands (Mizrahi et al. 2007) as well as singly. Double-crested cormorants are common at mid-size and large waterbodies, as well as larger marsh pools. They are breeding in large numbers (approximately 2,000 pairs) in the New York—New Jersey Harbor Estuary near the Meadowlands following a pattern of expansion elsewhere in the Hudson River estuary. In the Harbor Estuary they nest on floating navigation markers in open water (H. Carola, pers. obs.).

The pied-billed grebe has similar habitat affinities to the American coot (Muller and Storer 1999). The grebe nests on a floating platform in tall emergent vegetation in seasonal ponds ranging from fresh to brackish or in larger marshes. They feed opportunistically, mainly on small fish, crustaceans, and aquatic insects (Muller and Storer 1999). Kearny Marsh is a nesting and wintering area for pied-billed grebe (Kane and Githens 1997). The Meadowlands have long provided breeding habitat as pied-billed grebe was recorded breeding in Kearny in 1905 (Abbott 1907). Kingsland Impoundment also has breeding pied-billed grebe (NJTA 1986). The mudflats created during drawdown at Kingsland Impoundment are heavily used by grebes (Kane and Githens 1997).

Loons are regular winter visitors to the New York—New Jersey Harbor Estuary, including the Meadowlands, where they spend the colder months in brackish water that remains unfrozen.

Birds of Prey (Raptors)

Birds of prey include vultures, hawks, eagles, falcons, and owls. Twenty-five species of raptors use a variety of wetland and upland habitats in the Meadowlands for breeding, hunting, and roosting. The Meadowlands are an important wintering area for several of these species (Bosakowski 1983). Habitat characteristics favoring raptors in the Meadowlands include the extensive marshes and non-wooded uplands, little direct human disturbance, and an abundance of prey (e.g., fish, small birds, small rodents, depending upon the raptor species); formerly, the operating garbage landfills with their abundance of rodents were magnets for raptors. Species of hawks and owls commonly seen include the northern harrier, rough-legged hawk, red-tailed hawk, Cooper's hawk, American kestrel, short-eared owl, and long-eared owl

(Day et al. 1999, Mizrahi et al. 2007). The red-tailed hawk is the commonest raptor (Mizrahi et al. 2007). The bald eagle, northern harrier, northern goshawk, red-shouldered hawk, peregrine falcon, and short-eared owl are state Endangered species. The osprey, American kestrel, barred owl, and long-eared owl are listed as Threatened and sharp-shinned hawk, Cooper's hawk, broad-winged hawk, and barn owl are species of Special Concern.

The common reed marshes and landfills of the Meadowlands provide important habitat for populations of wintering raptors, especially the northern harrier, rough-legged hawk, red-tailed hawk, sharp-shinned hawk, osprey, American kestrel, short-eared owl, and long-eared owl (Bosakowski 1983, 1986, 1989). Red-tailed hawks are frequently seen perching in trees and watching for prey along the highways in the Meadowlands. In winter, northern harriers hunted predominantly in common reed marshes whereas rough-legged hawks hunted around landfills (Bosakowski 1983). Red-tailed hawks and short-eared owls also hunted in common reed marshes in the Meadowlands (Bosakowski 1989). A large communal roost of northern harriers was formerly located in a common reed marsh at Berry's Creek (Bosakowski 1983, Kane and Githens 1997). Bosakowski (1989) described evening emergence of short-eared owls from ground roosts in the marsh. They are crepuscular as well as nocturnal hunters. Perches in or adjacent to open areas are used by short-eared owls and red-tailed hawks for hunting. American kestrel uses grasslands and forb meadows on reclaimed landfills. Several other species are regularly observed including the merlin, Cooper's hawk, sharp-shinned hawk, and red-shouldered hawk (USACE 2000, eBird 2021). At gas capture plants associated with closed landfills, kestrels and other raptors have been scorched flying through the methane flares.

In addition to marshes, the areas around landfills were hunted by wintering short-eared owls (Bosakowski 1983, NJTA 1986). Studies of long-eared and short-eared owl pellets collected from roost sites near Berry's Creek in the Meadowlands were conducted by Anderson (1977) and Bosakowski (1982). Anderson (1977) identified meadow vole (*Microtus pennsylvanicus*) as the primary food source for long-eared owl followed by brown rat, house mouse, and small birds. Bosakowski (1982) found that house mouse was the major prey for long-eared owl. This unusual phenomenon was explained as likely due to snow cover making it difficult to capture meadow voles. Bosakowski identified house mouse, birds, and brown rat as the main prey items in short-eared owl pellets. Roost sites preferred by short-eared owls were large, planted conifers adjacent to the open expanses of marsh and field where they foraged (Bosakowski 1989). Short-eared owls have also roosted on the ground in common reed stands (Bosakowski 1986). Winter irruptions of snowy owls from the Far North occur periodically, most recently in 2017–18 (Carola 2018b); this has been a regular occurrence in more recent years.

Several raptor species, including the northern harrier, osprey, bald eagle, red-tailed hawk, peregrine falcon, great horned owl, and barn owl, breed in the Meadowlands. Northern harrier nests on the ground, in tall, dense herbaceous vegetation (Smith et al. 2011). The densest populations of harrier are typically associated with large undisturbed habitats dominated by dense growth (Apfelbaum and Seelbach 1983, Toland 1986, Kantrud and Higgins 1992). They nest in a variety of vegetation types, including common reed (Apfelbaum and Seelbach 1983, NJTA 1986, Kiviat 2019). One of only two pairs of northern harriers confirmed nesting in the urban core of New York City has nested at the Berry's Creek Marsh in the Meadowlands (NJTA 1986, Kane and Githens 1997, Day et al. 1999). This represents a remnant population of the 1–4 pairs that once nested here (Kane 1974, Day et al. 1999). In addition, territorial pairs or family groups of harriers have been observed at the Kane Natural Area and probably nested there in some years (Kane and Githens 1997; H. Carola, pers. comm.). Harriers forage in marshes and grasslands throughout the Meadowlands.

In 2007, a pair of ospreys successfully fledged two young in the Meadowlands for the first time in at least 50 years (Vos-Wein 2007). The nest was on a platform at the Public Service Electric and Gas (PSE&G) Hudson Generating Station on the Hackensack River in Jersey City. The pair fledged young at the same nest site the following year (Carola 2008a). Osprey have continued to increase as a breeding species in the Meadowlands and in 2019 there were 9 active nests in the Meadowlands District (Carola 2019b).

Beginning in approximately 2000, bald eagles have been seen fairly regularly in the Meadowlands. The bald eagle is now present year-round on the Hudson River estuary, is seen frequently, and there are 33 active nests. Widespread occurrence of bald eagles in the Meadowlands has been reported during winter (Carola 2007, 2008a, d). Bald eagles nested in Ridgefield Park and at Kearny Point in 2019 (Carola 2019b). Overwintering and migrating adults and immatures are regularly seen (Carola 2019a, c).

Recently, peregrine falcon pairs nested on the Route 3 westbound bridge, the NJ Turnpike Western Spur Bridge between Carlstadt and Ridgefield, the Eastern Spur Bridge between Secaucus and Kearny, and the quarry face at Laurel Hill (Carola 2019b). Peregrines are seen often enough in the Meadowlands, and suitable foraging habitat and prey populations are available, to indicate that some Meadowlands marshes and landfills are probably feeding sites for migrant peregrines, nonbreeding immatures, and peregrines breeding near the Meadowlands (Wander and Wander 1995).

The abandoned industrial buildings in the Meadowlands provide breeding sites for barn owls. Inasmuch as barn owl is doing well as a breeding bird in New York City, it is likely that the breeding population in the Meadowlands will remain stable or increase. However, recent nests on the Route 3 eastbound bridge, the HX Draw (Jackknife) Bridge in Rutherford, and the trash transfer station in North Arlington have not been active for the past three years (Carola 2019a). The disappearance of breeding barn owls is suspected due to widespread use of anticoagulant rodenticides which can cause secondary poisoning when predators or scavengers consume poisoned rodents.

Forested areas are important for several species. Red-tailed hawks very likely breed in forests in the Meadowlands. Great horned owls may also be using these forests as nest sites; they have been observed at Schmidt's Woods, Overpeck Preserve, and Teaneck Creek Conservancy (Carola 2008a, b). In suburban areas of northwestern New Jersey, presence of several species of forest raptors, including the red-shouldered hawk, northern goshawk, and barred owl, was positively correlated with larger forest patch size (>1000 ha; Bosakowski and Smith 1997). In the same study, presence of red-tailed hawk and great horned owl was positively associated with smaller forest patch size.

Galliform Birds

Three species of Galliformes, ring-necked pheasant, wild turkey, and northern bobwhite, occur in the Meadowlands. The ring-necked pheasant, an introduced Asian species, was widespread and fairly common in the Meadowlands, although the population has apparently decreased during the past several decades. Pheasants spend a lot of time in common reed stands but are believed to nest in other habitats such as upland meadows. They typically nest on the ground in tall herbaceous vegetation (Giudice and Ratti 2020). Thus, it is possible that they are nesting in drier stands of common reed. In recent decades there has been a dramatic decline of self-sustaining pheasant populations across their (introduced) North American range. In view of the fates of pheasant populations in most areas, the status of this species in the Meadowlands is striking. It was unusual in the 2000s to spend a day afield in the Meadowlands without seeing, hearing, or finding tracks of a pheasant; however, in the last decade they have declined. We posit that pheasants have thrived in the Meadowlands because of a combination of mild winter climate, plentiful food, and abundant cover, including that provided by common reed stands, which provides shelter from weather and predators. There may be an ecological lesson in the Meadowlands for managing pheasants elsewhere.

Wild turkeys have been observed in the Meadowlands over the past couple of decades (Mizrahi et al. 2007, Carola 2008a; E. Kiviat, pers. obs.). We expect this species to increase in abundance in the near future, if not already, as it is in some other urban areas nearby.

Passerines, Doves, Cuckoos, Woodpeckers, and Kingfisher

This group (a total of 156 species) accounts for a large proportion of the bird species in the Meadowlands. The Meadowlands provide habitat for 139 species of *passerines* (perching birds, most of them *songbirds*), as well as 2 doves, 7 woodpeckers, 1 swift, up to 2 goatsuckers, and other small birds during the breeding, spring and fall migrations, and winter seasons (see Bird List in Appendix 2). The most common songbirds are the red-winged blackbird, European starling, song sparrow, and American robin (Seigel et al. 2005, Mizrahi et al. 2007). Among the species found in the Meadowlands, the sedge wren, loggerhead shrike, Henslow's sparrow, and vesper sparrow are Endangered and red-headed woodpecker, savannah sparrow, grasshopper sparrow, horned lark and bobolink are Threatened in New Jersey. The least flycatcher, winter wren, veery, gray-cheeked thrush, blue-headed vireo, northern parula, black-throated blue warbler, black-throated green warbler, blackburnian warbler, cerulean warbler, worm-eating warbler, Kentucky warbler, Canada warbler, hooded warbler, and eastern meadowlark are species of Special Concern.

The richness of small birds is enhanced by the presence of a diversity of habitats in the Meadowlands. Marsh-upland edges and rights-of-way with mixed vegetation including vines, sumacs, and other abundant upland shrubs provide late fall, winter, and early spring food. Many forbs and grasses provide potential fall-winter-spring food for sparrows and other *granivorous* (seed-eating) birds. Shrub thickets and reed stands offer dense escape cover and shelter from the weather, and sparrows eat the seeds of both reed and knotweed (*Polygonum cuspidatum s.l.*; E. Kiviat, pers. obs.). There may be reduced pressure from nest predators such as snakes, chipmunks, and raptors because some of these species are rare or absent in the Meadowlands and others may have difficulty reaching some of the bird habitats due to highways, wetlands, and canals.

A few species of songbird require or prefer wetland habitats for breeding. The species most commonly found nesting in a variety of freshwater wetlands and salt marshes, including common reed marshes, are the red-winged blackbird, marsh wren, common yellowthroat, willow flycatcher, and swamp sparrow (NJTA 1986, Kane and Githens 1997, USACE 2000, Mizrahi et al. 2007). Most of these common wetland species are found in even small patches of habitat throughout their range (Benoit and Askins 2002). The red-winged blackbird, the most abundant and widespread passerine in the Meadowlands, is the songbird most commonly found nesting in common reed (NJTA 1986, Mizrahi et al. 2007). The marsh wren, also very common in reed marshes but even more so in *Spartina* marshes in the Meadowlands, nests in tall vegetation in a variety of marshland habitats throughout its range (Kroodsma and

Verner 2013). This species eats mainly adult insects and spiders. The swamp sparrow uses a wide variety of wetland habitats throughout its range (Reinert and Golet 1979, Greenberg and Droege 1990). Its simple habitat requirements are shallow standing water or wet soil, low, dense vegetation, and elevated perches for singing (Reinert and Golet 1979).

Quinn (1997) suggested that the sedge wren may be returning as a breeding species in the Sawmill Creek WMA and in Kearny Marsh. However, this species is extremely rare and there is only one recent observation, at Harrier Meadow (eBird 2021). Meadows of bluejoint grass (*Calamagrostis canadensis)* interspersed with reed on a portion of the Kane Natural Area most closely approached the requisite short grass or sedge habitat for this species, but the bluejoint is gone (E. Kiviat, pers. obs.).

Two other wetland-associated breeders, seaside sparrow and saltmarsh sparrow, were once more frequent breeders in the Meadowlands but are now rare. Observations of seaside sparrows are much less frequent at sites in the Meadowlands than saltmarsh sparrows (eBird 2021). The seaside sparrow is a salt marsh habitat specialist (Post and Greenlaw 2018). In high quality salt marsh habitat, this species can reach high densities during the breeding season and is a useful indicator of the health of the salt marsh (Post and Greenlaw 2018). They nest above spring high tides in a variety of vegetation including shrubs and common reed but they prefer short grass meadows with saltmeadow cordgrass (*Spartina patens*). They forage for a variety of invertebrates in open mudflats, along ditches, and in smooth cordgrass.

While uncommon, saltmarsh sparrows have been recorded throughout the Meadowlands in recent years (Mizrahi et al. 2007, eBird 2021). The saltmarsh sparrow is found in a wider range of fresh and saltwater habitats (Greenlaw et al. 2018), though their breeding and foraging range greatly overlaps with seaside sparrow. Along the Atlantic coast they prefer salt marshes, especially at the ecotone where salt marshes dominated by smooth cordgrass, saltmeadow cordgrass and black rush (*Juncus gerardii*) are bordered by cattail, common reed, and marsh-elder (*Iva frutescens*).

Benoit and Askins (2002) found seaside and saltmarsh sparrows were negatively affected by reed invading the short grass meadows of the high salt marsh in Connecticut. Their low abundance in smaller marshes indicates that both species are area sensitive (Benoit and Askins 2002). Rising sea level is thought to threaten these species in Connecticut salt marshes because active nests are more likely to be flooded (Bayard and Elphick 2011).

Uplands are often overlooked in conservation planning despite their value in supporting many bird species (and other biota) in the Meadowlands. Upland areas of shrubs and trees along ditches and canals and in upland meadows support many breeding species. Some are the northern flicker, downy woodpecker, eastern kingbird, eastern phoebe, tree swallow, fish crow, American robin, brown thrasher, gray catbird, northern mockingbird, yellow warbler, warbling vireo, common grackle, northern oriole, indigo bunting, blue grosbeak (rare), American goldfinch, song sparrow, savannah sparrow, and (nonbreeding) Lincoln's sparrow (NJTA 1986, Wargo 1989, Kane and Githens 1997). The savannah sparrow, a Threatened grassland species, was found at a Moonachie site during spring, summer, and fall surveys (USACE 2000). Savannah and grasshopper sparrows have been reported in the Meadowlands in grassland habitats on capped landfills (Fallon 2019b). Development projects threaten these habitats, despite the inherent hazards of landfills for construction. Migrating bobolinks (at Kingsland Landfill) and savannah sparrows (at Sawmill Landfill) declined significantly from 2004 to 2005 (Mizrahi et al. 2007). This decline may be due to changes in habitat composition at the landfill sites or to timing of migration events, or to large-scale population declines throughout their ranges.

During fall migration, large numbers of songbirds and other small birds are found in upland habitats throughout the Meadowlands. In addition to the year-round residents, some of the small birds present in fall include the yellow-bellied sapsucker, purple martin, bank swallow, hermit thrush, winter wren, American pipit, ruby-crowned kinglet, golden-crowned kinglet, bobolink, and a long list of wood warblers (Kane and Githens 1997). Seewagen and Newhouse (2018) found the former Erie Landfill site was used by an abundance of grassland and shrubland songbirds during autumn stopover. They did not observe strong trends of increasing body mass during stopover there, nor did they find evidence that the site was an energy sink in which birds lose mass. Possibly the site, which was dominated by mugwort, served other purposes such as providing resting areas and reducing predator risk during migration. Native shrub fruits in at least some cases have been shown to have higher nutritional content (fats and energy) than fruits of nonnative plants during fall migration (e.g., Smith et al. 2013). However, in areas where invasive fleshy-fruited vines and shrubs are common, as is the case in the Meadowlands, removal of exotic invasive species should take place when and where native replacements are available (Bricklin et al. 2016). Historically, large non-breeding roosts of European starlings (Kalmbach and Gabrielson 1921) and swallows (tree, bank, barn, cliff, and northern rough-winged; Chapman 1900) were reported from the Meadowlands. Roosting congregations of the European starling, blackbirds, and swallows still occur in reed stands (H. Carola, pers. comm.) and perhaps groves of trees.

The remnant lowland forests at Teterboro Airport, Losen Slote Creek Park in Little Ferry, and Schmidt's Woods in Secaucus are important as stopover habitat for many songbirds, including Neotropical migrants, as well as nesting habitat for several resident and Neotropical forest songbirds. For example, Losen Slote Creek Park contains forest habitat that is used for breeding by the downy woodpecker, American crow, blue jay, and tufted titmouse (Kane and Githens 1997). In addition, the hairy woodpecker, red-bellied woodpecker, black-capped chickadee, American robin, wood thrush, gray catbird, common grackle, and northern cardinal were present in the forests at Teterboro Airport during the breeding season in 2001 (K. MacDonald, pers. obs.) and are potentially nesting in other forest remnants. Surveys during migration in 1995 at Losen Slote found 44 species of migrant songbirds, including 5 flycatchers, ruby-crowned kinglet, blue-gray gnatcatcher, 5 thrushes, cedar waxwing, 3 vireos, 18 warblers, orchard and northern orioles, rose-breasted grosbeak, bobolink, 4 sparrows, and American goldfinch (Kane and Githens 1997). Other small forests along the upper Hackensack River estuary are probably similarly used.

The Meadowlands have historically provided habitat for several other songbird species that are rare in the urban core areas. North of the Vince Lombardi Service Area of the Turnpike, a large sandy area formerly provided nesting habitat for horned lark (NJTA 1986). The common raven began nesting in 2006–2010 on the abandoned quarry face on the north side of Laurel Hill in Secaucus (Figure 7.2; Carola 2006a, Mizrahi et al. 2007; E. Kiviat, pers. obs.). After an absence, a breeding pair nested again at Laurel Hill in spring 2019 but the three chicks were taken by peregrine falcons (Carola 2019b). The Meadowlands are far from the typical breeding range of ravens in the more mountainous regions of northwestern New Jersey. However, the presence of ravens in the Meadowlands is not surprising given that ravens nest in highly variable habitats including rock cliffs, quarries, woodlands, built structures, and urban areas throughout their range (Boarman and Heinrich 2020).

Figure 7.2 Abandoned mine face, Laurel Hill, used by nesting common ravens and peregrine falcons. The cliffs reach 35–45 m in height. Photo by Erik Kiviat.

The belted kingfisher has been found at a number of sites in the Meadowlands, but its nesting habitat has not been reported. This species typically constructs nests in earthen banks close to water but will also use a variety of habitats away from water (Kelly et al. 2020). Eroding landfill side-slopes, and higher water-control or containment berms, are potential nesting habitats, as are various dirt piles and portions of the scarp at the western edge of the Meadowlands. The common nighthawk was "uncommon" in spring, summer, and fall in the Sports Complex area (McCormick & Associates 1978); it was not stated whether these were breeding birds or late and early migrants. They continue to show up as observations on eBird (2021). We have seen no clear evidence of breeding nighthawks. There is a historical reference to eastern whip-poor-wills in the Meadowlands (Cruickshank 1942). Notable species that winter in the Meadowlands include the horned lark and American pipit (NJTA 1986, Mizrahi et al. 2007).

Monk parakeet, a species that began colonizing the region in the early part of the twenty-first century, is becoming more widespread in the Meadowlands (eBird 2021; H. Carola, pers. comm.; E. Kiviat, pers. obs.). Escaped monk parakeets, native to South America, have become established locally in many states (Burgio et al. 2016). It is the only species of parrot that builds stick nest structures containing many nest holes and there are typically nest structures located close together in an area; thus, the monk parakeets are communal and colonial (Burgio et al. 2016). Burger and Gochfeld (2009) studied monk parakeet nest site selection and nesting behavior in Edgewater, New Jersey, a small city near the Meadowlands. Most parakeet nest structures were located on utility poles, street overpasses, and in trees. Parakeets tended to rebuild nest structures immediately after removal by utility companies. There seems to be affinity, or at least tolerance, for the noisy and conspicuous monk parakeets among residents in the municipalities where they are found.

Conclusions

Despite the urban context, the Meadowlands support a high diversity of New Jersey's birds relative to most other places of comparable land area in the state and comparable to other estuaries; although, this is not true for all taxa in the Meadowlands. There are inherent characteristics of birds along with biogeographic and ecological influences imposed at several different spatial scales that support the diversity of birds we see in the Meadowlands today. It is useful to address these intrinsic and extrinsic variables so that we might gain an understanding of bird diversity patterns in the Meadowlands. Some of these variables include: birds are a species-rich group of vertebrates, with the most species after fishes; habitat diversity in the Meadowlands favors bird diversity, especially the mixture of upland and wetland habitats in a sheltered estuary; the coastal situation of the Meadowlands and the climate of the region favor a wide variety of birds and other taxa; and global influences may have a larger impact on birds than most other taxa because many species migrate long distances between wintering and breeding grounds.

Other important drivers of birds' success in the Meadowlands are that they are potentially more resilient to past habitat loss, degradation, and loss of connectivity than most other vertebrate taxa for several reasons. Because of their ability to fly, birds can overcome barriers in the landscape that might limit other taxa. This might, for example, allow them to colonize locales that were once unsuitable when habitat conditions are once again suitable. This makes birds as a group relatively good indicators of present habitat conditions. Furthermore, because many birds are migratory and often travel thousands of miles from nesting to wintering grounds, they are adapted to survival in challenging new habitats. In fact, passerines (especially *corvids*, the jays, crows, and raven) are among the most intelligent of birds (Emery 2006), a trait that can help them successfully maneuver through new terrain and adapt to new situations and utilize human-dominated habitats. About 25% of the bird species in North America are *synanthropic* (human-associated) to some extent (Johnston 2001).

Birds are good targets for biodiversity conservation planning and management in the Meadowlands for several reasons. There are many species and groups that are rare or declining that rely upon habitat types in the Meadowlands. Also, in many cases we understand what is needed to improve their protection. Expenditures of millions of dollars in wetland mitigation monies are often justified by the predicted increase in particular bird species. Habitat restoration projects aimed at eradicating common reed often posit the overall benefits to birds. However, many birds use reed-dominated habitats for nesting, foraging, and roosting, thus marsh management should provide reed habitats as well as cordgrass habitats. Finally, non-marsh habitats in the Meadowlands are important to many birds including species of conservation concern, and these forests, shrublands, upland grasslands, and built habitats have been overlooked in much of the policy and management of the Meadowlands (see Chapter 3). Habitat management recommendations for birds and other wildlife are summarized in the Conclusion chapter. The most important actions for bird conservation are protection of greenspace habitats and their management for birds (e.g., maintenance of large areas of forest, shrubland, and meadows), as well as reducing chemical hazards that include rodenticides, herbicides, and contaminants in soils and sediments, water, and air.

Chapter 8

Reptiles and Amphibians

The United States are an important center of herpetofaunal species diversity, especially for turtles and salamanders (Stein et al. 2000). In general, the herpetofauna of the northeastern states has been well-studied. Large-scale urbanization of the U.S. is wreaking havoc on herpetofaunal diversity (Mitchell and Brown 2008). Many species are sensitive to urbanization, agriculture, forestry, disease, habitat fragmentation, road mortality, or pollution, and many require particular hydrological or soil conditions. In addition to coping with anthropogenic stressors, amphibians and reptiles must find suitable temperatures, moisture levels, food, mates, and safe *hibernacula* (overwintering places), as well as avoiding predation. Aquatic eggs and larvae of most amphibians, and the terrestrial eggs and terrestrial or aquatic juveniles of reptiles, are exposed to contaminants and other adverse factors that may compound threats to a species.

The conservation of frogs is of intense concern worldwide; it is less publicized that many salamanders, turtles, and snakes are also highly threatened (e.g., Gibbons et al. 2000). Amphibians and reptiles, although seemingly requiring relatively small amounts of food on a year-round basis, may play important roles in ecosystems, such as controlling litter fauna and decomposition, serving as important prey for larger animals, disturbing marsh sediments, or controlling waterfowl populations (Coulter 1957, Burton and Likens 1975, Kiviat 1980, Wyman 1998). Moreover, these animals are aesthetically, culturally, and scientifically important. Thus, despite a *depauperate* (impoverished) herpetofaunal assemblage in the Meadowlands, its relationships to the urban-industrial environment are worthy of documentation and analysis.

The reptiles and amphibians of the Meadowlands contrast starkly with the birds: the herpetofauna is poorly represented and little studied, although a number of species are New Jersey Species of Greatest Conservation Need (SGCN). The birds are species-rich, well studied, and also with many species of conservation importance. The proportion of available herpetofauna in the surrounding region that has been documented in the Meadowlands is small

Table 8.1. Reptiles and amphibians of the New Jersey Meadowlands.

Common name	Scientific name	Habitat[a]	Abundance[b]	Last found	References[c]
SALAMANDERS					
Red-spotted newt (eastern newt)	Notophthalmus viridescens	FN	X	1960	10
Red-backed salamander	Plethodon cinereus	U woods, aban-doned quarries	LC	2009	10, 14
Northern red salamander $	Pseudotriton ruber	Springs	R[d] P	Ca. 1994	16
FROGS					
Northern cricket frog*	Acris crepitans	FN; FN, U edges	X[d]	< 2001	7
American toad	Anaxyrus americanus	F or sB non-wooded; U	R	2006	7, 14
Fowler's toad $	Anaxyrus fowleri	F or sB non-wooded; U	X[d]	1960s	7, 10, 13
Northern gray treefrog[e]	Hyla versicolor[e]	FN; FN, U	X	< 2001	2, 7
Bullfrog	Lithobates catesbeianus	Permanent F or sB	R	2009	7, 11, 12, 14
Green frog	Lithobates clamitans	F or sB; moist U	R	2004	1, 6, 11, 14
Pickerel frog*	Lithobates palustris	FN; moist U	X	< 1998	6
Atlantic Coast leopard frog[f] $	Lithobates kauffeldi	F or sB; moist U	LC[d]	2014	1, 2, 6, 10, 14, 15
Wood frog	Lithobates sylvaticus	FN; U	R P	2010	14
Spring peeper	Pseudacris crucifer	FN; F & U; P	X	2017	2, 7, 14, 17
New Jersey chorus frog*	Pseudacris kalmi[e]	FN; U	X[d]	< 2001	7
Eastern spadefoot $	Scaphiopus holbrookii	FN; U	X[d]	1936	10
TURTLES					
Snapping turtle	Chelydra serpentina	F or B	LC	2019	1, 2, 3, 4, 6, 9, 11, 14
Painted turtle $	Chrysemys picta	F or B	LC[d]	2019	1, 3, 6, 9, 14, 19
Spotted turtle $	Clemmys guttata	F or B; U?	X[d]	< 2001	1, 2, 7, 10

Common name	Scientific name	Habitat	Status	Date	References
Eastern mud turtle $	Kinosternon subrubrum	F or B; U?	R?[d]	<1998	1, 6
Diamond-backed terrapin $	Malaclemys terrapin	BT	LC[d]	2004	1, 3, 5, 6, 8, 9, 14, 15, 19
Musk turtle (stinkpot)	Sternotherus odoratus	Permanent F or sB	R?	2000s	7, 19
Eastern box turtle (woodland box turtle) $	Terrapene carolina	U; F or sB marsh?	R[d]	2006	6 (pp. 22, 41), 11, 14
Red-eared slider	Trachemys scripta elegans	FN; sB?	LC	2018	14, 19
Yellow-bellied slider	Trachemys scripta scripta	FN	R	2009	14
LIZARDS					
Five-lined skink*	Plestiodon fasciatus	Rocky U; swamps?	X[d]	<2000	7
SNAKES					
Black racer*$	Coluber constrictor	U	X[d]	<2000	7
Northern ring-necked snake $	Diadophis punctatus	U	R?[d]	2007?	17
Eastern hog-nosed snake $	Heterodon platirhinos	Sandy U	X[d]	1980s	7
Milk snake	Lampropeltis triangulum	U	R?	2006	2, 9, 10, 14, 15
Smooth green snake*$	Opheodrys vernalis	U; F edges?	X	<2001	7
Northern water snake	Nerodia sipedon	F, sB; U edges	R	2000s?	1, 2, 6, 9
Black rat snake (eastern rat snake) $	Pantherophis alleghaniensis (Elaphe obsoleta)	U	X[d]	<1998	6, 18
Brown snake	Storeria dekayi	U	U?[d]	2005	2, 10, 14, 15
Eastern ribbon snake $	Thamnophis sauritus	F, U	X[d]	<2001	2, 7
Eastern garter snake	Thamnophis sirtalis	U, F edges	U	2006	1, 2, 9, 10, 14, 15

Note: Scientific names follow CNAH (2020) and common names attempt to conserve traditional usage in the northeastern U.S. All species are native except the two slider turtles.

* Indicates unverified reports; $ indicates Species of Greatest Conservation Need (SGCN; NJDEP 2018a).

[a] Refers to habitat generally in the area of northern New Jersey. For frogs, the first habitat is breeding habitat, the second is nonbreeding habitat; F = freshwater wetlands, pools, and ponds; B = brackish water wetlands and ponds; sB = slightly brackish water; N = nontidal only; T = tidal only; U = uplands.

[b] LC = locally common, U = uncommon, R = rare, X = possibly extirpated (these assessments refer only to the Meadowlands), P = peripheral areas or just outside the Core Meadowlands.

c 1. McCormick & Associates 1978; 2. NJTA 1986; 3. Kraus and Bragin 1989; 4. Albers et al. 1986; 5. Kane and Githens 1997; 6. Quinn 1997; 7. USACE 2000; 8. Urffer 2002; 9. Mohn no date; 10. American Museum of Natural History unpublished specimen data as of 31 December 2018; 11. Arnold 2008; 12. Steve Covacci; 13. Richard Kane; 14. Erik Kiviat; 15. Kristi MacDonald; 16. Steve Quinn; 17. Joan Hansen and Don Smith; 18. Kyle Spendiff; 19. Jason Tesauro. Numbers 12 through 19 refer to personal communications.

d New Jersey Species of Greatest Conservation Need (NJDEP 2018a)

e The references do not identify the species; they would have been New Jersey chorus frog (*Pseudacris kalmi*) and northern gray treefrog based on geographic ranges (Schwartz and Golden 2002). The chorus frog has historically been considered as the upland chorus frog (*Pseudacris feriarum*) but based on recent genetic analysis from other portions of the state is referable to the New Jersey chorus frog (Lemmon et al. 2007).

f Reported as "*Rana pipiens*"; that taxon was subsequently split into *R. pipiens* (now *Lithobates pipiens*), the northern leopard frog, with range beginning about 150 km north of the Meadowlands, and *R. sphenocephala utricularia* (= *R. utricularia*, now *Lithobates sphenocephalus utricularius*), the southern leopard frog, which was thought to be distributed from Orange County, New York, southwards, and throughout New Jersey (see Klemens et al. 1987, Schwartz and Golden 2002). However, a recently described species of leopard frog has been identified genetically, acoustically, and morphologically (Newman et al. 2012, Feinberg et al. 2014) and is the only species of leopard frog in northern New Jersey (Schlesinger et al. 2018). The Atlantic Coast leopard frog was confirmed in the Meadowlands in 2012–2014 (Feinberg 2015), and is the entity identified as southern leopard frog in Kiviat (2011). In addition to the recent documentation of several localities in the core Meadowlands, there are many old records of what was presumably the Atlantic Coast leopard frog in or near the Meadowlands in Pace (1974) and among specimens in the American Museum of Natural History.

(Table 8.1), and many reported species are unverified, rare, or apparently extirpated. A few species (red-backed salamander, eastern box turtle) persist at the ecological and geographic edges of the Meadowlands region. However, two SGCN, diamond-backed terrapin and Atlantic Coast leopard frog, are apparently doing well at a few Meadowlands sites.

HABITATS OF THE MEADOWLANDS HERPETOFAUNA

Spotted salamanders and wood frogs need intermittent woodland pools for successful breeding and large patches of upland forest to live in; this habitat complex and its amphibians are lost as cities develop. Diamond-backed terrapins and snapping turtles, however, spend most of their time in the water and tolerate poor water quality. As long as they can survive migration to and from upland nesting areas and among ponds or tidal marshes, these two species may thrive in urban preserves. Many of the herpetofauna are small and inconspicuous. And unlike fish, birds, and a few plants, reptiles and amphibians have often been ignored during decisions about land use and management in the Meadowlands.

John Quinn (1997) mentioned that his brother Steve, as a child (probably in the early 1960s), found salamanders at the edge of the Core Meadowlands north of Overpeck Creek. In 2008 Steve Quinn showed Erik the spring in Ridgefield Park where he used to find the salamanders. They were northern red salamanders (*Pseudotriton ruber*), last seen there in the early 1990s. We did not find a salamander but the habitat looked suitable. The location was above normal tidal influence and buffered by undeveloped land that may be sufficient to protect wetland quality. Springs and seeps are often scarce on the landscape and are important features for biodiversity support.

In 2005, at the long-abandoned Granton Quarry in North Bergen, Erik found several adult and juvenile red-backed salamanders (*Plethodon cinereus*). They were beneath rocks, boards, and trash, some only the 20 m width of the railroad embankment from the marshes of the Core Meadowlands. The red-backed salamander can thrive in relatively dry upland habitats, and sometimes persists in urban parks. The portion of Granton Quarry that was not paved or built on covered ca. 0.8 ha; it was bordered by commercial buildings and shopping plazas on three sides, and by the railroad and marshes on the fourth (western) side. The undeveloped patch is characterized by bare rock and pockets of poor soil, with sparse woody and herbaceous plants and a small patch of woodland. Though not the ideal conditions for a salamander population, there is evidently enough moisture, soil organic matter, and invertebrate prey for the redback, the most urban-tolerant salamander of the northeastern states.

On an early-summer visit, following wet weather, to the mature forest and swamp of the South Woods of Teterboro Airport, we turned tens of logs and pieces of trash but found no salamanders. Nor did we find salamanders in mid-spring 2005 under suitable-looking cover objects in the Teaneck Creek Conservancy preserve not far from the locality in Quinn (1997). At Laurel Hill, however, in 2006 and 2009, we found redbacks at several locations on the hill and on the quarry floor at the foot of the hill, suggesting a substantial population. In 2009, we found red-backed salamanders in the abandoned quarry at the western edge of the Core Meadowlands in North Arlington, and in 2017 under rock debris in an abandoned quarry at Long Swamp in Palisades Park. It is remarkable that the red-backed salamander, albeit surprisingly common in rural woodlots and relatively tolerant of dry soil and human activities, has survived in these quarried, graffiti-painted, and development-encroached enclaves. Quarrying and abandonment at all four sites has provided habitats with abundant rock fragments and sufficient woody vegetation for the redback, whereas much of the rest of the Meadowlands is essentially rock-free. This exemplifies how special geological and cultural features can support particular species: a phenomenon not unique to cities, but perhaps more prominent there. Redbacks also occur in the old forest of Borg's Woods in Hackensack (Kiviat, pers. obs.) outside the Core Meadowlands, which does not have a history of quarrying.

Salamanders, and other sensitive organisms, are evidently distributed irregularly in and around the Meadowlands in a mosaic of environmental conditions and accidents of human history. Relationships are obscured because of taxonomic biases and the paucity of study effort for amphibians, as well as small mammals, most invertebrates, lichens, mosses, and wildflowers, in the Meadowlands. Gibbons et al. (1997) showed for the 800 km² Savannah River Site, South Carolina, that several decades of skilled field effort may be necessary to discover uncommon species. The herpetofauna is especially prone to this challenge due to the cryptic and secretive nature of many species such as those that spend most of their lives in the soil. This underlines the importance of natural history field work and synthesis for any region with intensifying land use.

WHICH SPECIES HAVE BEEN FOUND IN THE STUDY REGION?

The Meadowlands are in and around an urban estuary. Although North American herpetologists have conducted little research in tide-affected habitats, many species use those habitats and some even visit or reside in brackish waters, whereas other species are sensitive to low levels of salinity and are restricted to freshwater tidal habitats or are not found in estuaries at all. The amphibians and reptiles that use estuaries show distinct patterns of behavior related to tidal fluctuation (Kiviat 1989). Snapping turtles studied in a freshwater tidal marsh of the Hudson River followed the tide in and out of shallow pools and creeks, foraging in flooded areas, and moving to deeper water, burying in the mud, or hiding in muskrat burrows at lower tide stages (Kiviat 1980). The basking behavior of painted turtles and red-bellied turtles on the Patuxent River estuary in Maryland was shaped by tidal fluctuation as well as time of day and season (Swarth and Kiviat 2009 and unpublished data).

Various documents associated with development, remediation, and other projects in the Meadowlands mention amphibian and reptile species for which we have not found verification (i.e., no specimen, photograph, or expert identification). Several such documents were among sources of Meadowlands locality records mentioned by Mohn (no date). Because some documents list numerous unverified species, we suspect that preparers might have listed species solely on the basis of range maps in field guides, that is, potential species rather than documented locality records.

Historic Species, Peripheral Species, and the Species Pool

In general, herpetofaunas are species-poor in urban areas, although a few species tolerate urbanization and there are notable instances of native species thriving in cities (e.g., Neill 1950, Schlauch 1978, Klemens 1985, Mitchell and Brown 2008). We expected low herpetofaunal diversity in the Meadowlands, considering the toxic contamination, habitat fragmentation, urbanization, and brackish flooding. Few herpetofaunal surveys have been conducted and available data are limited to small areas; detailed biological studies have rarely been performed. Contaminants in tissues of the diamond-backed terrapin and snapping turtle have been studied (Albers et al. 1986, McIntyre 2000, Sherwood 2017).

Table 8.1 shows 15 species of the herpetofauna documented in the Core Meadowlands during the past twenty years; 3 species found during this period but only in peripheral parts of the Greater Meadowlands; and another 15 species known only before 2000. The overall pool of native herpetofauna in the surrounding region of northeastern New Jersey is 48; the 14 additional species very likely occurred in the study region at one time, but we have found no documentation. Certain species, such as spring peeper and red salamander, that have recently been found only in peripheral areas, would have occurred in the Core Meadowlands earlier. Several of the Meadowlands species are Species of Greatest Conservation Need (SGCN) in New Jersey (Table 8.1); there is a strong association between SGCN listing and our Possibly Extirpated category.

Several factors make the herpetofauna difficult to study. Many amphibians and reptiles are small and inconspicuous, spending most of their time in dense vegetation, burrowing, hiding in leaf litter or beneath logs and rocks, in the soil, or in water. Some of the local species are difficult to identify. American toad and Fowler's toad are easily confused and also hybridize (Conant and Collins 1991). Leopard frog and pickerel frog can be mistaken for each other—especially the juveniles—and even the calls can be confused; as well, the Atlantic Coast leopard frog and wood frog can be confused acoustically. Garter snake and ribbon snake can be hard to distinguish unless they are in the hand or viewed at close range through binoculars. Mud turtle and musk turtle may be misidentified in the field. Despite good intentions, field naturalists make mistakes, and data shown in Table 8.1 may reflect errors of identification or the recording of localities, as well as unwarranted assumptions that species occur everywhere within the generalized range maps in field guides. Such errors can result in the reporting of species that do not occur in the Meadowlands or overlooking species that do occur. Some locality records are vague and the animals may have been found in or outside the Meadowlands (e.g., the Jersey City newt, Table 8.1). Furthermore, many accurate observations undoubtedly escaped our notice, because as is typical of such situations, most data are anecdotal, unpublished, or unavailable. We suspect that biologists who have conducted consulting projects and other work in the Meadowlands have made many observations of the herpetofauna without reporting their data in print.

Frogs

Many species of frogs are undergoing non-cyclical population declines, and some other species that do not seem to be declining have a high prevalence of disease or malformations (Green 1997). Whether frogs as a group are distinctly worse off than, for example, freshwater mussels, crayfishes, stoneflies, freshwater fishes, turtles, snails, or even birds (Stein et al. 2000) is unclear; nonetheless, the global plight of frogs must be taken seriously. Because factors prominent in the Meadowlands, including urbanization, chemical pollution, habitat fragmentation, nonnative invasive animals, and hydrological alterations to aquatic habitats are associated with threats to a number of frogs elsewhere, we have to look closely at the status of frogs in this region.

Eleven species of frogs occur in northeastern New Jersey (approximately Bergen and Hudson counties, based on range maps in Conant and Collins [1991] and Schwartz and Golden [2002]), and all 11 have been reported in the Meadowlands. Only one species, the green frog, has been widely reported in the Meadowlands, and it is rarely observed. Quinn (1997) opined that the species was less common in the Meadowlands than previously. Frog-call surveys

Figure 8.1 Stormwater pond breeding habitat of Atlantic Coast leopard frog. Photo by Erik Kiviat.

in spring 2006 at the Upper Penhorn Marsh in East Bergen, and Teterboro Airport in Teterboro and Moonachie (Figure 8.1), revealed choruses of leopard frogs, but no green frogs (Kiviat 2011). The erroneous identification of these frogs as southern leopard frogs is described in Table 8.1 footnote f.

The Atlantic Coast leopard frog (Figure 8.2) occurs from Connecticut to North Carolina, within 40 km of a coast or estuary and at 0–208 m elevation, mostly in large marshes or ponds (Schlesinger et al. 2018). Its Meadowlands habitats include a shallow, seasonally flooded, ca. 0.1 ha stormwater pond; a 65 ha, freshwater, common reed marsh with large channels and pools; and a sprawling complex of freshwater reed marsh with large ponds, totaling 6+ ha. All three habitats are surrounded by developed areas with parking lots, highways, railroads, buildings, industry, and forest fragments.

In 2008, we were shown a bullfrog population in a marshy railroad ditch at the edge of Overpeck County Park on the east side of Overpeck Creek. Bullfrogs possibly spend part of the year in the ditch and part in Overpeck Creek, perhaps as salinity or another factor changes. In late July 2009, we heard bullfrogs calling in a side channel of Overpeck Creek in the municipality of Palisades Park, and in May 2017 we found bullfrogs in Teaneck Creek and ditches in the northern side of Overpeck Creek Park. In twenty years, we have observed American toad twice. New Jersey chorus frog and northern cricket frog are SGCN. These are among the species historically reported

Figure 8.2 Atlantic Coast leopard frog (male with paired vocal sacs during early spring breeding season). Photo by Erik Kiviat.

from the Meadowlands but with no photographic or specimen documentation and no recent records. Possibly several other species are rare or localized enough in the Meadowlands that they are easily overlooked, but the Meadowlands frog fauna is poor in species and in abundances. Mathewson (1955) found spring peeper, wood frog, green frog, pickerel frog, and leopard frog (then considered northern leopard frog but now known to be Atlantic Coast leopard frog) "common," and northern cricket frog, gray treefrog, bullfrog, and Fowler's toad "not common"; eastern spadefoot "rare," and did not mention American toad, for Staten Island.

Frogs are sensitive to chemical pollutants, perhaps more so than many animals because of the permeable skin of adults and their aquatic larvae. Several pollutants that are common in the Meadowlands potentially affect frogs. Several local species of frogs also require complementary breeding and nonbreeding habitats: an aquatic habitat for mating, egg-laying, and development of eggs and larvae, and a terrestrial habitat for the *metamorphs* (recently transformed froglets) and adults. Larvae (tadpoles) take anywhere from a few weeks to two years or more to develop and metamorphose, and the larval habitat must remain flooded to a suitable depth with water of acceptable quality during that entire time. Most frogs are sensitive to salinity, although a few species (e.g., American toad, Fowler's toad, Atlantic Coast leopard frog) can tolerate slightly brackish water. Other species, such as pickerel frog, northern cricket frog, and perhaps bullfrog seem to require stable water levels and rarely if ever spend time in tidal waters. Tidal fluctuation, even without salinity intrusion, evidently is stressful to most northeastern frog species. Some frogs undergo mass migrations to and from breeding pools, and if a road with high traffic volume crosses the migration path, large segments of a population may perish. Certain species require breeding habitats free of predatory fishes for the tadpoles to thrive. Many frog species are now known to be vulnerable to diseases and parasites. Finally, frogs require a stable, cool, but not deeply freezing microhabitat for overwintering, sufficient moisture during the active season, and as adults a degree of refuge from predation by fish, birds, and other animals. The Meadowlands seem a hazardous place for frogs.

Green frog, bullfrog, Atlantic Coast leopard frog, Fowler's toad, and American toad are more tolerant of environmental conditions associated with urbanization, and we would expect these species to survive best in the Meadowlands. The fresh, nontidal, aquatic habitats have not been well-studied. For example, Mehrhof Pond, a large, claypit pond bordered by woody vegetation on the west side of the Hackensack River in Little Ferry and South Hackensack, a pond evidently created by inadvertent diking of tidal marsh at Meadowlark Marsh near Bellman's Creek in North Bergen, several pools and ponds near Laurel Hill, and various stormwater management ponds, might support species such as spring peeper and gray treefrog, in addition to the urbanization-tolerant species just mentioned. Spring peeper still occurs in Long Swamp in Palisades Park, just outside the Core Meadowlands, and wood frog occurs similarly in Borg's Woods, Hackensack (Kiviat, pers. obs.).

Salamanders

Of 13 species of salamander known in northeastern New Jersey, we have found records of 3 species in the Meadowlands. The stories of the red salamander and red-backed salamander were told above. An old specimen of the red-spotted newt (Table 8.1) was collected at an unknown location in Jersey City. Thus, salamanders have fared as poorly as frogs in the Meadowlands, where they are potentially sensitive to all the same factors. Mathewson (1955) stated that two-lined salamander was "fairly common," northern dusky salamander "common," red-spotted newt "not common," and four-toed salamander, spotted salamander, and marbled salamander extirpated on Staten Island. Mathewson considered the northern red salamander likely to be extirpated in the near future due to the loss of clear cold springs and streams. The decline of salamanders in New York City during the past two centuries was chronicled by Pehek (2007). Some salamanders (e.g., spotted salamander) use larger landscapes than most frogs, and are more readily harmed by habitat fragmentation and road mortality, as well as the loss of specialized habitats and habitat complexes (Miller and Klemens 2004). Those species associated with streams, mature forests, and woodland (vernal) pools, are also highly dependent on the physical, chemical, and biological structures of the substrate. Some salamanders spend most of their lives burrowing in the soil, or using burrows made by other animals. Logging, scouring and siltation of streams, and acidification of soils cause loss of the more sensitive species. The near absence of natural upland soils in the Meadowlands is probably an impediment to the establishment and persistence of salamander populations. Urban forests and swamps experience more extreme hydrological fluctuations than those in non-urban environments, and forest fragments are subject to greater lateral light penetration, higher temperatures, and drying of the soil than occur in larger forest blocks. Urban forests and swamps (Teterboro Airport Woods is an exception) may lack the large decaying downed wood present in more natural forests that is so important to salamanders and other small animals on the forest floor. Urban soils are highly variable (Sukopp et al. 1979, Gilbert 1989), and upland urban soils tend to be droughty. Calcium from demolition debris and weathering of concrete is often abundant in soils and this may be favorable to reptiles and amphibians because most species are intolerant of acidity.

The near absence of salamanders from the Meadowlands is similar to the situation on Jekyll Island, a Georgia Sea Island in the southeastern U.S. Jekyll, apart from the salt marshes, encompasses ca. 18 km^2. Several herpetofaunal surveys and data compilations have been conducted during the past 35 years (Anderson 1976, Sandifer et al. 1980, Schwab and Shelton 1981, Kiviat 1982b and unpublished data, Schoettle 1996, Frick 1998, Georgia Sea Turtle Center unpublished data) and only one salamander has been found (red-spotted newt). Woodlands contain ample logs and bark slabs on well-developed soil that look like good cover objects for salamanders. Although the Georgia Sea Islands in general are not rich in salamanders, several species have been found on islands other than Jekyll (Sandifer et al. 1980). Do Jekyll Island and the Meadowlands have common features that help explain the scarcity of salamanders in both places? Jekyll is one of the more developed Georgia Sea Islands—much of it is suburbanized, and it underwent extensive drainage of freshwater wetlands that probably resulted in salinity intrusion into formerly fresh habitats. Several heavily trafficked roads crisscross the island. And pesticides appear to have been applied heavily in residential areas for mosquito control and on the island's golf courses. Furthermore, both areas had an agricultural history until several decades ago. Jekyll, however, differs from the Meadowlands in having larger blocks of greenspace with more natural soils and vegetation, as well as being subtropical—both factors that would tend to increase the species richness of the herpetofauna. Jekyll is also an island (although connected to the mainland by a long-standing causeway), which acts to decrease the number of species present. Overall, it seems that salinity intrusion, roads, pesticides or other contaminants, and agricultural conversion of habitats, have contributed to today's scarcity of salamanders in the Meadowlands as well as on Jekyll. Karraker (2008) identified salinity from road salt as an important mortality factor for amphibian larvae at concentrations as low as 1 ppt chloride (equivalent to about 2 ppt salinity, a modest level indeed for either road salt or natural estuarine salinity). Salinities during early spring 2013 following Hurricane Sandy (October 2012) exceeded this level in some of the ponds where leopard frogs were calling (Feinberg 2015).

Turtles

Turtles are the best represented herpetofauna. Of 10 species occurring in northeastern New Jersey, 8 have been reported in the Meadowlands. The diamond-backed terrapin, a brackish tidal marsh specialist, abounds in Saw Mill Creek and nearby areas; we saw at least 30 during a four-hour canoe trip on Saw Mill Creek on 24 August 1999. Painted turtles can be seen in a number of places where salinity is not too high (the painted turtle is a Species of Greatest Conservation Need in New Jersey despite being common and widespread there). Snapping turtles are widespread in the Meadowlands; they were abundant and conspicuous in the upper Hackensack River estuary (between New Bridge Crossing and the Oradell Dam) on an ebbing tide, 1 July 2004. A population of the nonnative red-eared slider is widespread in nontidal waters. Kiviat saw several large adults in Losen Slote in 2003 (prior to tide gate failure), some of which had lost their red head markings (sliders darken with age) and thus were a challenge to identify through binoculars. This species, native to the southeastern U.S., was introduced to our region by the release of store-bought pets and became common in urban habitats in New Jersey compared to other environments of the state (Duchak and Holzapfel 2011). In 2006 there were small juveniles in a pond at the solid waste baling facility in North Arlington that may have been hatched in the wild or were released pets. All the red-eared sliders we have seen have been at nontidal waters, including Kearny Marsh West, Upper Penhorn Marsh, and Losen Slote. There is a population of red-eared sliders in the freshwater tidal Denning's Point Bay on the Hudson River at Beacon, New York, thus this turtle may also occur in tidal waters of the Meadowlands.

The yellow-bellied slider (the other subspecies of *Trachemys scripta*, native to the southeastern U.S.) has also been liberated widely in the northeastern states and is established as an introduced species in New York (Gibbs et al. 2007). In June 2009, we found a well-marked yellow-bellied slider in Indian Lake in Little Ferry.

The musk turtle is a small, inconspicuous species reported from upper Penhorn Creek. Musk turtles could be present in many other waters of the Meadowlands where they would likely escape notice due to cryptic coloration and highly aquatic behavior. The mud turtle and spotted turtle potentially inhabit a variety of shallow waters and wetland pools, creeks, or ditches, and small populations might go undocumented. Although mud turtle is not a listed species in New Jersey, it is rare in northeastern New Jersey and is Endangered in New York where the last few populations persist on Long Island and Staten Island. The box turtle was reported as common just outside the Overpeck Creek marshes in 1957 (S.C. Quinn, quoted in Quinn 1997). In June 2006, J.G. Barbour and Erik Kiviat found an adult male box turtle at the Bergen County Utilities Authority site in Little Ferry. A box turtle was photographed in Little Ferry in 2014 by Don Torino and another in DeKorte Park by Chris Takacs (meadowblog.net). Box turtle populations are believed viable in some urban areas but not others (Mitchell et al. 2008). We found no records of wood turtle from the Meadowlands; however, this species would once have occurred in freshwater tributaries and probably in the upper Hackensack River estuary (there are small populations in the fresh-tidal Hudson River in and near tidal swamps [Kiviat and Barbour 1996]). Mathewson (1955) considered spotted turtle "not common," wood turtle "not rare," box turtle "common," and mud turtle and musk turtle of undetermined abundance on Staten Island.

The diamond-backed terrapin is the most charismatic of the Meadowlands herpetofauna. The terrapin population has increased over the past thirty years (Quinn 1997, HMDC 1999), possibly in response to improvement in water quality though this resurgence could simply be part of a coast-wide recovery from the ubiquitous overharvest for food from the colonial period well into the mid-1900s. Terrapin populations in some East Coast marshes are affected by mortality in abandoned crab pots (Grosse et al. 2009); blue crabs are fished very little in the Meadowlands and the absence of crab pots may be helpful for terrapin population viability. Diamond-backed terrapins are now common enough that their propensity to climb up on the New Jersey Turnpike shoulders to nest is a matter of public concern, and fences were built at four locations to steer terrapins under the highway to other nesting areas, or at least keep them off the pavement (Urffer 2002). Although the fences reduce highway mortality, terrapins may be blocked from some of the best nesting habitats (Jason Tesauro, Jason Tesauro Consulting, pers. comm.). Terrapins in the Meadowlands also nest on the spoil banks of old mosquito ditches, and on railroad verges. Despite hotspots of heavy metal contamination in the Meadowlands, metals levels in the livers of terrapins in Saw Mill Creek did not reflect that contamination (McIntyre 2000), although mercury contamination may be a concern (Sherwood 2017). Yet not all turtle species may have been as fortunate as the diamond-backed terrapin in surviving, or recovering from, long-term loss caused by harvest for food and pets as well as road mortality and the ubiquitous predation on eggs and hatchlings by raccoons, striped skunks, and other animals that benefit from human activities. Ner and Burke (2008) documented a large reproducing population of terrapins in Jamaica Bay (New York City) with a high rate of depredation of nests by raccoons, rats, and gulls.

Although several species of sea turtles occur along the New Jersey and New York coasts, and some are known to enter sheltered brackish waters, we have not found reports of sea turtles in the Hackensack River estuary (nor does the Marine Mammal Stranding Center of New Jersey have any such records). It is likely that sea turtles enter Newark Bay and the lower Hackensack River estuary occasionally. The species frequenting the New Jersey coast are the green turtle, Kemp's ridley, hawksbill, loggerhead, and leatherback (Anonymous 2006).

Lizards

A single species, the five-lined skink, is known in northeastern New Jersey and reported from the Meadowlands. Skinks, including young individuals with their striking bright blue tails, are regularly seen in some of the rocky, sun-exposed, crest environments of the nearby Hudson River Palisades. The five-lined skink could occur on the natural and mined rocky habitats of Laurel Hill and elsewhere in the Meadowlands, and possibly in less-rocky, moister habitats with suitable basking sites. Railroads might serve as dispersal corridors between greenspaces. The eastern fence lizard is known from central New Jersey not far south of the Meadowlands (Schwartz and Golden 2002), on three small mountainous areas of the east side of the Hudson River in New York, and at a location in Brooklyn, New York, where it was probably introduced (Feinberg 2004). The Italian wall lizard, introduced to the U.S., is established locally in western Long Island, New York City including Staten Island, Westchester County, and Philadelphia (Conant and Collins 1991, Burke and Deichsel 2008, Mendyk and Adragna 2014, Lambert et al. 2016). The wall lizard, skink, and fence lizard should be watched for in the Greater Meadowlands.

Snakes

Although 9 of 15 available species have been reported in the Meadowlands, evidently only 3 species are now seen regularly: brown snake, garter snake, and milk snake. The northern brown snake is well-known in urban habitats (Gaul 2008), such as trash piles in vacant lots, which may support high population densities (Schlauch 1978). The brown snake also does well in some of the rural areas on the glacial lake clays and glacial tills of northwestern Dutchess County, New York (Kiviat, pers. obs.), thus it is interesting that 2 of the 3 individuals we have seen in the Meadowlands were on the clayey soils near the Hackensack River in Moonachie and River Edge. One brown snake was above the soil climbing in coarse meadow forbs. The garter snakes we have seen in the Meadowlands have been on old fill of various kinds, close to existing wetlands. Among other prey, the eastern garter snake might feed substantially on earthworms (Hamilton 1951), and the brown snake eats introduced slugs (Gaul 2008). Large earthworms are common on fill in the Meadowlands, as are introduced slugs such as the large leopard slug (*Limax maximus*; E. Kiviat, pers. obs.). Because the milk snake specializes on small mammal prey, we wonder if the decline of rats and mice that followed the closing of the garbage landfills and that is believed to have reduced populations of wintering birds of prey, also affected the milk snake population. Quinn (1997) mentioned the "common . . . blacksnake," referring to either the black racer or black rat snake, as occurring historically at Laurel Hill. There have been two recent, unverified, reports of black snakes (species unidentified) at DeKorte Park in Lyndhurst.

Quinn (1997) related tales of abundant and large water snakes at Laurel Hill in the 1800s and earlier, and also suggested that the area supported venomous snakes in the time of the early Dutch settlers. Those would have been the timber rattlesnake (*Crotalus horridus*) and northern copperhead (*Agkistrodon contortrix*) which were likely widespread in northern New Jersey then, and which even today occupy similar landscapes in eastern New York. The water snake is now rare enough in the Meadowlands that we have never seen one. Snakes as a group appear to be sensitive to environmental contaminants (Campbell and Campbell 2002). We can only speculate that the pervasive contamination by metals and organochlorines, as well as the high traffic volume on many local roads, have affected the water snake and perhaps other snakes in the Meadowlands. Mathewson (1955) considered the water snake, brown snake, and garter snake "common" on Staten Island; black racer, milk snake, and ring-necked snake "not common"; and worm snake, hog-nosed snake, and ribbon snake extirpated or very rare.

FACTORS AFFECTING THE HERPETOFAUNA: WHAT ARE THE CAUSES OF THE OBSERVED PATTERNS?

Several reptiles and amphibians reported to occur in the Meadowlands lack documentation. Reports of some species may have been based on field guide range maps, misidentifications, old records, or occurrences near but outside the Meadowlands. These species are: spotted turtle, five-lined skink, ribbon snake, black racer, smooth green snake, northern cricket frog, spring peeper, gray treefrog, and New Jersey chorus frog. It is likely that all these species, plus others known from northeastern New Jersey, were present in the Meadowlands two centuries ago, when mature swamp forests, bogs, fens, springs, streams, and dry rocky woodlands covered large areas. The interesting question is not so much whether other species were present, but how long they survived in the Meadowlands and whether they will occur here again either by spontaneous re-establishment or human-assisted restoration.

In field work since 2000 (> 200 person-days in the field during the growing season), we have seen only 13 species of amphibians and reptiles, one of which was the introduced slider. Excepting the diamond-backed terrapin, snapping turtle, red-eared slider, red-backed salamander, and Atlantic Coast leopard frog, we found very few individuals of each species. This contrasts with the Meadowlands bird fauna which is rich in wetland, aquatic, and woodland-fragment species. Why is the Meadowlands herpetofauna *depauperate* (poor in species and low in abundance of most species) whereas large segments of the bird fauna are the opposite? This could be due to greater sensitivity of amphibians and reptiles to environmental contaminants, soil structure, and other factors, but could also be due to the greater ability of birds to safely cross fragmented and developed landscapes. Furthermore, the good fortunes of some of the raptors, herons, wild turkey, and other birds, as well as mid-size carnivores such as raccoon and house cat, may result in greater predation on the smaller or younger amphibians and reptiles. The herpetofauna, raccoons, and many birds tend to concentrate around small pools and the edges of larger waterways, increasing the potential for predatory interactions. Killing of snakes by people and persistent road mortality have undoubtedly also been factors in the decline of this taxon.

In the freshwater tidal Hudson River and Connecticut River, herpetofaunal diversity increases from lower to higher wetland elevations. Supratidal pools and marshes, located from Mean High Water (MHW) level to 1 m vertically above MHW, are ecologically transitional between the tidal marshes and vernal pools (intermittent woodland pools). These supratidal pools are subject to flooding by the higher high tides, spring floods, and storm surges, and support species such as spotted salamander that are not found in the regularly flooded intertidal zone (Kiviat and Stevens 2001). In brackish tidal wetlands, habitats apparently subject to groundwater discharge and consequent dilution with fresh water, may support species not found in the more brackish areas, such as breeding American toad (Kiviat and Stapleton 1983). Finally, wetland species such as four-toed salamander may occur at the upland-wetland edge at or just above the MHW elevation where forest litter and tidal wrack both accumulate (E. Kiviat, pers. obs.). The presence of development, water control structures, or fill, and as a result the lack of natural soils and vegetation at and near MHW in the Meadowlands, may eliminate these high elevation but tide-affected herpetofaunal habitats. Nonetheless, remaining natural or semi-natural habitats of these types may be important to amphibians and reptiles, and field work is needed in areas such as the low-salinity tidal marshes along the upper Hackensack River estuary. It is noteworthy that the red-backed salamander and the box turtle have been found mostly at the edges of the Core Meadowlands; some species may be unable to penetrate into the interior areas due to salinity or other factors. Many of the amphibians and reptiles may be squeezed between brackish waters and urban-industrial development, and sea level rise will intensify this stress.

A plausible historical scenario explaining low herpetofaunal diversity in the Meadowlands includes rapid development of agricultural and wild lands for residential, industrial, transportation, and waste disposal use during the first several decades of the 1900s, accompanied by extensive industrial contamination, and starting in the 1950s heavy use of DDT for mosquito control. Once these factors had eliminated or reduced populations of sensitive species, the highways, developed areas, polluted waterways, and post-Oradell Dam salinity inhibited dispersal of many species into the Meadowlands from surrounding, less-developed areas. This process is still active; for example, the extensive infill development on vacant lands (e.g., just north of Laurel Hill Park) and the development of former garbage landfills (e.g., the Lyndhurst warehouse project) are eliminating the oldfield-like habitats of the milk snake, garter snake, and brown snake and many other animals. Contemplated flood protection works could be adverse or favorable for the herpetofauna, depending on design; a possible system of high dikes would alter much habitat (Massachusetts Institute of Technology 2014). Although the currently proposed engineering is relatively low key (AECOM 2018), improved drainage ditches, "green infrastructure," and associated wetland mitigation could further degrade remnant swamp forests and other herpetofaunal habitats. Currently widespread use of rodenticide bait stations outside buildings may pose a hazard to rodent-feeding snakes such as milk snake and rat snake (see Lohr and Davis 2018).

Urban trash, which is so prevalent in the Meadowlands, can have adverse effects on herpetofauna. One curious example from México is that horned lizards (*Phrynosoma cornutum*) climb into dumped automobile tires lying flat on the ground but are unable to climb back out (Gatica-Colima et al. 2016); this could happen to other small animals. Negative effects of urbanization on the herpetofauna are obvious (French et al. 2018); urban habitats can also favor some species. Gravid northern water snakes used piles of scrap metal and other artificial materials for thermoregulation in an urban area of Pennsylvania (Pattishall and Cundall 2009). Great Basin gopher snakes hibernated in human-altered, steep, rocky slopes in a developed area (Williams et al. 2015). In both cases, there were thermal advantages to urban habitats. The lesser siren (a highly aquatic salamander) spent a cool, dry period aggregated in the hollows of buried concrete blocks at a Texas site (Powell et al. 2015). The wall lizard, an introduced European species, has colonized urban environments in New York and Ohio, exploiting warm microhabitats and dispersal corridors along rivers and railroads (Hedeen and Hedeen 1999, Lambert et al. 2016). Clearly, warm urban microhabitats and probably the urban heat island overall can favor herpetofauna that tolerate other urban conditions, and artificial or altered habitats can sometimes satisfy herpetofaunal requirements of moisture and temperature. Herpetofaunal use of artificial (refuse) cover objects, in place of logs and rocks, is common in urban and non-urban habitats (Baxter-Gilbert et al. 2013) where thermal advantages are a factor as well as moisture retention.

Although a few ecologically specialized species thrive in cities, some of them in association with cliff-like features or heat-conserving microhabitats that are common in urban areas, most urban wildlife appears to be ecologically generalized. The northern brown snake (see above) and the snapping turtle are good examples. Surasinghe (2013) referred to *disturbance-tolerators* and *disturbance-avoiders*, analogous to our *urban-tolerant* and *urban-sensitive* groups; stream salamanders are disturbance-avoiders. Which organisms have affinities for the urban environment depends on both requirements and tolerances of those species. The extremes of habitat fragmentation, salinization, contamination, and hydrological alteration that distinguish the Meadowlands region and many other urban or industrial regions, along with factors such as road mortality and noise, shape the urban biota but not always in predictable ways.

CONSERVATION, REMEDIATION, AND MANAGEMENT OF HABITATS

If we are correct in suspecting contaminant impacts, remediation of contamination hotspots in the freshwater or upland portions of the Meadowlands might benefit the herpetofauna and should be designed to provide habitats for amphibians and reptiles. Croteau et al. (2008) recommended reducing contaminant loads in amphibian breeding habitats. Remediation projects, of course, would have to be conducted with minimal remobilization of contaminants from earthmoving activities, and contaminant levels in downstream waters might be elevated temporarily. Ongoing and planned habitat enhancement or development projects should also consider reptiles and amphibians. Diked, formerly tidal wetlands, if fresh enough, are much better herpetofaunal habitats than the brackish tidal wetlands created from these impoundments in many mitigation efforts. Although we are skeptical in general about large-scale manipulations of wetlands, if such projects are undertaken, they should consider amphibians and reptiles (and other fauna and flora) as well as fishes and birds. For example, islands in marshes could be designed for turtle nesting habitat if capped with a minimum of 30 cm of sandy or gravelly, low-fertility soil and managed perpetually for sparse, low-growing vegetation. Constructed turtle nesting habitats could also serve as potential nesting areas for killdeer, spotted sandpiper, least tern, common nighthawk, and horned lark, as long as suitable soil and vegetation are maintained. Freshwater ponds or impoundments, as well as quiet areas of tidal creeks and pools, should have stumps and logs (either *in situ* or imported) for underwater hiding and foraging sites, and abovewater basking sites. Plastic netting that serves to protect plantings and stabilize soils can also entangle and kill snakes (Zappalorti and Mitchell 2008, Kapfer and Paloski 2011; E. Kiviat, pers. obs.) and probably other animals; its use on habitat development sites should be avoided and netting at existing Meadowlands sites such as Eastern and Western Brackish Marshes and Skeetkill should be removed. Landfill closure projects and development of recreation areas and other upland open space reserves, if of sufficient size, could be used for amphibian and reptile habitat development. This could include, for example, creation of shallow, fishless ponds for amphibian breeding, surrounded by substantial upland areas with friable soils of at least moderate depth and sufficient soil organic matter; both ponds and uplands should be forested and have ample logs and leaf litter. Such ponds in a forest matrix would provide sites for the potential re-introduction of wood frog, spotted salamander, and spotted turtle. Alternatively, constructed ponds in non-forested areas would serve as habitat for American toad, snapping turtle, painted turtle, northern water snake, and possibly species not recently recorded in the Meadowlands such as red-spotted newt and eastern hog-nosed snake. Ostergaard et al. (2008) found that stormwater ponds in Washington attracted breeding amphibians more when there was forest nearby, and less with increasing extent of impervious surfaces within 1 km. Retention of cover suitable for thermoregulation, and maintenance of an unobstructed southern exposure can improve habitat for snakes (Ovaska and Engelstoft 2008).

Habitat management projects require perpetual maintenance to allow created habitats to remain suitable for their intended users—for example, turtle nesting habitats must be treated non-chemically, such as by tilling, every few years to inhibit development of shading vegetation (see Dowling et al. 2010).

If large enough habitat areas can be enhanced or constructed, for example on closed landfill or on out-of-play areas of golf courses, it might make sense to translocate selected amphibian or reptile species to attempt to establish breeding populations. This was done successfully with the box turtle and painted turtle at Floyd Bennett Field, part of Gateway National Recreation Area in Brooklyn (Cook and Pinnock 1987, Cook 1996). Existing upland and wetland forests should be strictly preserved, with no loss of area to other land uses. Upland buffer zones adjoining wetland breeding areas of the Atlantic Coast leopard frog may need to be several hundred meters wide for long-term population persistence (see Schlesinger et al. 2018), notwithstanding that this species currently breeds where buffer zones are minimal (Kiviat 2011). Habitat creation can and should be designed to serve not only the herpetofauna but also other groups of native organisms including vascular plants, mosses, lichens, clam shrimps, dragonflies, butterflies, moths, bees, and birds.

One of the most important measures to ensure conservation of the Meadowlands herpetofauna is to conduct thorough field surveys in advance of planning for any preservation, restoration, mitigation, or development project. All present and future parks and preserves (e.g., Laurel Hill, Richard P. Kane Natural Area, Teterboro Airport Woods) need to be surveyed and the survey results considered in management planning. Good herpetological surveys take a high level of expertise and lots of time and must be conducted at particular seasons and times of day. For example, because certain species of turtles (e.g., musk turtle, mud turtle, spotted turtle) are secretive, live-trapping or other special techniques must be used, trapping periods and numbers of traps must be adequate, and weather conditions must be appropriate. Surveys for calling frogs must be performed at the appropriate calling season for each species, the proper time of day (usually between dusk and midnight), and under conditions of mild temperatures and low wind-speed. Snakes are usually hard to find and require much effort, although special techniques are helpful for certain species. The best herpetological survey work spans multiple years; often a species is detected in one year and not another, and this is especially true of rare species present at low population densities (see Gibbons et al. 1997). Of course, statewide rare species are among the most important to document for conservation purposes. Well-performed surveys, repeated at intervals, would provide useful indicators of biodiversity trends and environmental quality in the Meadowlands as well as on the larger scale of the region. Surveys must carefully document species and numbers in order to be most useful.

Public and private entities are establishing nature trails, leading boat trips, and providing other opportunities for ecotourists, students, and recreationists to see the Meadowlands biota. It only makes sense to invest in thorough, regularly repeated biological surveys, and appropriate ecological studies, of the lesser-known components of that biota. Many millions of dollars are being spent on wetland mitigation projects and development projects that greatly change the environment. For a small fraction of those dollars, as part of the current efforts to plan ecological restoration of the Meadowlands and to preserve undeveloped lands, the reptile and amphibian fauna of the Meadowlands could be thoroughly surveyed and a Meadowlands-wide herpetofaunal management plan prepared. This effort would help ensure that species such as the diamond-backed terrapin continue to thrive and be viewed by wildlife watchers. It would also solve some of the mysteries concerning the impacts of urbanization on herpetofaunas that would inform environmental planning in the vast regions of the U.S. that will become urban during the next few decades.

As is generally true for other organisms, the diversity of the Meadowlands herpetofauna depends on the protection of large blocks of habitat (e.g., Upper Penhorn Marsh and the Little Snake Hill freshwater ponds for Atlantic Coast leopard frog), certain special habitats or habitat features (rock debris in abandoned quarries for red-backed salamander, coarse woody debris in forests, intermittent pools wherever they occur), multiple habitat units connected by movement corridors (the Penhorn Creek corridor between Upper Penhorn Marsh and Little Snake Hill), and regulatory control over the footprints of development, improvement, and mitigation projects to prevent gratuitous destruction of herpetofaunal habitats small or large. Leopard frogs elsewhere commonly move some distance into non-wetland habitats to forage in summer. At the Meadowlands leopard frog sites there is little space for this behavior and it is possible that strong selective pressure (road mortality) may have driven adaptation to remain in the wetlands.

Chapter 9

Fishes of the Meadowlands and Adjacent Waters

Robert E. Schmidt[1]

The lower portion of the tidal Hackensack River is the major watercourse in the Core Meadowlands. It is surrounded by extensive tidal and nontidal marshes with a long history of hydrologic modifications. The Hackensack River estuary had an important commercial fishery historically and catch-and-release sport fishing is common today. Fishes such as striped bass are iconic denizens, and the mummichog is a key ecological component of tidal marshes. Like all estuaries, the fish community of the Meadowlands is a mixture of resident estuarine species, *diadromous* (migrating between salt water and freshwater for spawning) species or seasonal migrants, freshwater species with a variable degree of salinity tolerance, and marine strays. The relative proportion of species in these categories at any given location is related to the position of the study area in the landscape and the history of anthropogenic effects on the physiography and water quality of the estuary. This chapter summarizes publicly available information on the fish fauna, emphasizing the main Hackensack River channel in the Meadowlands District (which includes much of the Core Meadowlands), supplemented by personal communications from A. Brett Bragin (New Jersey Sports and Exposition Authority) and a small number of my personal observations. I have attempted to place the ichthyofauna of the Core Meadowlands in the larger landscape of the Greater Meadowlands and the New York City estuarine delta by comparing the fish fauna upstream of the area in the Hackensack River and downstream of the area in Newark Bay and the Raritan Bay—New York Harbor complex. I also compare the Meadowlands fauna to that of the Hudson River.

SALINITY

Given suitable temperatures and tolerable dissolved oxygen levels, salinity is the main abiotic factor that influences fish distribution in an estuary. Salinities in the Hackensack River in the Meadowlands ranged from < 1 to 23 parts-per-thousand (ppt) in 2001–2003 (seawater being 33–38 ppt; Bragin et al. 2005). Thus, this area is best classified as mesohaline.

Low salinity water is found farther inland and there is a relatively steep salinity gradient. Highest salinities are seen in winter and the lowest in spring and early summer probably due to dilution by freshwater runoff.

Fishes in essentially freshwater families (Cyprinidae, Leuciscidae, Centrarchidae, Percidae) and gizzard shad (see Annotated List of Fishes [below] for scientific names) primarily occur in the low salinity areas whereas the marine species, for example, lookdown and striped searobin, are restricted to the high salinity areas near the mouth of the river. The most consistently abundant species in the Meadowlands District are mummichog and white perch, both *euryhaline* (tolerating a wide range of salinity) species, and both widely distributed in the river.

RESIDENT FISHES

Interactions between tides and basin morphometry in estuaries result in a long retention time and therefore estuaries accumulate nutrients and detritus making them one of the most productive aquatic ecosystems. Unfortunately, this same system concentrates toxins and other pollutants accumulated from the watershed. High productivity should be attractive to fishes wanting to participate in the food web but the stressors of twice daily fluctuations of salinity, temperature, dissolved oxygen, turbidity, and strong currents are not well tolerated by many species. Waldman et al. (2006) listed 8 resident estuarine fishes and 17 permanent or seasonal marine fishes from the Hudson River out of a total of 214 (now 235) fishes known from the watershed. The relatively small number of resident species is a testament to the difficulty for fishes to persist in such a fluctuating environment.

The Hackensack River estuary is much smaller than the Hudson estuary and has been sampled much less intensively (especially since 1974 when fish studies burgeoned in the Hudson [Klauda et al. 1988]). At present, 19 of the 25 Hudson estuary resident or permanent/seasonal marine fishes have been documented from the Meadowlands District. Some of the other species listed (Table 9.1) may well be found in the Hackensack estuary with further

Table 9.1. Estuarine and marine fishes of the Hackensack River estuary and Hudson River estuary.

Common Name	Scientific Name	Hudson	Hackensack
Shortnose sturgeon (PS)	*Acipenser brevirostrum*	+	-
Bay anchovy (R)	*Anchoa mitchilli*	1	+
Atlantic menhaden (PS)	*Brevoortia tyrannus*	+	+
Naked goby (PS)	*Gobiosoma bosc*	+	+
Striped mullet (PS)	*Mugil cephalus*	+	+
White mullet (PS)	*Mugil curema*	+	+
Rough silverside (R)	*Membras martinica*	+	+
Inland silverside (R)	*Menidia beryllina*	+	+
Atlantic needlefish (PS)	*Strongylura marina*	+	-
Mummichog (R)	*Fundulus heteroclitus*	+	+
Spotfin killifish (R)	*Fundulus luciae*	+	-
Crevalle jack (PS)	*Caranx hippos*	+	+
Summer flounder (PS)	*Paralichthys dentatus*	+	+
Winter flounder (PS)	*Pseudopleuronectes americanus*	6	+
Hogchoker (R)	*Trinectes maculatus*	2	+
Lined seahorse (PS)	*Hippocampus erectus*	+	-
Northern pipefish (PS)	*Syngnathus fuscus*	+	+
American sand lance (PS)	*Ammodytes americanus*	12	+
Bluefish (PS)	*Pomatomus saltatrix*	+	+
Fourspine stickleback (R)	*Apeltes quadracus*	+	-
Threespine stickleback (R)	*Gasterosteus aculeatus*	+	+
Grubby (PS)	*Myoxocephalus aenaeus*	+	+
White perch (R)	*Morone americana*	4	+
Weakfish (PS)	*Cynoscion regalis*	5	+
Spot (PS)	*Leiostomus xanthurus*	+	+

Note: List of resident estuarine fishes (R) and permanent or seasonal marine fishes (PS) in the Hudson River (Waldman et al. 2006) compared to the Meadowlands region (PSE&G 1998, Neuman et al. 2004, Bragin et al. 2019, Brett Bragin pers. comm.) of the Hackensack River (+ = present, - = absent). Numbers in the "Hudson" column are the rank in abundance of that species in trawl surveys from the lower Hudson River and Raritan Bay (Woodhead 1987). List is in phylogenetic order following Nelson et al. (2016).

sampling. It is unlikely that some species, shortnose sturgeon for instance, could establish simply because the Hackensack estuary is too small for that species.

Woodhead (1987) listed 12 species he considered dominant (total of 94% of the catch) from a trawl survey of the lower Hudson River and Raritan Bay. Six of those species were considered residents in the Hudson by Waldman et al. (2006) and all six have been documented from the Meadowlands. Likewise, Will and Houston (1992) listed the 12 most common fishes from trawl and gillnet studies in Newark Bay. All of those species were recorded from the Meadowlands. Woodhead's (1987) and Will and Houston's (1992) ranking of species by abundance is very different from the Meadowlands surveys, but it is not possible to determine how much habitat size has affected the abundances of the fishes involved and therefore these differences cannot be related to differences in water quality, habitat availability, or other abiotic correlates. There could also be subtle differences in sampling gear or deployment which could alter the apparent abundances of various species.

In general, the Core Meadowlands portion of the Hackensack estuary seems to support a substantial subset of the resident fishes in the Hudson estuary. There is no obvious pollutant, modification, or lack of habitat (except for shortnose sturgeon mentioned above) in the Meadowlands District that would explain the differences noted.

THE ROLE OF MUMMICHOG IN THE MEADOWLANDS ECOSYSTEM

The mummichog was the most abundant species in surveys of the ichthyofauna of the Meadowlands (Bragin et al. 2005). A casual visit to the area revealed huge numbers of this species congregating in the tidal shallows. You can see similar sights in any shallow tidal marshland from Florida to Nova Scotia, especially in areas of moderate salinity. There has been considerable research on the role of this species in tidal marsh ecosystems.

Sweeney et al. (1998) reported densities of about 14 mummichog/m^2 from population estimates and areal estimates of the habitat. Schmidt (1993) reported densities three times as high from seine collections in a Hudson River tidal marsh. Valiela et al. (1977) calculated the productivity of mummichog as 160 kg dry weight/ha, one of the highest productivity estimates for a natural fish population. Mummichog is inefficiently captured in seines (Allen et al. 1992), the primary gear collecting them in the Meadowlands District (Bragin et al. 2005). This means that the abundance estimates for mummichog in the Meadowlands are probably low due to gear inefficiency, perhaps by as much as an order of magnitude below the true abundance (Allen et al. 1992).

Mummichogs follow the edge of the rising tide from channels up onto marsh surfaces and can either remain on the surface in isolated pools or (most likely) retreat to the channels as the tide ebbs (Valiela et al. 1977). In summer months, individuals may have restricted home ranges: 36 m of shoreline (Lotrich 1975), or 450–650 m (Sweeney et al. 1998). They may abandon the shallow marshlands in winter (Chidester 1920) or winter in pools on the marsh surface (Smith and Able 1994).

While on the marsh surface, mummichogs eat a variety of foods. Studies on diet of mummichog (Vince et al. 1976, Kneib et al. 1980, Abraham 1985, Joyce and Weisberg 1986, Allen et al. 1994) listed a wide variety of animal foods, such as gammarid amphipods, copepods, dipteran larvae, arachnids, snails, and decapod crustaceans. Variation in the animal component of the diet is site-specific and can change seasonally (Valiela et al. 1977, Werme 1981). All authors also listed detritus and algae as substantial food items for this species. Stable isotope analyses (Kneib et al. 1980, Hughes and Sherr 1983, Griffin and Valiela 2001) demonstrated the role of mummichog as a primary or secondary detritus consumer with larger (older) mummichogs feeding on one trophic level higher than younger individuals (i.e., consuming animals rather than detritus and algae).

Mummichog, in turn, is preyed upon by a wide variety of organisms. Any large predatory fish will consume mummichog. In the Meadowlands, American eel, bluefish, and striped bass would be expected to feed on mummichog. Wading birds and waterfowl, which are present in abundance in the Meadowlands District, are likely eating mummichog, especially the herons, egrets, and ducks (Abraham 1985).

Mummichog may have a large role in transferring energy from plant detrital productivity of the marshlands to the nekton and the marsh-foraging avifauna in the Hackensack River. This ecological role of mummichog has been described as significant in other marsh systems. In the Meadowlands, the grass shrimp (*Palaemonetes pugio*) probably contributes a similar service to the ecosystem but only further study would allow us to determine whether grass shrimp or mummichog have larger roles in energy transfer.

MIGRATORY FISHES AND THEIR SPAWNING HABITS

I divided the migratory fishes in the area into three categories based on major characteristics of their life histories. Because they are migratory, only part of their life cycle is spent in the Meadowlands District.

Catadromous fishes are those that spend their juvenile life in low salinity water but migrate to the sea to spawn. There is one catadromous fish in the Meadowlands, the American eel. American eel collected in the Meadowlands are the juvenile stage, often called yellow eels. American eel may spend up to twenty-five years as yellow eels and this is the stage where the majority of growth occurs. The individuals in brackish water are most probably males and therefore would mature earlier than females. When sexually mature (then called silver eels), they change color and morphology and emigrate to the Sargasso Sea—southeast of Bermuda—to spawn. The adults are replaced by small (5–6 cm) transparent *glass eels* that have drifted from the Sargasso Sea in the Gulf Stream. Spawning takes place annually with a different cohort each year, and the adults die after spawning.

Anadromous fishes are those species that spawn in freshwater but spend most of their time in the ocean. There are records of 5 anadromous species

in the Meadowlands District and 2 other species that may or may not be considered anadromous.

Three species of anadromous herrings (Clupeidae) are recorded from the
Hackensack River. Alewife and blueback herring are probably spawning
in the Hackensack although the runs appear to be small. These species are
known to spawn in the Raritan River and the Hudson River as well. The
third herring, American shad, is represented by juveniles in the Meadowlands
although adults are seen in the Raritan River and the Hudson so they may
occur in the Hackensack as well. Spawning of American shad has not been
recorded recently in the Hackensack River. In all anadromous herrings, the
eggs are laid in freshwater and development occurs in fresh and brackish
water. Estuaries therefore are important nursery habitats for these species.
The juveniles caught in the Meadowlands District may have been spawned
upstream (e.g., in the upper Hackensack River estuary) or may have been produced in the nearby Hudson River (in the case of American shad, certainly).

Rainbow smelt is a small anadromous species that spawns in shallow freshwater, often in small tributary streams. Rainbow smelt are known from the
Hackensack, Passaic, and Raritan rivers but whether spawning occurred in
those areas is not known—in the late 1900s the most southern spawning population may have been in the Hudson River. The Hudson River population of
rainbow smelt crashed in 1995 and may now be extirpated from that drainage
(Daniels et al. 2005). (Rainbow smelt was at its southern range margin in the
Hudson, and possibly was affected by climate warming.) A similar fate seems
to be occurring in the Long Island Sound tributaries in western Connecticut.
Rainbow smelt was not collected in the most recent surveys (Bragin et al.
2019) and may have disappeared from the Meadowlands District.

The two temperate basses (Moronidae) native to the East Coast are seen
in the Meadowlands District—striped bass and white perch. The former is an
anadromous species with a large spawning population in the Hudson River.
Outside of the Hudson and Chesapeake Bay few spawning runs are known
for striped bass in the northeastern United States, and there is no evidence
that spawning occurs in the Hackensack River. The striped bass collected in
the Meadowlands District were mostly juveniles. White perch is anadromous
in the northern end of its range but members of the population in the Hudson
River are considered to be euryhaline residents. It is likely that white perch
in the Meadowlands District are spawning there and the adults are resident.

Very much like white perch, Atlantic tomcod is anadromous in the northern
end of its range. A large spawning population is present in the Hudson River
and the species is known from the Passaic River as well. Historically, spawning occurred as far south as Delaware Bay, but recent evidence suggests that
the Hudson estuary now has the southernmost spawning population (Daniels
et al. 2005). A recent population crash in the Hudson estuary suggests its

population of this essentially northern species is under considerable stress and that global warming may be implicated in the retraction of its spawning range. The Atlantic tomcod collected in the Meadowlands District are probably adults (spawning occurs at nine months old in most individuals in the Hudson) and spawning did occur in the late 1980s (Kraus and Bragin 1990) and late 1990s (PSE&G 1998). It is unlikely that spawning is occurring in the Hackensack River today.

A third type of migration is characterized by the inshore movement of young marine fishes spawned offshore. Juvenile bluefish, crevalle jack, and weakfish collected in the Meadowlands represent the end of the inshore movement for these species. These fishes use the brackish waters to feed and grow through the summer and they leave the estuary in late fall. This can be a significant event in the life cycle of these important marine fishes.

FRESHWATER FISHES UPSTREAM OF THE MEADOWLANDS

Fishes inhabiting the Hackensack River upstream of the Meadowlands are a potential source of species for the study area. Records provided here are from Arndt (2004) and Carlson et al. (2016). There are 22 species within the watershed (Table 9.2) that were not reported from the Meadowlands District. Of these species, 14 were recorded only from the headwaters near the New York State border. Urbanization, stream modifications, and pollution upstream of the Meadowlands have probably eliminated these species from most of the drainage. Many of these *headwater* species may never have occurred in the Meadowlands in any case because they may prefer small rocky streams and/ or have little salt tolerance. The other species (not listed as "headwaters" in Table 9.2) may be present in the Meadowlands but have simply not been documented.

FISHES RECORDED FROM NEWARK BAY

The fish community of Newark Bay is or can be a source of individuals entering the tidal Hackensack River. Sixty-three species were documented by recent publicly available surveys (Table 9.3), 42 of which were recorded from the Meadowlands District. In comparison, 61 of the 63 were present in the Hudson River and Raritan Bay (Table 9.3—note the discussion of chub mackerel). Therefore, the fishes in Newark Bay have a greater effect on the ichthyofauna of the Meadowlands (67% of the marine fishes in Newark Bay were observed in the Meadowlands) than the freshwater fishes from the

Table 9.2. Fishes from the Hackensack River not reported from the Meadowlands District.

American brook lamprey (*Lampetra appendix*)-headwaters
Atlantic sturgeon (*Acipenser oxyrhinchus*)-Hackensack/Passaic junction
Hickory shad (*Alosa mediocris*)-one record from middle Hackensack
Satinfin shiner (*Cyprinella analostana*)-upstream of Meadowlands
Common shiner (*Luxilus cornutus*)-headwaters
Spottail shiner (*Notropis hudsonius*)-headwaters
Blacknose dace (*Rhinichthys atratulus*)-upstream of Meadowlands
Longnose dace (*Rhinichthys cataractae*)-headwaters
Creek chub (*Semotilus atromaculatus*)-headwaters
Fallfish (*Semotilus corporalis*)-headwaters
White sucker (*Catostomus commersonii*)-upstream of Greater Meadowlands
Creek chubsucker (*Erimyzon oblongus*)-headwaters
Black bullhead (*Ameiurus melas*)-exotic species, one record upstream
Tadpole madtom (*Noturus gyrinus*)-headwaters
Redfin pickerel (*Esox americanus*)-headwaters
Chain pickerel (*Esox niger*)-headwaters
Brown trout (*Salmo trutta*)-exotic species, headwaters
Brook trout (*Salvelinus fontinalis*)-headwaters
Mud sunfish (*Acantharcus pomotis*)-headwaters
Bluespotted sunfish (*Enneacanthus gloriosus*)-headwaters (in New York)
Smallmouth bass (*Micropterus dolomieu*)-exotic species, headwaters
Tessellated darter (*Etheostoma olmstedi*)-upstream of Greater Meadowlands

Note: Data from Bean 1903, Arndt 2004, Feltes 2005, Carlson et al. 2016, and Bragin pers. comm.

watershed (44% of the freshwater fishes in the Hackensack watershed were observed in the Meadowlands). The freshwaters in the Meadowlands are less attractive to fishes than the saline habitats due to a combination of inappropriate habitats, anthropogenic modifications, and poor water quality.

The list of species found in Newark Bay (Table 9.3) is probably incomplete despite recent efforts. For instance, Schmidt and Wright (2018) reported the speckled worm eel (*Myrophis punctatus*) from the nearby Arthur Kill. This species is almost impossible to collect in conventional fish sampling gear and could easily be present in Newark Bay and the Core Meadowlands. Also, sheepshead minnow (*Cyprinodon variegatus*) has not been collected in the Meadowlands or Newark Bay despite the presence of appropriate habitat. The sheepshead minnow is a common and widely distributed cyprinodontid found in shallow, moderately saline waters and may be present in habitats that have been poorly sampled (Bragin, pers. comm.). Feather blenny (*Hypsoblennius hentz*) is established in lower New York Harbor and Raritan Bay (Schmidt and Moccio 2013, and unpublished data) and is likely to be present in the study area. Newark Bay and surrounding waters are dynamic systems with some species appearing occasionally or rarely.

Table 9.3. Fishes of Newark Bay (southern end of the Greater Meadowlands).

Common Name	Scientific Name	Hudson	Hackensack
Little skate (PS)	*Leucoraja erinacea*	+	-
Atlantic sturgeon (PS)	*Acipenser oxyrhinchus*	+	-
Conger eel (PS)	*Conger oceanica*	+	+
American eel (R)	*Anguilla rostrata*	+	+
Striped anchovy (PS)	*Anchoa hepsetis*	+	+
Bay anchovy (R)	*A. mitchilli*	1	+
Blueback herring (PS)	*Alosa aestivalis*	+	+
Alewife (PS)	*A. pseudoharengus*	+	+
American shad (PS)	*A. sapidissima*	+	+
Atlantic menhaden (PS)	*Brevoortia tyrannus*	+	+
Atlantic herring (PS)	*Clupea harengus*	+	+
Gizzard shad (R)	*Dorosoma cepedianum*	+	+
Rainbow smelt (PS)	*Osmerus mordax*	+	+
Atlantic tomcod (R)	*Microgadus tomcod*	+	+
Silver hake (PS)[a]	*Merluccius bilinearis*	+	-
Red hake (PS)	*Urophycis chuss*	+	-
White hake (PS)[a]	*U. tenuis*	+	-
Spotted hake (PS)	*U. regia*	+	+
Oyster toadfish (PS)	*Opsanus tau*	+	-
Naked goby (PS)	*Gobiosoma bosc*	+	+
Striped mullet (PS)	*Mugil cephalus*	+	+
White mullet (PS)	*M. curema*	+	+
Atlantic silverside (PS)	*Menidia menidia*	+	+
Atlantic needlefish (PS)	*Strongylura marina*	+	-
Mummichog (R)	*Fundulus heteroclitus*	+	+
Striped killifish (R)	*F. majalis*	+	+
Crevalle jack (PS)	*Caranx hippos*	+	+
Atlantic moonfish (PS)	*Selene setapinnis*	+	+
Lookdown (PS)	*S. vomer*	+	-
Windowpane (PS)	*Scophthalmus aquosus*	+	+
Smallmouth flounder (PS)	*Etropus microstomus*	+	-
Smooth flounder (PS)[a]	*Liopsetta putnami*	-	-
Summer flounder (PS)	*Paralichthys dentatus*	+	+
Fourspot flounder (PS)	*Hippoglossoides oblongus*	+	-
Winter flounder (PS)	*Pseudopleuronectes americanus*	6	+
Hogchoker (R)	*Trinectes maculatus*	2	+
Lined seahorse (PS)	*Hippocampus erectus*	+	-
Northern pipefish (PS)	*Syngnathus fuscus*	+	+
Chub mackerel (PS)[b]	*Scomber colias*	-	-
Butterfish (PS)	*Peprilus triacanthus*	+	+
American sand lance (PS)	*Ammodytes americanus*	12	+
Tautog (PS)	*Tautoga onitis*	+	+
Cunner (PS)	*Tautogalabrus adspersus*	+	+
Black seabass (PS)	*Centropristes striata*	+	-
Bluefish (PS)	*Pomatomus saltatrix*	+	+

Northern searobin (PS)	*Prionous carolinus*	+	-
Striped searobin (PS)	*P. evolans*	+	+
Rock gunnel (PS)	*Pholis gunnellus*	+	+
Threespine stickleback (R)	*Gasterosteus aculeatus*	+	+
Grubby (PS)	*Myoxocephalus aenaeus*	+	+
Longhorn sculpin (PS)	*M. octodecimspinosus*	+	-
White perch (R)	*Morone americana*	4	+
Striped bass (R)	*M. saxatilis*	+	+
Silver perch (PS)	*Bairdiella chysoura*	+	-
Weakfish (PS)	*Cynoscion regalis*	5	+
Spot (PS)	*Leiostomus xanthurus*	+	+
Northern kingfish (PS)	*Menticirrhus saxatilis*	+	-
Atlantic croaker (PS)	*Micropogonias undulatus*	+	+
Pumpkinseed (R)[a]	*Lepomis gibbosus*	+	+
Pinfish (PS)[a]	*Lagodon rhomboides*	+	-
Scup (PS)	*Stenotomus chrysops*	+	+
Planehead filefish (PS)	*Stephanolepis hispidus*	+	-
Northern puffer (PS)	*Sphoeroides maculatus*	+	+

Note: List of resident estuarine fishes (R) and permanent or seasonal marine fishes (PS) collected in Newark Bay (Will and Houston 1992, USACE 1997, Normandeau 2017) compared to the Meadowlands area (PSE&G 1998, Neuman et al. 2004, Bragin et al. 2019, Bragin pers. comm.) of the Hackensack River (+ = present, - = absent) and in the Hudson River (Waldman et al. 2006). Numbers in the "Hudson" column are the rank in abundance of that species in trawl surveys from the lower Hudson River and Raritan Bay (Woodhead 1987). Arrangement is phylogenetic following Nelson et al. (2016).

[a] Reported in 1987–88, but not in more recent surveys

[b] Although not recorded from the Hudson River or Raritan Bay, some specimens have probably been mis-identified as Atlantic mackerel, *Scomber scombrus*.

POTENTIAL FOR RARE AND ENDANGERED FISHES IN THE MEADOWLANDS

The only fish likely to occur in the Meadowlands that is listed as rare or endangered by the State of New Jersey is the federally endangered short-nose sturgeon (NJDEP 2012). This species has not been reported from the Meadowlands District although it is possible that individuals stray into this area from the Hudson River population.

The Meadowlands District does host the eastern mudminnow (*Umbra pyg-maea*), a member of a regionally rare fauna—the New Jersey Pine Barrens assemblage of fishes found in acidic tannin-stained waters. Most members of this assemblage have narrow habitat preferences and would never survive in the Meadowlands District even if it were pristine. However, eastern mudmin-now is a widespread species tolerant of poor water quality and is found in the southern Hudson River watershed and in the Wallkill drainage in New York.

The spotfin killifish (*Fundulus luciae*) is present in the Meadowlands District. Yozzo and Ottman (2003) documented this species from the Hudson estuary and nearby locales. Spotfin killifish habitat is small marsh pools in the upper intertidal zone. They have been found in *Phragmites* reedbeds (Yozzo and Ottman 2003). This species is known from areas both north and south of the Meadowlands District including the Hackensack River estuary at Lincoln Park in Jersey City (Yozzo and Ottman 2003, Arndt 2004). It has been often overlooked because it is superficially similar to the very abundant mummichog and the two species can be found in the same pools. They are common in New Jersey (Able et al. 1983) and would probably not be eligible for listing, however.

CONTAMINANTS

Some aquatic environments in the Meadowlands have very high concentrations of mercury in the sediments. This is particularly true of Berry's Creek. Mercury can bioaccumulate through the food web, especially in the methyl mercury configuration. There is voluminous literature on the effects of mercury on mammals (including humans) and birds that feed on contaminated fishes. Little study has been done on the effect of high mercury concentrations on the fishes themselves. For instance, Webber and Haines (2003) reported that fish tissue concentrations of 518 ng/g (nanograms per gram) caused behavioral changes in golden shiner but no differences in growth rate or mortality among several treatments. They suggested that the behavioral changes may make golden shiner more vulnerable to predation.

The mercury concentrations that caused behavioral changes (Webber and Haines 2003) were comparable to the levels of mercury found in white perch in the Meadowlands District. Weis and Ashley (2007) reported that 32% of their white perch samples exceeded 470 ng/g of mercury (compare to 518 ng/g in Webber and Haines 2003). Although these concentrations caused behavioral changes in golden shiner, white perch is only very distantly related to that species and may have a different tolerance for mercury.

Polychlorinated biphenyls (PCBs) are bioaccumulating compounds with an affinity for fatty tissues. Some of the common effects of PCBs (Safe 1994, Monosson 2000) on fishes include feminization, reproductive dysfunction, and enzyme inhibition. Severe effects occur in piscivorous predators such as eagles (Elliott et al. 1996) and mink (Heaton et al. 1995). White perch in the Meadowlands District contained > 1 ppm PCB (Weis and Ashley 2007). This level is certainly high enough to have some physiological effect on white perch, but there are no studies available on that topic. Based on Weis and Ashley's (2007) data, one would expect a similar concentration in other resident invertebrate feeders like mummichogs and the sunfishes. The impacts of contaminants on Meadowlands biodiversity will be explored further in a future book.

REMEDIATION AND HABITAT QUALITY

Bragin et al. (2019) described significant changes in the Meadowlands District fish population compared to a similar study done in 1987–88 (Kraus and Bragin 1989). Among the biotic changes were increases in abundance of game and forage fishes, both considered to be improvements. Concomitantly, they also described positive changes in dissolved oxygen levels and temperatures in the Hackensack River, cessation of several point sources of pollutants, and remediation efforts in several habitats.

Efforts to improve environmental quality in the Hackensack River are laudable and have shown some success. More needs to be done. Care should be taken in interpreting the changes in the fish community documented by Bragin et al. (2019) because sampling effort varied among the studies. There is no information on what annual variation may have occurred from 1990–2015 and thus determining whether the fish community has improved is not possible. Put another way, fish species composition in samples from low salinity tidal wetlands and shallows tends to vary considerably, thus we are not sure of the meaning of differences in samples several years apart.

FURTHER RESEARCH

Fish sampling in the Meadowlands District has primarily been done in the open estuary with less effort in the brackish tidal wetlands. The ichthyofauna of the upland freshwater habitats has been considerably less well documented, including ponds and ditches. There may be some interesting species occurrences still to be seen in these shallow marginal habitats such as the spotfin killifish (discussed above) and others like the ninespine stickleback recorded by Learn et al. (2004) and not seen elsewhere.

The role of the Hackensack estuary as spawning habitat for fishes has also not been documented thoroughly. I suggest that sampling the *ichthyoplankton* (fish eggs and larvae) in the area would elucidate which species are spawning in the Meadowlands District. At this point in our understanding of the Hackensack estuary, ichthyoplankton data are critical to understanding the habitat quality of the estuary and planning for future alterations and management.

SUMMARY

The Hackensack River estuary in the Core Meadowlands has a fish fauna generally typical of estuaries in the northeastern United States. The size of this estuary and the relatively small sampling effort that has been made so far contributes to the relatively short list of species documented from the area. The fish fauna contains species in each of several categories that one expects to see in an estuary: diadromous (anadromous and catadromous) species, euryhaline residents, freshwater and saltwater strays, and marine species using the estuary as a nursery. The estuary is providing nursery habitat for herrings, striped bass, bluefish, and crevalle jack but there are few data on the abundance of these species. The freshwater fishes in the watershed have been negatively affected by water quality and modifications to the watershed not necessarily related to the Meadowlands *per se*. There is a substantial lack of data on which species may be spawning in the Meadowlands District which should be addressed in future projects especially since dredging, wetland alterations, contaminant hotspot remediation, and alteration of hydrologic engineering structures may affect spawning and nursery habitats.

Continued improvements in water quality, and removal of contaminants, are the most important management actions to improve the environment of the fish fauna. Large-scale developments, including flood control engineering and shipping channel deepening, must be assessed carefully to predict their impacts on fishes. Currently, very little consumption of Hackensack River fish is advised (NJDEP 2021) because of contamination. This is unlikely to change soon; nonetheless, the fish fauna remains important ecologically and culturally, and the estuary may be contributing to fish populations in the coastal waters as well as remaining viable as a potential future food source.

ANNOTATED LIST OF FISHES DOCUMENTED FROM THE MEADOWLANDS

The designation *SGCN* is appended to those species listed by the State of New Jersey as Species of Greatest Conservation Need (NJDEP 2018a). Ecological comments are general unless specifically ascribed to the Meadowlands.

Lepisosteidae (gars)

Atractosteus spathula (alligator gar)-Introduced, Freshwater. Only one collected. This individual was probably an aquarium release.

Congridae (conger eels)

Conger oceanica (conger eel)-Native, Marine. This species is primarily found in high salinity waters. The one individual documented from the Meadowlands was probably a stray from Newark Bay.

Anguillidae (freshwater eels)

Anguilla rostrata (American eel)-Native, Catadromous. Juveniles inhabit the area and have a wide range of environmental tolerance. SGCN

Engraulidae (anchovies)

Anchoa hepsetis (striped anchovy)-Native, Marine. This species is an occasional visitor from high salinity waters.

Anchoa mitchilli (bay anchovy)-Native, Marine. This species is probably spawning in the Meadowlands. It can tolerate very low salinities.

Clupeidae (herrings)

Alosa aestivalis (blueback herring)-Native, Anadromous. Both juveniles and adults were collected in the Meadowlands. There is probably some spawning in the Hackensack River. SGCN

Alosa pseudoharengus (alewife)-Native, Anadromous. Both juveniles and adults were collected in the Meadowlands. There is probably some spawning in the Hackensack River. SGCN

Alosa sapidissima (American shad)-Native, Anadromous. No adults were documented and spawning in this area is unlikely. Juveniles are using the Meadowlands as a nursery. SGCN

Brevoortia tyrannus (menhaden)-Native, Marine. Young of year, juveniles, and adults were present in the Meadowlands. This species can tolerate low salinities. SGCN

Clupea harengus (Atlantic herring)-Native, Marine. Larvae and small juveniles are present in late winter. SGCN

Dorosoma cepedianum (gizzard shad)-Native, Freshwater. This species is a relatively recent immigrant to the northeastern United States. It can tolerate high salinities.

Cyprinidae (minnows and carps)

Carassius auratus (goldfish)-Introduced, Freshwater. This species has a very high tolerance for poor water quality.

Cyprinus carpio (common carp)-Introduced, Freshwater. This species has a very high tolerance for poor water quality.

Xenocyprididae (East Asian minnows)

Ctenopharyngodon idella (grass carp)-Introduced, Freshwater. This species was reported from the Meadowlands or vicinity (AECOM 2018). Unverified report.

Hypophthalmichthys nobilis (bighead carp)-Introduced, Freshwater. This species was reported from the Meadowlands or vicinity (AECOM 2018). Unverified report.

Leuciscidae (minnows)

Notemigonus crysoleucas (golden shiner)-Native, Freshwater. Adults can be found in moderate salinities.

Pimephales promelas (fathead minnow)-Introduced, Freshwater. Found in Harrier Meadow, is tolerant of poor water quality.

Ictaluridae (North American catfishes)

Ameiurus catus (white catfish)-Native, Freshwater. This large catfish has a moderate tolerance for poor quality water and a high salinity tolerance. SGCN

Ameiurus natalis (yellow bullhead)-Native, Freshwater. Neuman et al. (2004) reported this species. This catfish is rarely found in brackish water. SGCN

Ameiurus nebulosus (brown bullhead)-Native, Freshwater. This catfish also has a high salinity tolerance.

Umbridae (mudminnows)

Umbra pygmaea (eastern mudminnow)-Native, Freshwater. This small species prefers quiet, muddy freshwater habitat.

Osmeridae (smelts)

Osmerus mordax (rainbow smelt)-Native, Anadromous. This species is probably extirpated from the Meadowlands in a general range contraction.

Gadidae (cods)

Microgadus tomcod (Atlantic tomcod)-Native, Euryhaline. Juveniles and adults were documented from the Meadowlands but there has probably been no spawning recently. SGCN

Phycidae (hakes)

Urophycis regia (spotted hake)-Native, Marine. This species is known to occur in moderate salinities.

Batrachoididae (toadfishes)

Opsanus tau (oyster toadfish)-Native, Marine. This benthic species rarely enters brackish water. Only one specimen was collected. SGCN

Gobiidae (gobies)

Ctenogobius shufeldti (American freshwater goby)-Native, Euryhaline. Documented by Schmidt and Bragin (2021). This record is a range extension of a semitropical species. Specimen examined.

Gobiosoma bosc (naked goby)-Native, Euryhaline. This species is difficult to catch because it likes to hide in crevices.

Gobiosoma ginsburgi (seaboard goby)-Native, Marine. This species is also difficult to catch and prefers more saline water than the naked goby.

Mugilidae (mullets)

Mugil cephalus (striped mullet)-Native, Marine. This species can tolerate low salinity water.

Mugil curema (white mullet)-Native, Marine. A specimen was recently reported from the study area.

Atherinopsidae (New World silversides)

Membras martinica (rough silverside)-Native, Marine. A few larvae have been reported from the study area, probably as strays.

Menidia beryllina (inland silverside)-Native, Euryhaline. This species probably spawns in the Meadowlands.

Menidia menidia (Atlantic silverside)-Native, Euryhaline. This species probably spawns in the Meadowlands. SGCN

Fundulidae (killifishes)

Fundulus diaphanus (banded killifish)-Native, Freshwater. This record is based on a photograph by E. Kiviat from Teaneck Creek (the habitat is shown in Figure 9.1).

Fundulus heteroclitus (mummichog)-Native, Euryhaline. This is the most abundant resident fish in the Meadowlands. SGCN

Fundulus luciae (spotfin killifish)-Native, Euryhaline. This species prefers shallow pools in heavily vegetated brackish marshes. It is very similar in appearance to the mummichog. SGCN

Fundulus majalis (striped killifish)-Native, Marine. This species will tolerate moderate salinities.

Poeciliidae (livebearers)

Gambusia affinis (western mosquitofish)-Introduced, Freshwater (Figure 9.2). This species was probably imported for mosquito control and is now self-perpetuating. Eastern mosquitofish (*Gambusia holbrooki*) was reported from Harrier Meadow (Feltes 2005). I doubt this identification because all of the identifiable mosquitofish I have examined from New York and New Jersey have been *G. afffinis* (western mosquitofish) and Arndt (2004) likewise never identified eastern mosquitofish from New Jersey (and questioned

Figure 9.1 Perennial stream (Teaneck Creek), banded killifish habitat. Photo by Erik Kiviat.

Figure 9.2 Supratidal channel (Kingsland Creek), western mosquitofish habitat. Railroad at far right. Photo by Erik Kiviat.

whether it was native to the state). For a definitive identification it is necessary to examine the bony structures of the gonopodium of a mature male
under a microscope (Rauchenberger 1989).

Carangidae (jacks)

Caranx hippos (crevalle jack)-Native, Marine. Juveniles use the Meadowlands
as a nursery in summer and fall.

Selene setapinnis (moonfish)-Native, Marine. One record from the
Meadowlands, a stray from the south.

Selene vomer (lookdown)-Native, Marine. This species occasionally strays
into the Meadowlands.

Scophthalmidae (turbots)

Scophthalmus aquosus (windowpane)-Native, Marine. Juveniles are frequently found in moderate salinity.

Paralichthyidae (flukes)

Paralichthys dentatus (summer flounder)-Native, Marine. Juveniles are frequently found in moderate salinity. SGCN

Pleuronectidae (right-eyed flounders)

Etropus microstomus (smallmouth flounder)-Native, Marine. Juveniles stray into estuaries on occasion.

    Pseudopleuronectes americanus (winter flounder)-Native, Euryhaline. Juveniles use the Meadowlands as a nursery. SGCN

Achiridae (soles)

Trinectes maculatus (hogchoker)-Native, Euryhaline. This species may be spawning in the Meadowlands.

Syngnathidae (pipefishes and seahorses)

Syngnathus fuscus (northern pipefish)-Native, Euryhaline. This species is most commonly found among submersed vegetation.

Stromateidae (butterfishes)

Peprilus triacanthus (butterfish)-Native, Marine. Individuals occasionally stray into estuaries. SGCN

Ammodytidae (sand lances)

Ammodytes americanus (sand lance)-Native, Marine. A few larvae were collected in the study area. SGCN

Labridae (wrasses)

Tautogalabrus adspersus (cunner)-Native, Marine. This is generally a common species in shallow, rocky, high salinity habitats. SGCN

    Tautoga onitis (tautog)-Native, Marine. In general, abundant in high salinity rocky habitat. SGCN

Centrarchidae (sunfishes)

Lepomis cyanellus (green sunfish)-Introduced, Freshwater. Moderate salinity tolerance.

Lepomis gibbosus (pumpkinseed)-Native, Freshwater. Moderate salinity tolerance.

Lepomis gulosus (warmouth)-Introduced, Freshwater. Unverified report from the Meadowlands (AECOM 2018).

Lepomis macrochirus (bluegill)-Introduced, Freshwater. Moderate salinity tolerance.

Lepomis hybrid-Unverified report in AECOM (2018).

Micropterus salmoides (largemouth bass)-Introduced, Freshwater. Moderate salinity tolerance.

Pomoxis nigromaculatus (black crappie)-Introduced, Freshwater. Moderate salinity tolerance.

Percidae (perches)

Perca flavescens (yellow perch)-Native, Freshwater. Moderate salinity tolerance.

Pomatomidae (bluefishes)

Pomatomus saltatrix (bluefish)-Native, Marine. Young are using the Meadowlands as a nursery. SGCN

Triglidae (searobins)

Prionotus carolinus (northern searobin)-Native, Marine. This species is less common than the striped searobin. Only one specimen was collected in the Meadowlands. SGCN

Prionotus evolans (striped searobin)-Native, Marine. Juveniles tolerate moderate salinity and migrate into brackish areas in their first year. SGCN

Pholidae (gunnels)

Pholis gunnellus (rock gunnel)-Native, Marine. This benthic species often burrows under rocks and can use intertidal habitat.

Gasterosteidae (sticklebacks)

Gasterosteus aculeatus (threespine stickleback)-Native, Euryhaline. This species is most commonly found among submersed vegetation.

Pungitius pungitius (ninespine stickleback)-Native, Euryhaline. This species is uncommon and prefers higher salinities than threespine stickleback.

Cottidae (sculpins)

Myoxocephalus aenaeus (grubby)-Native, Marine. This is a common inshore sculpin.

Moronidae (temperate basses)

Morone americana (white perch)-Native, Euryhaline. Juveniles and adults were documented from the Meadowlands and they are probably reproducing there.

Morone saxatilis (striped bass)-Native, Anadromous. There is no spawning of this species in the Meadowlands, but juveniles are using the area as a nursery.

Sciaenidae (drums or croakers)

Cynoscion regalis (weakfish)-Native, Marine. Young often seek moderate salinities in their first summer. SGCN

Leiostomus xanthurus (spot)-Native, Marine. This species can tolerate low salinity water.

Micropogonius undulatus (Atlantic croaker)-Native, Marine. This species appears in estuaries in years of high abundance.

Pogonias cromis (black drum)-Native, Marine. Anglers reported catching this species in winter 2014–2015 (Carola 2015). SGCN

Sparidae (scups)

Stenotomus chrysops (scup)-Native, Marine. Probably an occasional visitor in areas with high salinity. SGCN

Tetraodontidae (puffers)

Sphoeroides maculatus (northern puffer)-Native, Marine. A few larvae were identified from the study area, probably present as strays. SGCN

NOTES

1. Associate Director, Hudsonia

Chapter 10

Invertebrates

Although most people only notice invertebrates that are impressively beautiful (large butterflies and moths, fireflies, or colorful beetles) or a nuisance (biting flies, wasps, cockroaches, carpenter ants, garden and farm pests), invertebrates have an enormous role in nature. Nonetheless, few surveys of invertebrates are conducted for land use planning or environmental monitoring. The larger *benthic* (bottom-dwelling) aquatic invertebrates, and the *nektonic* (free-swimming) estuarine crustaceans are exceptions. Many taxa, because of their small size, large numbers of species, and similarity of appearances, are poorly studied. The paucity of training and funding for taxonomy exacerbates this situation.

The invertebrates are a vast and varied group of organisms, and urban areas support important invertebrate diversity (Jones and Leather 2013). There is an urgent concern about declines of insect pollinators and of insects in general. We highlight a few taxa here; our selection is based on existing information about some invertebrates (e.g., the estuarine nekton and the butterflies) as well as on our knowledge of a few forms found in nearby rural areas that appear to be absent from the Meadowlands. Certain taxa are as informative by their absence as others are by their presence.

Some invertebrates are aquatic or estuarine; these kinds live in the water part or all of the time. Aquatic and estuarine invertebrates include mosquitoes, chironomid midges, and dragonflies that have aquatic immatures (larvae), although the larvae of some species live in seeps, wheel ruts, or water-filled treeholes. Other taxa include the polychaetes (bristle worms), mollusks, many amphipods, crayfish, and crabs that spend all or nearly all their lives in the water. Clam shrimps are also aquatic invertebrates because their active swimming and feeding stage occurs in the water, although there is a dormant *resting egg* stage that survives in the sediment of dried-up pools. Terrestrial invertebrates, in contrast, do not have aquatic life stages, and include butterflies, most moths, many beetles, and grasshoppers, bees, ants, centipedes, most spiders, and terrestrial earthworms. Invertebrates range in size from microscopic mites to the blue crab with a leg span larger than a person's hand. *Macroinvertebrates*, addressed here, are those invertebrate animals that are large enough to be readily visible to the naked eye.

Although many invertebrates are small or concealed, some can be abundant (large populations or high density per unit area). Invertebrates, in many cases, are also functionally important—their activities drive or influence ecological processes such as decomposition of dead plant matter, grazing living plant matter, formation and conditioning of soils, predation on smaller animals, food supply for larger animals including most birds, amphibians, and fishes, or transmission of diseases of plants and animals. Mosquitoes and other biting flies are nuisances and disease vectors to humans, whereas the blue crab and American oyster are important sources of high quality human food.

As with other taxa, urban environments filter the invertebrate fauna; some species can disperse across urban landscapes and find favorable habitats, whereas others cannot. Urban gardens, parks, and unbuilt lots can support many species. Even brownfields can provide temporary habitat that supports invertebrate diversity between industrial abandonment and redevelopment (e.g., Kattwinkel et al. 2009). The Meadowlands host a rich urban assemblage of invertebrates with interesting patterns of species abundance and habitat affinity.

CLAM SHRIMPS (CRUSTACEA: SPINICAUDATA)

The clam shrimps are an unusual group of crustaceans with bivalve-like shells enclosing their arthropod appendages. Superficially, especially in their environment, clam shrimps look like fingernail clams (Mollusca: Sphaeriidae) but swim and therefore move much faster than clams. Most known extant clam shrimps live in temporary waters although a few species live in lakes in other world regions. Triassic fossil clam shrimp in the genus *Cyzicus* have been found in Granton Quarry at the edge of the Meadowlands. Clam shrimps belong to three crustacean orders, but only one, the Spinicaudata, has been documented in the Meadowlands.

Mattox's clam shrimp *Cyzicus gynecius* (= *Caenestheriella gynecia*; Figure 10.1) was described as a new species in the 1930s and it is now known from perhaps as few as 15 extant localities in several states. As far as has been documented, Mattox's clam shrimp has only been found in temporary pools on dirt roads and trails. We found this species for the first time in New Jersey in 2001 on a 1 km stretch of the Paterson Lateral Gas Pipeline service road where it crosses the east end of the Richard P. Kane Natural Area in South Hackensack and Carlstadt (Schmidt and Kiviat 2007). There were about 40 pools in the habitat complex on the road. ORV and maintenance vehicle

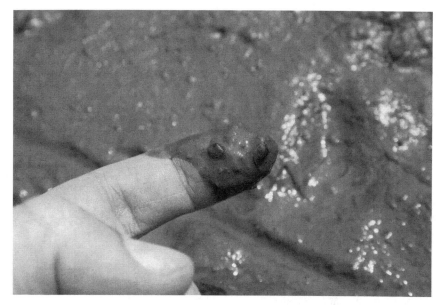

Figure 10.1 Mattox's clam shrimp (*Cyzicus gynecius*), a globally-rare inhabitant of a threatened habitat, temporary pools. **Photo by Erik Kiviat.**

traffic apparently created and maintained the pools (Figure 10.2). The pools were about 2–5 m across and 10–15 centimeters deep with a shallow layer of fine mineral sediment constituting the pool bottoms. These pools contained water most of the time although we saw them dry or nearly so on a few occasions during droughts. In July 2008, mean water temperature of several pools was 28.9 C, mean dissolved oxygen was 6.8 mg/L (87% saturation), pH was 6.9, and salinity was 0.3 ppt (Orridge et al. 2009). We have seen ostracods (Ostracoda), damselfly larvae, egg-laying dragonflies, snapping turtles, and tracks of raccoon and birds in the pools, and muskrats taking plant cuttings into the pools to eat. Few vascular plants grew in the pools.

In 2010, the entire clam shrimp pool complex was destroyed when the road was rebuilt during construction of the Kane Natural Area Wetland Mitigation Bank. Although we looked at many similar rain pools on ORV trails, dirt roads, and clay flats elsewhere in the Meadowlands, even just across the Turnpike from the Kane area, we found no other clam shrimp until 2013. We then found a population of Mattox's clam shrimp in similar rain pools on the Jersey City Aqueduct service road in Lyndhurst, 6 km southwest of the Kane site. Two years later and on subsequent visits we failed to detect clam shrimp, and the road was resurfaced in 2019 resulting in destruction of much of the rain pool habitat.

Figure 10.2 Off-road vehicles and Mattox's clam shrimp habitat on gas pipeline service road. Photo by Erik Kiviat.

During a 2019 search for Mattox's clam shrimp in Lyndhurst, we found a clam shrimp of the genus *Eulimnadia*, in a similar rain pool habitat on a nearby dirt road. (*Ovigerous* or egg-bearing females are necessary for species identification in *Eulimnadia,* and we have yet to collect them.) The presence of these two taxa underlines the biodiversity importance of rain pools on roads and the general potential for clam shrimp conservation in the Meadowlands. This habitat, rain-fed intermittent pools on roads and trails, has been overlooked by northeastern ecologists, and we expect other rare or unusual species of invertebrates or algae to occur there. There is worldwide concern about conservation of clam shrimps, and they are at risk of local extirpation or global extinction because of loss and degradation of their temporary aquatic habitats, but there has been little conservation action (Gołdyn et al. 2007, Boven et al. 2008, Rogers 2021). Documentation of the distribution and habitat affinities of these unusual organisms is the first step toward their conservation.

AMPHIPODS (AMPHIPODA) AND
OPOSSUM SHRIMP (MYSIDA)

The opossum shrimp (*Neomysis americana*) is a small (12 mm) decapod crustacean that lives in open estuarine and marine waters; it is common on the East Coast (Gosner 1978). This species has a wide salinity tolerance, favors deeper waters, stays near the bottom during the day and migrates up in the water column at night (PSE&G 1998). Opossum shrimps were present in small numbers in Ponar grab (bottom sediment) samples in the 2002 Meadowlands benthic invertebrate study (Bragin et al. 2009).

A scud (*Gammarus daiberi*) was one of the four most abundant macroinvertebrates in the Bragin et al. (2009) study and was considered pollution-tolerant. Scuds are important fish food organisms. A corophiid amphipod and an *Unciola* amphipod were recorded in the diet of Meadowlands fish (Neuman et al. 2004).

CRABS, CRAYFISH, AND GRASS SHRIMP (DECAPODA)

Crabs and crayfish are decapod crustaceans whose adult stages are larger than the shrimps and scuds just discussed. North American crabs are overwhelmingly marine or estuarine. Crabs have *planktonic* (tiny, drifting) larvae that can be carried long distances by water currents.

Five native crabs occur in the Meadowlands. The blue crab (*Callinectes sapidus*) is the largest species and was an important commercial and subsistence fishery species until recent decades. Blue crabs accumulate metals such as cadmium and are now banned from fishing in the Meadowlands. Blue crabs are very mobile and range from strongly brackish water to tidal fresh water in the Hudson River; they would be expected anywhere in the tidal habitats of the Meadowlands.

Two fiddler crabs, the mud fiddler (*Uca pugnax*) and the red-jointed fiddler (*Uca minax*), occur in the tidal marshes. The mud fiddler lives in moderately brackish water, whereas the red-jointed fiddler lives in slightly brackish water. Fiddler crabs burrow among the underground parts of emergent marsh plants and play an important role in allowing water and air to move through the marsh soil.

The white-fingered mud crab (*Rhithropanopeus harrisii*) is a brackish water species associated with hard surfaces such as pilings and oyster reefs (Gosner 1978). It appeared in samples of fish and benthos (Brett Bragin, pers. comm.). The marsh crab (*Sesarma reticulatum*) was found once on the high salt marsh at Riverbend (*ibid.*); its abundance and distribution in the Meadowlands are unclear. The Meadowlands are within the geographic ranges of several other native crabs (Weis 2012) as yet not found there.

The U.S. is a global hotspot for crayfish diversity (Stein et al. 2000). Most of this diversity is outside the northeastern states, and only six species of crayfish are known from New Jersey (Francois 1959, Fetzner 2019). A single individual of the northern crayfish (*Orconectes virilis*) has been found in the Meadowlands, by Brett Bragin (pers. comm.). It is a stream species (Hobbs and Hall 1974) with a large geographic range in the eastern U.S. The northern crayfish may have been introduced to New Jersey (Anonymous 2010).

Crayfishes generally are intolerant of salinity and water pollution (Hobbs and Hall 1974), including many of the problems affecting waters of the Meadowlands, thus the study area may lack suitable habitat. It is likely that individuals occasionally disperse or wash downstream from the nontidal Hackensack River above the Oradell Dam or perhaps a tributary below the dam. There has been no sampling specifically for crayfish in the Meadowlands; therefore, populations could exist unnoticed in fresher waterbodies (e.g., Mehrhof Pond, Willow Lake, Wolf Creek, Coles Brook, or the spring system west of Harrier Meadow). In the freshwater tidal Hudson River, crayfish (especially the spiny-cheeked crayfish [*Orconectes limosus*]) are uncommon to moderately common in rocky tidal stream mouths and water-chestnut (*Trapa natans*) beds. There is very little water-chestnut in the Meadowlands, although dense beds of floating marsh pennywort (*Hydrocotyle ranunculoides*) bear some ecological similarity to water-chestnut beds. And rocky tidal stream mouths are few.

The dagger-blade grass shrimp (*Palaemonetes pugio*) is an important estuarine invertebrate in the Meadowlands. When both were available, the dagger-blade grass shrimp did not show a preference for smooth cordgrass (*Spartina alterniflora*) or common reed (*Phragmites australis* ssp. *australis*) habitat and detrital food in the Meadowlands (Weis 2000, Weis and Weis 2000, Weis et al. 2002). This species was found in fish diets in the Meadowlands (Neuman et al. 2004).

MILLIPEDES AND CENTIPEDES (MYRIAPODA)

These terrestrial invertebrates are fairly common beneath cover objects (logs, bark slabs, rocks, refuse) and in plant litter in the Meadowlands, but no taxonomic survey has been conducted. A recently published taxonomic catalog of North American centipedes (Mercurio 2010) will facilitate such study. We found the house centipede (*Scutigera coleoptrata*) beneath cover objects outdoors on at least five occasions in the months of April, June, and July, 2008–2018. In the Hudson Valley this species seems to occur only within buildings.

House centipedes do not survive the winter outdoors in Pennsylvania (Jacobs 2013). Possibly the warm, humid microclimate in much of the Meadowlands is more favorable, or perhaps there is less predation pressure than in rural environments.

In Europe, certain millipedes occur in urban areas north of their typical geographic ranges (Vilisics et al. 2012). In Olomouc (Czech Republic), most millipede species were positively associated with less-artificial habitats that featured greater plant litter and canopy cover (Riedel et al. 2009). In Budapest and vicinity, millipedes were diverse, due to diversity of anthropogenic and natural habitats and evidently importation of a few species with horticultural materials (Korsós et al. 2002).

Notable by their apparent absence from the Meadowlands are the xysto-desmid millipedes. The Xystodesmidae are large (to about 6 x 1 cm), broad, *depressed* (flattened top-to-bottom) millipedes, and many species have bright yellow, orange, or red markings on the dorsum. Several species occur in the Hudson Valley, mostly in calcareous uplands. Xystodesmids may indicate non-urban conditions. We do not know what prevents their occurrence in the Meadowlands, although many xystodesmid species have restricted geo-graphic ranges and may be ecologically specialized. Xystodesmidae were not found in any of the three studies of European cities cited above.

SPRINGTAILS (COLLEMBOLA)

Springtails are small terrestrial and aquatic invertebrates (often 1–2 mm long) that are variously associated with the soil, leaf litter, tree bark, ponds, wet-lands, and flowers. Springtails were formerly considered a primitive order of insects, the Collembola. They are now considered a separate taxon (a class or subclass) related to the insects.

Springtails, second only to the mites, are typically the most abundant inver-tebrates in surface soils and litter in many habitats of the northeastern U.S. A study of invertebrates in the then-freshwater, nontidal marsh of the Kane Natural Area found these two taxa most abundant in the litter (Grossmueller 2001), as did a study of invertebrates in common reed and cattail (*Typha*) litter in a freshwater tidal marsh of the Hudson River (Kiviat and Talmage 2006).

HARVESTMEN (OPILIONES)

Harvestmen are arachnids, related to mites, but are much larger; they have what appears to be a single body segment with long, articulated legs. In our experience, harvestmen are very scarce in the Meadowlands and we do not have species identifications. A few harvestman species of the British Isles have an affinity for urban habitats (Sankey 1988). Species found in urban Sofia, Bulgaria, were those that tolerated warm, dry habitats (Mitov and Stoyanov 2004). In urban greenspaces of Toledo, Ohio, harvestman abundance and species richness were positively correlated with woody plant richness and negatively correlated with forb richness (Philpott et al. 2014). Concurrent with establishment of nonnative harvestmen, some native species declined and others increased in Danish cities (Toft 2018). Perhaps most relevant to the scarcity of harvestmen in the Meadowlands is the finding of Zolotarev and Nesterkov (2015) that harvestmen declined with proximity to a copper smelter, presumably due to deposition of heavy metals in the soil and litter habitat.

TICKS AND MITES (ACARI)

Hard ticks belong to the family Ixodidae and include most or all ticks of medical importance in the U.S. Our field experience indicates that the hard ticks that bite humans are rare in the Meadowlands. In two decades of field work, we have found few ticks on our bodies or clothes. These were black-legged ticks (deer ticks, *Ixodes scapularis*) and wood ticks (American dog ticks; *Dermacentor variabilis*), from tree- or shrub-dominated areas. The rabbit tick (*Haemaphysalis leporispalustris*) has been collected from eastern cottontail (*Sylvilagus floridanus*) at Teterboro (Occi et al. 2019). We have not seen the Lone Star tick (*Amblyomma mericanum*) in the Meadowlands; this is the third species of medical importance that occurs in northern New Jersey, after the black-legged tick and wood tick. Seven other ixodids have been collected in New Jersey (*ibid.*) and some of those undoubtedly occur, or will occur, in the Meadowlands. The nonnative long-horned tick (*Haemaphysalis longicornis*) has recently been documented in Bergen County (Beard et al. 2018; exact locality not reported), and should be watched for in the Meadowlands.

The scarcity of hard ticks in the Meadowlands may be related to the limited distribution of forests, low numbers of white-tailed deer, and the predominance of deeply-flooding soils. Inasmuch as the white-tailed deer population is increasing, we anticipate expansion of the black-legged tick population as well. The correlation of black-legged tick populations with deer populations was emphasized by Telford (1993). The white-footed mouse, which is the most important host for black-legged tick larvae and is also important in the tick life cycle, is likely widespread in the Meadowlands.

We have found no information on Acari other than the hard ticks in the Meadowlands, with one exception. Unidentified mites were the most abundant invertebrates in litter samples from marshes in the Kane Natural Area

(Grossmueller 2001), and in a freshwater tidal marsh of the Hudson River (Kiviat and Talmage 2006). This is expected, given the abundance of mites in litter and soils in many ecosystems.

SPIDERS (ORDER ARANEAE)

The spiders are a diverse and important group of predaceous invertebrates (certain spiders also eat pollen, nectar, and other plant materials [Nyffeler et al. 2016]). Almost all spiders are exclusively terrestrial; a few species, including the fishing spiders (*Dolomedes*) and the long-jawed orb weavers (*Tetragnatha*), use water edges and water surfaces for foraging and other activities. There seem to be many species of spiders in the Meadowlands, but there has been no taxonomic or ecological survey.

Some spiders are known to have narrow affinities for microhabitats, such as the bark of particular tree species (Kefyn Catley, Western Carolina University, pers. comm.). This raises the possibility that, like certain bees and birds, particular spiders might be associated with common reed leaves, tassels, or stubble in the Meadowlands, or other common Meadowlands microhabitats. In the Meadowlands and the Hudson Valley, a large funnel-web spider (*Agelenopsis pennsylvanica*) builds its web among the bases of robust knotweed (*Polygonum cuspidatum s.l.*) stems and uses the hollow top of a broken dead stem as a shelter (this spider is presumably not restricted to knotweed stands).

The black widow (*Latrodectus* sp.), a spider potentially dangerous to humans, occurs on the New Jersey Palisades and in the Meadowlands (Sullivan 1998b, 25). It should be looked for on the dry rocky portions of Laurel Hill and beneath boards or other debris on dry wetland fill or landfill cover. We have not seen this spider despite turning many cover objects.

APHIDS AND SCALE INSECTS (HEMIPTERA, SUBORDER STERNORRHYNCHA)

Aphids, scale insects, and related insects (all terrestrial) constitute part of the insect order Homoptera. Aphids and scales are individually small and often inconspicuous, but they are diverse and widely distributed. Many species have distinctive host plant associations. Aphids and scales feed on plant liquids obtained by piercing and sucking mouthparts. They are capable of doing considerable damage to their hosts, both by direct feeding effects and by transmission of plant diseases. Species feeding on agricultural crops or ornamentals may have substantial economic impact. Angalet et al. (1979)

identified 26 species of aphids during their study of lady beetles in the East Rutherford area (see below). Common reed is the host plant of the mealy plum aphid (*Hyalopterus pruni*), the most abundant aphid in the Meadowlands (Angalet et al. 1979). Reed is also host of the reed scale (*Chaetococcus phragmitis*), likely the most abundant scale insect in the Meadowlands. Reed scale and mealy plum aphid do not have an obvious impact on their reed host. Mealy plum aphid alternates seasonally between reed and woody plants in the genus *Prunus* (cherries, plums, etc.). No alternate host of this aphid in the Meadowlands has been determined. However, black cherry (*Prunus serotina*) is a common Meadowlands tree, and it and other *Prunus* species may serve as the winter hosts of the aphid.

The adult stage of the reed scale resembles a tiny brown pancake, reaching a length of about 4 mm, and lives in the tight space between the leaf sheath and the lower *culm internode* (aerial stem segment) of the reed shoot. The insect secretes a waxy material that leaves a footprint-like feature when the insect dies or is removed. Reed scale seems to be abundant just above flooding levels in wet reedbeds, and perhaps is absent from dry reedbeds (e.g., those on well-drained landfill cover). The immature reed scales (the *crawler* stage) are able to move from one reed shoot to another.

This abundant insect is capable of reaching a high biomass. In one year in Tivoli North Bay on the Hudson River reed scale constituted 74% of early spring and 60% of early summer insect biomass associated with common reed (Krause et al. 1997). In the same study, reed scale biomass sometimes exceeded 1g dry weight per m^2 of marsh. Although this may seem trivial, it is a lot of biomass for an insect. If reed scale existed at this level in a 100 ha reedbed, the total dry biomass of scale insects would equal one metric ton!

Reed scale is often preyed upon by songbirds. Black-capped chickadee is the most frequently observed predator of reed scale in the northeastern states. We have seen red-winged blackbirds feeding on this insect in the Meadowlands. Because songbirds require a fat-rich diet to survive cold winter weather, and insects provide this nutrition for many birds, it seems likely that the relatively large (for a scale insect), abundant, and easily found reed scale could provide this benefit to other bird species. Moreover, birds foraging inconspicuously in the lower levels of many reedbeds may benefit from reduced exposure to wind and predators due to the density of the reeds.

We have not seen birds eating mealy plum aphids, although they seem a likely food for marsh wren or wetland-frequenting warblers (sedge warblers in England fed on mealy plum aphids to fatten for the fall migration [Bibby et al. 1976]). However, lady beetles (and surely other insects and spiders) eat the aphids. Mealy plum aphids vary considerably in numbers from year to year. They seemed to be absent or scarce in 2019, for example. They reach peak adult population size in about July; in some years, live reed leaves may each support hundreds of aphids. Live and dead aphids fall to the ground where they presumably provide food to ground-dwelling insectivores.

Nothing is known of other Hemiptera, including true bugs (suborder Heteroptera) and cicadas and hoppers (suborder Auchenorrhyncha), in the Meadowlands.

DRAGONFLIES AND DAMSELFLIES (ODONATA)

Odonates are colorful, fast flying, predatory insects with aquatic larvae and complex adult social behavior. Odonates have attracted much scientific study, amateur observation, and folklore. The past several years have seen publication of excellent field guides to the odonates of the northeastern states (e.g., Nikula et al. 2003, Barlow et al. 2009). Now that non-specialists can identify many of these insects (with practice, and in some cases specimen collection), dragonflies and damselflies will be monitored and used as environmental indicators more. The Meadowlands have a moderately rich odonate fauna (Table 10.1) for an urban area but lack most of the many species that are specialized for development in flowing fresh waters. The Meadowlands, so far as known, support a moderate number of odonate species compared to the state as a whole, 30 of 182 species. Besides the conspicuous lack of unpolluted streams, the Meadowlands have few freshwater peatlands (the nontidal Upper Penhorn Marsh and formerly tidal Kearny Marsh West), woodland seeps, or other specialized habitats important for odonate diversity.

In 2006, Hudsonia tested a simple method of transect survey for adult odonates and butterflies (Kiviat 2007). We established five strip transects, each 1 km long and divided in ten 100 m segments. Transects were located near fresh or slightly brackish water in areas sheltered from the wind. We surveyed each transect once quantitatively at the end of June when most odonates are in the more easily seen and identified adult stage.

We identified 22 species on the transect counts. The five transects seemed homogeneous for species richness and total numbers of individuals of dragonflies, but not for richness or numbers of damselflies. Damselflies were positively correlated with measures of woody plant abundance and diversity, and dragonflies were negatively correlated with the same measures. Yet the damselflies we counted tended to be in openings rather than in dense woody vegetation. Possibly woody thickets provide damselflies with a degree of refuge from wind and predators. Damselflies are weaker fliers and generally do not venture far from suitable freshwater larval habitats, thus may be less evenly distributed across the Meadowlands landscape.

Although Needham's skimmer dragonfly is considered Special Concern by the New Jersey Odonata Survey, it is rated as locally common in New

Table 10.1 Dragonflies and damselflies of the New Jersey Meadowlands.

Common name	Scientific name	Statewide status
Coenagrionidae		
Blue-fronted dancer	*Argia apicalis*	Common
Powdered dancer	*Argia moesta*	Common
Azure bluet	*Enallagma aspersum*	Common
Familiar bluet	*Enallagma civile*	Common
Big bluet*	*Enallagma durum*	Locally common
Orange bluet	*Enallagma signatum*	Common
Citrine forktail	*Ischnura hastata*	Common
Fragile forktail	*Ischnura posita*	Common
Rambur's forktail	*Ischnura ramburii*	Locally common
Eastern forktail	*Ischnura verticalis*	Common
Lestidae		
Slender spreadwing	*Lestes rectangularis*	Common
Aeshnidae		
Common green darner	*Anax junius*	Common
Swamp darner[a]	*Epiaeschna heros*	Common
Libellulidae		
Halloween pennant	*Celithemis eponina*	Common
Banded pennant	*Celithemis fasciata*	Locally common
Eastern pondhawk	*Erythemis simplicicollis*	Common
Seaside dragonlet	*Erythrodiplax berenice*	Locally abundant
Dot-tailed whiteface	*Leucorrhinia intacta*	Common
Widow skimmer	*Libellula luctuosa*	Common
Needham's skimmer*	*Libellula needhami*	Locally common
Twelve-spotted skimmer	*Libellula pulchella*	Common
Four-spotted skimmer*	*Libellula quadrimaculata*	Locally common
Painted skimmer	*Libellula semifasciata*	Common
Great blue skimmer	*Libellula vibrans*	Common
Blue dasher	*Pachydiplax longipennis*	Common
Spot-winged glider	*Pantala hymenaea*	Common
Eastern amberwing	*Perithemis tenera*	Common
Common whitetail	*Plathemis lydia*	Common
Meadowhawk	*Sympetrum*[b]	
Autumn meadowhawk[a]	*Sympetrum vicinum*	Common
Black saddlebags	*Tramea lacerata*	Common

Note: Scientific and common names, and statewide status, follow Barlow et al. (2009). Species sources: Kiviat 2007; Erik Kiviat, observations; meadowblog.net.

* Species of Concern (NJODES 2020).

[a] Photo-documented in https://meadowblog.net/2013/08/teaser-answered-28/.

[b] We do not have species-level identifications of most meadowhawks observed.

Jersey (Barlow et al. 2009) and seems widespread and fairly common in the Meadowlands. Just a short distance farther north and east in New York State, Needham's skimmer is rare and a Species of Greatest Conservation Need (SGCN). Other species of Special Concern are the four-spotted skimmer and big bluet. The only Meadowlands odonate (Table 10.1) that is listed as a Species of Greatest Conservation Need in New Jersey is the seaside dragonlet. We found seaside dragonlet populations near Berry's Creek in Rutherford and Lyndhurst.

On the fifty transect segments in our study, the occurrence of Needham's skimmer was positively associated with the presence of common reed (*Phragmites australis*), while blue dasher, another dragonfly, was negatively associated with reed. Needham's skimmer seems to preferentially perch on reed culms but we do not know what if any benefit this behavior has. The blue dasher is a smaller insect; possibly it avoids reeds because of predation by the larger Needham's skimmer. Vegetation structure is generally important to adult odonates (Clausnitzer et al. 2009).

Several species of the Meadowlands odonate fauna have been reported tolerant of brackish water: familiar bluet, big bluet, Rambur's forktail, eastern amberwing, Needham's skimmer, and especially seaside dragonlet (Nikula et al. 2003, Wilson 2008, Barlow et al. 2009). Certain species have been described as tolerant of polluted waters or degraded habitats: orange bluet, fragile forktail, blue dasher, eastern pondhawk, common whitetail, twelve-spotted skimmer, and dot-tailed whiteface (Nikula et al. 2003, Barlow et al. 2009). Probably due in part to these tolerances, the familiar bluet, fragile forktail, blue dasher, and Needham's skimmer are among the most common Meadowlands odonates.

The pond habitats of the Meadowlands, including stormwater ponds, claypit lakes, temporary rain pools, and tidal impoundments, all appear to be constructed ponds (i.e., human-created). (Possible exceptions are the small, apparently springfed pool southwest of Harrier Meadow in North Arlington, and the large, temporary, wet meadow pool west of the Losen Slote Creek Park parking lot in Little Ferry. The latter pool has a population of citrine forktail, a damselfly that seems to be rare in the Meadowlands.) Constructed ponds, if they have good water quality and especially if there is diverse submergent vegetation and fish are absent, can be good larval habitats for odonates and may even support rare species. For example, in a comparison of ponds on nature reserves and parks versus golf course ponds in Stockholm, Sweden, the IUCN Red-listed white-faced darter dragonfly was found only in the golf course ponds (Colding et al. 2009). Our single observation of a four-spotted skimmer at the Kane Natural Area on 12 June 2008 may represent an individual from Mehrhof Pond, as this is a species that breeds in lakes, swamps, and bogs (Barlow et al. 2009). The only individual of powdered dancer seen in the Meadowlands was near Mehrhof Pond. The powdered dancer breeds in streams and in lakes with wave-washed rocky shores (Lam 2004); the riprapped north shore of Mehrhof Pond presumably was suitable larval habitat. There is a single record of a swamp darner in August 2013 at

DeKorte Park where there doesn't seem to be appropriate swamp habitat; this individual may have been a migrant (see Barlow et al. 2009).

KATYDIDS, CRICKETS, AND GRASSHOPPERS (ORTHOPTERA)

These large, mostly herbivorous, sound-producing insects could be moderately diverse in the Meadowlands but have not been surveyed. Some species seem urban-tolerant. We noted a few species that are distinctive in appearance, calls, or diet.

The handsome meadow katydid (*Orchelimum pulchellum*) occurred in the reedbeds of Kingsland Impoundment, both at the edges of the mowed areas near the NJSEA buildings, and in the interior reedbeds along the boardwalk. Here this species feeds on reed leaves and is numerous. Its calls alternate long buzzes with a series of ticks. This is a common and widespread species in the eastern U.S. (Capinera et al. 2004).

We found the restless bush cricket (*Hapithus agitator*) on knotweed leaves at Van Buskirk Island in September 2008. This orthopteran has a distinctive dorsolateral yellow or cream stripe on either side. It is evidently near the northeastern extent of its geographic range in northern New Jersey (Capinera et al. 2004). It would be interesting to discover if this species, and many other insects found on knotweed, are eating it or using it for shelter, basking, or nesting.

The common true katydid (*Pterophylla camellifolia*) is familiar to many who live in rural areas as the insect whose call is transliterated as *Katy did, No she didn't*. This song originates from the forest canopy at night. We heard it only once, at Overpeck Creek in 2009. Although the species has large wings it cannot fly more than to flutter to the ground if disturbed from a tree perch (Capinera et al. 2004, Elliott and Hershberger 2007). Lack of flight and the need for forest presumably limit the distribution of common true katydid in the Meadowlands and other large urban areas.

MANTIDS (MANTODEA)

Three species of mantids occur in New Jersey: European mantid (*Mantis religiosa*), Chinese mantid (*Tenodera sinensis*), and Carolina mantid (*Stagmomantis carolina*). The European species is specifically referred to as the praying (or preying) mantis, but the other two species are also called praying mantis. The European and Chinese mantids were introduced to the U.S. in the 1890s (Milne and Milne 1980). These two insects, although nonnative species, are considered beneficial rather than harmful. The Carolina mantid is native to the U.S.

All three mantids are large insects that ambush other insects, spiders, and occasionally hummingbirds. The large mantid forelegs, held in a semi-folded position in front of the body, are used to catch and hold the prey. Mantids in the Meadowlands frequent flowering herbs such as goldenrods (*Solidago*), and other plants. The female mantid deposits her eggs on an herb stem or woody twig in a foam that hardens into a tan or brown mass. This structure, called the *ootheca*, apparently protects the overwintering eggs from weather and predators. Our three mantids create distinctive oothecae that can be used for species identification (Breland and Dobson 1947). We have often seen oothecae on common reed culms as well as other plants.

We have not specifically surveyed for mantids, but we have identified oothecae of the Chinese mantid at several Meadowlands locations. We also found an ootheca of the European mantid in 2001 in Carlstadt, and one of the Carolina mantid in the edge of the Meadowlands in North Arlington in 2006. This southern species may be favored by a warm microclimate in the Meadowlands.

LADY BEETLES (COLEOPTERA: COCCINELLIDAE)

The beetles (Coleoptera) are the second-most species-rich group of insects (after the Hymenoptera). Beetles range greatly in size, habitat affinities, diet, and other traits. A single group of beetles, the lady beetles (Coccinellidae, also called ladybugs or ladybirds), has received attention in the Meadowlands. We have often noticed adult lady beetles on mugwort (*Artemisia vulgaris*), common reed, and other plants in the Meadowlands where the beetles are presumably feeding on aphids and possibly pollen as well. Opportunistically collected specimens were identified as two nonnative species, the multi-colored lady beetle (*Harmonia axyridis*) and the seven-spotted lady beetle (*Coccinella septempunctata*).

The first confirmed location in the United States where the seven-spotted lady beetle had *naturalized* (established and reproduced) was in the Meadowlands, in East Rutherford, adjacent to a landfill (Angalet et al. 1979). This lady beetle had been introduced elsewhere in the U.S. for biological control of aphids in 1958 and 1973; it is not known if the beetle was intentionally introduced in the Meadowlands or dispersed there on its own. During their study, Angalet et al. identified 26 species of aphid and 17 species of lady beetle in the East Rutherford area. The abundance of common reed and mealy plum aphid in the Meadowlands (Kiviat, pers. obs.) presumably helped make this region suitable for the seven-spotted lady beetle. The commercial harvest of lady beetles from the Meadowlands for sale to gardeners and farmers for biological control of aphids, alluded to by Sullivan (1998b), may pertain to this species. Aphids and lady beetles were prominent in samples of inverte-brates from the marsh surface at the Empire Tract (now called Kane Natural Area) reported in USACE (2000). Many native lady beetle species have declined in the northeastern states during the past few decades, and it is not known if the Meadowlands have retained the diversity of species reported by Angalet et al. We collected three specimens of the seven-spotted lady beetle at different Meadowlands localities in 2004 and 2006 indicating persistence of this species.

BUTTERFLIES (LEPIDOPTERA, IN PART)

Butterflies have fascinated naturalists and the general public for ages. In recent years butterfly watching has been a rapidly growing avocation for many naturalists, gardeners, and birdwatchers. Butterflies are mostly approachable and observable with the naked eye, close-focusing binoculars, or a camera, and this interest can appeal to children and adults who might shy away from birdwatching or botanizing. Although the ecology of many northeastern butterflies has been well studied, the complete life histories of many species remain poorly known (e.g., the larval host plants have not been identified for some of the skippers). And more advanced students of butterfly science can discover much new knowledge by qualitative or quantitative sur-veying of regional faunas and observations of behavior. The North American Butterfly Association (NABA) administers annual Fourth of July Butterfly Counts in selected 24 km diameter circles. The Greenbrook Sanctuary Count includes the northern portion of the Greater Meadowlands south to Mehrhof Pond and Skeetkill, and the Staten Island Count includes the southern end of Newark Bay, but it is not clear how much these portions of our study region are covered during the counts.

Table 10.2 lists 66 species of butterflies known from recent or historic observations in the Meadowlands. The butterfly fauna of New Jersey num-bers 149 species (Gochfeld and Burger 1997). In Philadelphia, Pennsylvania, the John Heinz National Wildlife Refuge (also called Tinicum Marsh), an urban wetland complex ca. 125 km southwest of the Meadowlands, had 73 butterfly species (Shapiro 1970). On Staten Island, a suburban borough of New York City with fairly extensive wetlands and forests, just across the Kill van Kull from Newark Bay, 106 species of butterflies were reported (Shapiro and Shapiro 1973). Because of the different periods of observation, these numbers should be taken as only approximate comparisons.

Table 10.2. Butterflies of the New Jersey Meadowlands.

Common name	Scientific name	Specialization	Larval host plants
	Danaidae		
Queen[a]	*Danaus gilippus*	Medium generalist	Milkweeds (*Asclepias*)
Monarch*	*Danaus plexippus*	Generalist	Milkweeds
	Hesperiidae		
Delaware skipper	*Anatrytone logan*	Generalist	Switchgrass (*Panicum virgatum*), other native grasses
Least skipper	*Ancyloxypha numitor*	Generalist	Grasses (Poaceae)
Sachem[b]	*Atalopedes campestris*	Generalist	Weedy grasses
Silver-spotted skipper	*Epargyreus clarus*	Generalist	Black locust (*Robinia pseudoacacia*), other legumes
Wild indigo duskywing	*Erynnis baptisiae*	Medium generalist	Crown vetch (*Securigera varia*)
Horace's duskywing[b]	*Erynnis horatius*	Generalist	Oaks (*Quercus*)
Dreamy duskywing[c]	*Erynnis icelus*	Generalist	Poplars (*Populus*), willows (*Salix*), black locust
Juvenal's duskywing[d]	*Erynnis juvenalis*	Generalist	Oaks (*Quercus*)
Dun skipper[b]	*Euphyes vestris*	Generalist	Sedges (*Carex*)
Fiery skipper[b]	*Hylephila phyleus*	Generalist	Weedy grasses
Common sootywing[b]	*Pholisora catullus*	Generalist	Pigweed (*Chenopodium album*), amaranths (*Amaranthus*), etc.
Broad-winged skipper	*Poanes viator*	Medium generalist	Common reed (*Phragmites*)
Zabulon skipper	*Poanes zabulon*	Generalist	Grasses
Long dash	*Polites mystic*	Medium generalist	Bluegrass (*Poa*); other grasses?
Peck's skipper[b]	*Polites peckii*	Generalist	Grasses
Little glassywing	*Pompeius verna*	Generalist	Purpletop (*Tridens flavus*); other grasses?
European skipper	*Thymelicus lineola*	Generalist	Common weedy grasses
Long-tailed skipper[a]	*Urbanus proteus*	Generalist	Leguminous vines
Northern broken-dash	*Wallengrenia egeremet*	Generalist	Large panic grasses

Common name	Scientific name	Specialization	Larval host plants
Lycaenidae			
Juniper hairstreak[c]	*Callophrys gryneus*	Medium generalist	Eastern red cedar (*Juniperus virginiana*)
Red-banded hairstreak[b]	*Calycopis cecrops*	Medium generalist	Sumacs (*Rhus*), oaks, others
Spring azure[d]	*Celastrina ladon s.l.*	not rated	Viburnums (*Viburnum*), flowering dogwood (*Cornus florida*)
Summer azure	*Celastrina neglecta*[e]	not rated	Spireas (*Spiraea*)
Eastern tailed-blue	*Cupido comyntas*	Generalist	Legumes (Fabaceae)
Bronze copper*	*Lycaena hyllus*	Medium specialist	Docks (*Rumex*), possibly smartweeds (*Polygonum s.l.*)
White-M hairstreak	*Parrhasius m-album*	Medium generalist	Oaks
Edwards' hairstreak[c]	*Satyrium edwardsii*	Specialist	Certain oaks
Striped hairstreak[c]	*Satyrium liparops*	Medium generalist	Heath family (Ericaceae), rose family (Rosaceae), etc.
Gray hairstreak[b]	*Strymon melinus*	Generalist	Mallows (Malvaceae), legumes, others
Nymphalidae			
Hackberry emperor[b]	*Asterocampa celtis*	Medium generalist	Hackberry (*Celtis occidentalis*)
Tawny emperor[b]	*Asterocampa clyton*	Medium generalist	Hackberry
Harris' checkerspot[c]	*Chlosyne harrisii*	Medium specialist	Flat-topped white aster (*Doellingeria umbellata*)
Silvery checkerspot[c]	*Chlosyne nycteis*	Medium generalist	Aster family (Asteraceae)
Variegated fritillary[a,b]	*Euptoieta claudia*	Generalist	Violets (*Viola*), etc.
Common buckeye[b]	*Junonia coenia*	Generalist	Plantains (*Plantago*), figworts (Scrophulariaceae), vervains (*Verbena*), etc.
Viceroy[b]	*Limenitis archippus*	Medium generalist	Willows, poplars, etc.

Red-spotted purple[b]	*Limenitis arthemis*	Medium generalist	Various woody plants
Mourning cloak	*Nymphalis antiopa*	Generalist	Willows, elms (*Ulmus*), poplars, birches (*Betula*), hackberry
Compton tortoiseshell[c]*	*Nymphalis vau-album*	Medium generalist	Birches, poplars, willows
American snout	*Libytheana carinenta*	Medium generalist	Hackberry
Pearl crescent	*Phyciodes tharos*	Generalist	Asters
Eastern comma[b]	*Polygonia comma*	Generalist	Hops (*Humulus*), other nettles (Urticaceae), elms
Question mark[b]	*Polygonia interrogationis*	Generalist	Elms, hackberry, nettles
Aphrodite[c]	*Speyeria aphrodite*	Medium generalist	Violets
Great spangled fritillary	*Speyeria cybele*	Medium generalist	Violets
Red admiral	*Vanessa atalanta*	Generalist	Nettle family (Urticaceae)
Painted lady[f]	*Vanessa cardui*	Generalist	Many herbs
American lady[b]	*Vanessa virginiensis*	Generalist	Everlastings (*Gnaphalium*, *Anaphalis*), other Asteraceae
Papilionidae			
Pipevine swallowtail[a,b]	*Battus philenor*	Medium specialist	Native *Aristolochia* spp.
Zebra swallowtail[c]	*Eurytides marcellus*	Medium specialist	Pawpaw (*Asimina triloba*)
Giant swallowtail[a,b]	*Papilio cresphontes*	Medium generalist	Prickly-ash (*Zanthoxylum*), wafer-ash (*Ptelea*)
Eastern tiger swallowtail	*Papilio glaucus glaucus*	Generalist	Tuliptree (*Liriodendron*), cherries (*Prunus*), ash (*Fraxinus*)
Palamedes swallowtail[c]	*Papilio palamedes*	Medium specialist	Laurel family (Lauraceae)
Black swallowtail	*Papilio polyxenes*	Generalist	Parsley family (Apiaceae)
Spicebush swallowtail	*Papilio troilus*	Generalist	Spicebush (*Lindera*), sassafras (*Sassafras*)
Pieridae			
Clouded sulphur	*Colias philodice*	Generalist	Legumes

Common name	Scientific name	Specialization	Larval host plants
Orange sulphur	*Colias eurytheme*	Generalist	Legumes
Cloudless sulphur[a,b]	*Phoebis sennae*	Generalist	Wild sennas (*Cassia*)
Cabbage white	*Pieris rapae*	Generalist	Mustard family (Brassicaceae)
Satyridae			
Common wood nymph	*Cercyonis pegala*	Generalist	Grasses
Little wood satyr	*Megisto cymela*	Generalist	Grasses

Note: Degree of specialization, which integrates larval host plants, habitat, and other factors, from Cech and Tudor (2005); larval hosts from Gochfield and Burger (1997), Cech and Tudor (2005). Species sources: Gochfield and Burger (1997), Kiviat (2007 and personal observations), and the Meadowlands Nature Blog (https://meadowblog.net).

* New Jersey Species of Greatest Conservation Need. Host plants are those species likely to be used in the Meadowlands. Note that standardized butterfly names use the spelling *sulphur* whereas the general technical usage in American English for the chemical element is *sulfur*.

a Single recent record of a stray (disperser) apparently outside the geographic range where reproduction occurs.

b Source of record meadowblog.net. Only records documented with a photograph or observed by an experienced butterfly-watcher are presented here. In addition, many of the bolded species found on the 2006 survey have also been documented on the meadowblog. Common sootywing was also photographed by Kiviat (pers. obs.).

c Known only from historic records cited in Gochfield and Burger (1997) from "Newark," "Elizabeth," "West Hoboken," "Rutherford/Carlstadt," "Fort Lee," or "Jersey City." In some cases these were strays whose typical larval hosts did not occur in the region.

d Unverified report.

e The genus *Celastrina* may contain several cryptic species in the eastern states, and the species-level taxonomy is a matter of debate (see Pratt et al. 1994).

f https://www.inaturalist.org/observations/32706268.

In 2006, Hudsonia counted 21 butterfly species in a survey on five 1-km transects in the Core Meadowlands (Kiviat 2007). Butterfly numbers and species richness were not correlated with the abundance of nectar plant species along the transects, nor with the woody plant abundance or plant species richness on the transects. These analyses suggest either that the butterfly assemblage was controlled by factors we did not assess (e.g., predators or contaminants), or that the area was not saturated (not at carrying capacity) with butterflies. Both explanations are likely in an urban environment. Of course, absence or scarcity of host plants (e.g., violets for fritillaries), does limit the Meadowlands butterfly fauna.

Waste areas (or *waste grounds*), habitats (often temporary) where vegetation has been removed or greatly disturbed and where mineral soil that is often sparsely vegetated predominates, commonly support many butterfly host plants and nectar plants. Some Meadowlands waste areas are sparsely vegetated and others have developed thickets of dense shrubs, small trees, and herbs. These habitats, which some people find aesthetically displeasing, are extensive on wetland fill, landfill cover, former building sites, and brownfields in the Meadowlands. Waste grounds can be important butterfly habitats with a few native butterfly species seeming to favor waste grounds of the U.S. East Coast (Cech and Tudor 2005). Spiders, aphids, beetles, flies, and true bugs also may be diverse in these habitats, and ground-nesting bees and wasps likely occur. Wetland management projects, insofar as they favor opportunistic nectar plants (e.g., saltmarsh-fleabane [*Pluchea odorata*]), may have some short-term or even long-term benefit to butterflies and other flower-visiting insects, yet this remains to be studied. However, the conversion of waste areas to ornamental landscaping, lawns, buildings, and even certain kinds of native plantings eliminates an important urban habitat.

Nineteen species of skippers (Hesperiidae) have been found in the Meadowlands in recent years, a modest number for this species-rich family. Many of the skippers not currently known in the Meadowlands have host plant ranges limited to one or a few native grasses or sedges of particular habitats. Some of these host plants (such as tussock sedge [*Carex stricta*] and tall cordgrass [*Spartina cynosuroides*]) are rare in the Meadowlands, accounting for the absence of specialist butterflies that might otherwise be present. As we discuss in the Seed Plants chapter, sedge diversity and abundance are modest in the Meadowlands.

We included 12 butterfly species in Table 10.2 based solely on historic records compiled by Gochfeld and Burger (1997) (see Footnote C in Table 10.2). Naturalists of a century ago were often less specific in recording localities; however, we believe these species occurred in or very close to the Greater Meadowlands. Additional species were reported on Staten Island by Shapiro and Shapiro (1973) and could have occurred in the Meadowlands.

Of the 12 species known only from historic records, only 1 was rated as a Generalist and 11 were rated as Medium Generalists, Medium Specialists, or Specialists by Cech and Tudor (2005). In contrast, of the species observed in recent years, none is rated as a Specialist and only one (a stray) as a Medium Specialist. This comparison suggests that, as the Meadowlands have become more urbanized during the past century, and as specific habitats and plants have declined, the butterfly species of conservation concern have also declined. (It should be noted that the Cech and Tudor [2005] rating system blends consideration of larval host plants with habitat, thus its interpretation is not straightforward.) Urban fringes can support some rare butterflies if the larval host plants and suitable nectar flowers are present (Haddad 2019).

Some of the historic species were probably strays (usually from farther south, such as the long-tailed skipper). More southern species whose ranges are advancing northward with climate warming may colonize the Meadowlands in the future if suitable habitats and host plants are available. Some of the historic species were butterflies that have undergone larger-scale declines, such as the Harris' checkerspot. In still other cases, human influences may have eliminated a habitat or host plant from the Meadowlands and likewise the butterfly that depended on it. Eastern red cedar is now rare in the Meadowlands, and there is probably not enough host biomass to support the historically recorded juniper hairstreak. There are no Meadowlands records of Hessel's hairstreak, a butterfly whose larva eats only Atlantic white cedar, although this butterfly occurs elsewhere in the state. Hessel's hairstreak likely occurred in the Meadowlands until the extirpation of its larval host plant a century ago.

MOTHS (LEPIDOPTERA, IN PART)

Moths are much more species-rich and less well known than the butterflies, and generally harder to identify to species. Many moths are host plant specialists or habitat specialists. We have found no survey of Meadowlands moths. The diversity of potential larval host plants and the broad range of moisture levels in different habitats could support a diverse urban moth fauna. For example, the substantial population of floating marsh pennywort, a common southern plant that is rare in New Jersey, could support moths or other insects associated with that species. We have observed abundant evidence of an insect mining the leaves of this plant; Walsh et al. (2013) reported weevils, flies, moths, and other insects feeding on floating marsh pennywort.

The giant silk moths (Saturniidae) include the largest North American moths and many of the most colorful, yet these insects are often unnoticed by those who do not know where and when to look. Among the largest silk moths are the Cecropia moth and Polyphemus moth. Their larvae eat leaves of a wide variety of plants, mostly trees and shrubs such as dogwoods and maples. They also feed on purple loosestrife leaves. In the Hudson Valley, Cecropia and Polyphemus selected loosestrife even when it was near typical native host plants (Barbour and Kiviat 1997). Cecropia tolerates low density urban environments (Tuskes et al. 1996). Nonetheless, we have only seen Cecropia once, in the cocoon stage, near Teterboro Airport. All stages of the saturniids should be watched for on purple loosestrife and woody plants.

Henry's marsh moth (Noctuidae: *Acronicta insularis [=Simyra henrici]*) is a generalist feeder. We observed an outbreak of larvae defoliating common reed in October 2003 where reed mingled with bluejoint grass (*Calamagrostis canadensis*) in a small part of the Kane Natural Area. We also identified Henry's marsh moth in Kearny Marsh West in October 2013 and Upper Penhorn Marsh in July 2016, on common reed, but without obvious defoliation.

A survey of hawk moths (sphinx moths, Sphingidae) found 6 species in rural northern New Jersey study sites but none in Harrier Meadow of the Meadowlands (Tartaglia 2013). A snowberry clearwing moth (*Hemaris diffinis*) at butterfly bush (*Buddleja*) in July 2018 and other clearwing sphinx observations have been documented in meadowblog.net, and we photographed a Nessus sphinx (*Amphion floridensis*) at butterfly bush in DeKorte Park in June 2006. The absence from Harrier Meadow could be due to the taxonomic and temporal pattern of nectar plant occurrence or to the microclimate; butterfly bush is especially favorable for certain moths and butterflies and the plantings around the buildings in DeKorte Park are apparently in a warm location. Interestingly, urban odors have been found to interfere with flower-finding by a hawk moth (Riffell et al. 2014), and Harrier Meadow lies between two garbage landfills.

PHANTOM CRANE FLY (DIPTERA: PTYCHOPTERIDAE)

One of the most unusual and beautiful North American insects, the phantom crane fly (*Bittacomorpha clavipes*), is notable by its apparent absence from the Meadowlands. The phantom crane fly occurs in many calcareous wetlands with an organic soil layer in northern New Jersey, the Hudson Valley, and northwestern Connecticut (Jason Tesauro, pers. comm.; E. Kiviat, pers. obs.). Phantom crane flies have striking large black-and-white banded legs held like the spokes of a tiny wheel as the insect flies or drifts through the air and alights on leaves of sedges or other plants. This insect is evidently intolerant of urbanization, salinity, or another factor of the Meadowlands, and may be a negative indicator of urbanization.

MOSQUITOES (DIPTERA: CULICIDAE)

The *biting flies* are several groups of true flies (Diptera) that contain blood-feeding species, principally the mosquitoes, horse flies and deer flies, biting midges, and black flies. Only the females bite. Mosquitoes are arguably the best-known Meadowlands insects although the mealy plum aphid and chironomid midges, for example, are probably more abundant overall. Mosquitoes have attracted the attention of travelers and residents in the Meadowlands for centuries. Of the 63 mosquito species known from New Jersey (Anonymous 2019), 22 species have been found in the Meadowlands; twice that number have been found in Bergen and Hudson counties overall (Headlee 1945).

Trapping is often used to sample and monitor the adult mosquito assemblage. The most commonly used traps are light traps and carbon dioxide traps. The former uses a light that attracts night-flying species, and the latter emits CO_2 which is an attractant for females seeking a blood meal. The mosquito species most abundant in light trap and carbon dioxide trap catches in the Bergen County municipalities in and close to the Meadowlands are the unbanded saltmarsh mosquito (*Culex salinarius*), the northern house mosquito (*Culex pipiens*), and the white-dotted mosquito (*Culex restuans*). These three species are difficult to distinguish when in worn condition in traps, thus they may be combined in catch data. The three accounted for 74% of the light and CO_2 trap catches in 2019 with the unbanded saltmarsh mosquito the most abundant (Matthew Bickerton, Bergen County Department of Health Services, pers. comm.). The inland floodwater mosquito (*Aedes vexans*) and the brown saltmarsh mosquito (*Aedes cantator*) are next in abundance at 7–8% each, followed by the tiger mosquito (*Aedes albopictus*) (2%) and eastern saltmarsh mosquito (*Aedes sollicitans*) (1%). Fifteen additional species were present in the traps at < 1% each. It should be noted that some mosquitoes are not readily caught in these trap types.

The northern house mosquito is abundant in urban areas, where it uses a variety of larval habitats, including water-filled containers such as tires and cans, as well as surface waters polluted by sewage or other organic materials (Crans 2016). The northern house mosquito is a vector to humans of Saint Louis encephalitis virus and West Nile virus. The white-dotted mosquito also breeds in a variety of natural as well as polluted or artificial container habitats and is thought to be primarily a bird feeder (Crans 2014). The unbanded saltmarsh mosquito uses diverse brackish and freshwater larval habitats, and breeds most abundantly in impounded, formerly tidal, marshes (Crans no date). It bites both birds and humans and may also be a *bridge vector* (bird to human) for West Nile virus.

The golden saltmarsh mosquito and the brown saltmarsh mosquito are widespread, often abundant, mosquitoes on the East Coast. Their larval habitat is saline pools above Mean High Water on the high salt marsh that are flooded only by the higher high tides and are somewhat isolated from fish activity. The saltmarsh mosquitoes are infamous for their adult abundance in large, synchronized emergences. Their ability to fly long distances inland can cause a high level of nuisance to humans who are not even near the breeding location. Prior to human disturbances, the saltmarsh mosquitoes would have been abundant in the Meadowlands. The eastern saltmarsh mosquito is still common in a few areas (Bickerton, pers. comm.). A combination of habitat alterations, including filling of the salt meadow (high salt marsh), drainage from ditching, the deterioration of ditches, replacement of salt meadow grasses by common reed, marsh restoration, exposure to contaminants, and insecticides, have apparently suppressed the saltmarsh mosquitoes. Ironically, large populations of the saltmarsh mosquitoes are an indicator of an intact, high salt marsh ecosystem, although those mosquitoes are not desirable from a human viewpoint.

The cattail mosquito (*Coquilletidia perturbans*) is unusual in that the larvae obtain oxygen directly from the roots of cattails (*Typha*) and other aquatic plants in nontidal marshes rather than from the air (Batzer and Sjogren 1986). Cattail mosquito larvae are hard to sample because they do not appear dependably in mosquito dipper samples. Considering the current scarcity of cattails in the Meadowlands, this mosquito probably also lives on roots of common reed (as reported elsewhere, Johnson et al. [2017]) in flooded nontidal marshes such as Upper Penhorn Marsh or Kearny Marsh West. Cattail mosquitoes constituted 0.9% of the 2019 catch in light and CO_2 traps in the Bergen County areas (Bickerton, pers. comm.).

We believe that the following information is correct about mosquito control in the Meadowlands in recent years; however, specific details are difficult to obtain. Several techniques are used to control Meadowlands mosquitoes. All of these techniques (except breeding container removal) involve risk to biodiversity, with adulticides and water management probably the greatest threat, and the bacterial larvicides probably the least harmful.

Many parks and outdoor athletic facilities in the Meadowlands adjoin wetlands and have populations of biting flies (e.g., mosquitoes at Overpeck County Park in Palisades Park, and horse flies at Laurel Hill County Park in Secaucus).

Mosquito adulticides such as organophosphates, pyrethroids, or carbamates affect a wide spectrum of organisms, presenting hazards to wildlife (including non-target insects) and humans (e.g., Bordoni et al. 2019). Adulticides are used routinely in the Meadowlands and these materials are very broad spectrum in their toxicity to organisms. Resmethrin, a pyrethroid, and malathion, an organophosphate, are used. Pyrethroids are highly toxic to fish (e.g., Haya 1989). Resmethrin affects the glucocorticoid receptor in humans (Zhang et al. 2016) thus could be an endocrine disruptor in wild mammals.

Several kinds of larvicide are commercially available. Methoprene is an insect hormone analog that prevents maturation of mosquito larvae. There are also three bacterial larvicides and a fungal larvicide. It should be possible to control mosquitoes by using the bacterial larvicides *Bacillus thuringiensis israelensis* (Bti) and *Bacillus sphaericus* (Bs), which affect few taxa other than mosquitoes, but the extensive areas of degraded nontidal wetlands and the abundance of artificial container breeding habitats make mosquito management a challenge. (Nonetheless, bacterial larvicides can affect chironomid midges [e.g., Liber et al. 1998]. In a large area of the Camargue wetlands in France, Bti treatments reduced abundance of odonates, evidently due to suppression of the numbers of their chironomid prey [Jakob and Poulin 2016].) The larvicide methoprene is toxic to a wide range of insects (e.g., Hershey et al. [1998]) and crustaceans (Olmstead and LeBlanc 2001, Ghekiere et al. 2006).

Monomolecular films of organic chemicals or oils are used to coat mosquito breeding waters. These materials interfere with the gas exchange of mosquito larvae and other aquatic invertebrates that rise to the water surface film to breathe. At least two of the commercial preparations were toxic to nontarget invertebrates in experiments, including a tadpole shrimp (not found in the eastern U.S.; Su et al. 2014), a clam shrimp (*Eulimnadia*), a backswimmer (Notonectidae), a water boatman (Corixidae), and a beetle (Takahashi et al. 1984).

In the Meadowlands, the western mosquitofish (*Gambusia affinis*; see Chapter 9) has been widely stocked for larval control of mosquitoes and has established extensively. Mosquitofish can adversely affect populations of various aquatic invertebrates, fishes, and frogs *via* competition and predation (Pyke 2008).

Water management includes a broad range of hydrological, soil, and vegetation alterations, and can be adverse to biodiversity. We have seen attempts to drain a woodland amphibian breeding pool and temporary rain pool habitats of clam shrimps in the Meadowlands. However, emptying water from artificial containers, such as bird baths and clogged gutters, removing dumped vehicle tires, and cleaning up other dumped containers, is generally harmless to native biodiversity and can be beneficial for reducing populations of container-breeding vector mosquitoes such as the northern house mosquito. Artificial containers tend to accumulate with urbanization and are a worldwide public health problem; cemeteries and gardens, for example, can play important roles as providers of container habitats (Vezzani 2007, Townroe and Callaghan 2014).

Water management becomes problematic when wetlands are altered (Rey et al. 2012). Historically, tidal wetlands in the Meadowlands and many other East Coast regions were grid-ditched (e.g., Riverbend Marsh) or impounded (Saw Mill Creek marshes) for mosquito control. Grid ditching alone may have affected ≥ 90% of East Coast tidal wetlands (Rey et al. 2012). Although more strategic and effective water management methods are now used in many East Coast marshes, there are still community and ecosystem-level impacts (*ibid.*).

Mosquito management must continue, ideally prioritizing control of disease vectors over nuisance biters. More field research on biodiversity effects of various techniques is needed to allow optimization of disease management with non-target impacts. Newer, less-toxic pesticides may continue to reduce the non-target impacts of mosquito management. Certain fungi and actinomycete bacteria provide important opportunities for targeted biological control of mosquitoes (Hallmon et al. 2000, Scholte et al. 2004, Paulraj et al. 2016).

Do adult or larval mosquitoes provide important food for any other Meadowlands animals, and do predators help control mosquitoes? Although we have not seen data specifically from the Meadowlands, it is likely that aerial insectivores such as the swallows and adult dragonflies feed on adult mosquitoes, and that small fishes and other aquatic animals eat mosquito eggs, larvae, and pupae. It seems improbable that there would be enough dragonflies or swallows to have much impact where mosquitoes are abundant, except perhaps when large roosting or migrating swallow flocks and large crepuscular swarms of dragonflies are foraging. On the other hand, abundant mosquitofish and mummichog (*Fundulus heteroclitus*) may exert some control over mosquitoes in the larval stage.

BITING MIDGES (DIPTERA: CERATOPOGONIDAE)

Biting midges (Ceratopogonidae or *no-see-ums*) are small flies, 1–3 mm long, whose larvae generally occur in organic-rich moist and wet habitats (Marquardt et al. 2000) including the estuarine benthos, tidal marshes, cattle feedlots, and treeholes. Adult biting midges are very weak fliers. They bite in and near dense vegetation during the day, and in more open areas during calm twilight periods. Biting midges are small enough to pass through many common screens, and often crawl beneath clothing to bite. The bites can itch intensely and may produce small pustules. Biting midges are not known to transmit human disease in the northeastern states.

We have found virtually no information on biting midges, and no species identifications, from the Meadowlands, although we have occasionally encountered these insects. Biting midge larvae in benthic samples from brackish tidal wetlands were reported by several workers including Yuhas et al. (2005).

NON-BITING MIDGES (DIPTERA: CHIRONOMIDAE)

Chironomid midges are delicate-looking flies that superficially resemble mosquitoes but do not bite or feed as adults. Chironomids also lack scales on their wings, whereas mosquitoes have scales that are visible with a 10x hand lens. The worm-like aquatic larvae of chironomids live in the sediment surface or on the submerged surfaces of plants and nonliving objects in fresh and slightly brackish waters. Some chironomid larvae have a red blood pigment that transports oxygen efficiently, allowing the larvae to inhabit water with low levels of dissolved oxygen. Chironomid larvae are often one of the most abundant taxa in the macrobenthos of freshwater marshes (along with aquatic oligochaetes). There are many species of chironomids in different habitats and microhabitats, and they are difficult to identify in either the larval or adult stage. Chironomid larvae and adults are very important food for many fishes, birds, and other animals (Armitage et al. 1995).

Chironomids are abundant in some, perhaps many, of the fresh and slightly brackish Meadowlands marshes. At Kingsland Impoundment, for example, we have seen large numbers of emerged adults perching on shoreline vegetation. Chironomids are thought to be important in transferring energy, nutrients, and contaminants from Meadowlands wetlands to the terrestrial environment. Chironomids were found important in the diet of Meadowlands tree swallows (Kraus 1989). The midge *Chironomus decorus* was a common Meadowlands chironomid (Utberg and Sutherland 1982). Complaints about nuisance insects often involve chironomids, despite the fact that they do not bite (Armitage et al. 1995). In commercial and industrial areas of the Meadowlands where bright outdoor lights are on all night, huge numbers of chironomid adults are attracted and complaints are common (Gregory M. Williams, Hudson Regional Health Commission, pers. comm.). The Hudson County and Bergen County mosquito control agencies do not treat for chironomids, or biting flies other than mosquitoes. Hudson County recommends alternative lighting technology that is less attractive to chironomid midges.

BLACK FLIES (DIPTERA: SIMULIIDAE)

Black fly larvae and pupae require flowing fresh water with ample dissolved oxygen. They attach to underwater substrates such as rocks that are not thickly covered with algae, and are often abundant in small streams with rocky bottoms (Jamnback 1969, Pennak 1978, Marquardt et al. 2000). There is little of this habitat type in the Meadowlands today, although a stream such as Coles Brook in Hackensack, Wolf Creek in Ridgefield, or Van Saun Brook on the southern boundary of River Edge might have suitable habitat. Concrete water control structures may substitute for rocky streams (Marquardt et al. 2000). The AMNET (1993–2018) aquatic invertebrate data for a station near Brookside Cemetery in Englewood, above the Meadowlands, show no black fly larvae in 1993 and 5–16 larvae for the succeeding five samples at five-year intervals. Arnold (2008) reported black fly larvae in Teaneck Creek, a freshwater tributary of Overpeck Creek, but we have not encountered black flies in the Meadowlands. Black flies may migrate several kilometers from larval habitats in search of blood meals (Marquardt et al. 2000). Black flies can be an important nuisance in the northeastern states but are not known to vector human diseases here.

HORSE FLIES AND DEER FLIES
(DIPTERA: TABANIDAE)

While human-biting tabanids do not seem to be a widespread problem in the Meadowlands, they were common enough in certain places and times to constitute a local nuisance. One such occurrence was in late spring and early summer 2006 when we opportunistically collected two species of deer flies: *Chrysops fuliginosus* at Laurel Hill and *Chrysops atlanticus* at Kearny Marsh West.

Estuarine marshes in other areas of the East Coast sometimes support large numbers of human-biting tabanids. Several such species have been studied (Daiber 1982), including the salt marsh greenhead fly (*Tabanus nigrovittatus*), which is troublesome near coastal marshes elsewhere in New Jersey (Hansens and Race no date). Larvae live in the tidal marsh soil and wrack near Mean High Water and adult females may fly to nearby woods to seek a blood meal. Possibly water quality and habitats in the Meadowlands are not suitable to support pestiferous tabanid populations.

GOLDEN-BACKED SNIPE FLY
(DIPTERA: RHAGIONIDAE)

Chrysopilus thoracicus is a large (about 1 cm long), boldly-patterned fly. Some snipe flies are blood feeders, but this species is not known to bite humans although its diet and many other details of its ecology are poorly known. Golden-backed snipe flies were reported to frequent low vegetation

in woodlands of North Carolina (Cotinis 2020). The species is widely distributed in the eastern U.S. In the Hudson Valley the golden-backed snipe fly
is typically seen in or near calcareous wet meadows where it is occasionally
common. We have seen this species once in the Meadowlands, in June 2008
close to the serpentinite dump in the Kane Natural Area.

BEES, WASPS, AND ANTS (HYMENOPTERA)

Various types of Hymenoptera provide a range of ecological services. Bees
are among the most important pollinators, and ants are important in soil formation, decomposition, and protection of vascular plants from herbivores.
Parasitoid wasps are extraordinarily diverse, likely outnumbering even beetles in numbers of species (Forbes et al. 2018). Many species of Hymenoptera
occur in urban areas, and gardens in some cities play an important role in bee
conservation. Ant assemblages in two midwestern cities differed according to
habitat types, with forested areas, vacant lots, and community gardens each
contributing distinctly to the diversity of urban ants (Uno et al. 2010).

A 2.4 ha study area on an inactive garbage landfill in Kearny yielded 49
species of bees (Yurlina 1998). Seventy-seven bee species were found at four
locations in the Meadowlands (reported as 78 in O'Neill 2010); see Table
10.3. Almost all of the Meadowlands bees are native species. Yurlina's study
area was upland meadow and young woodland, surrounded by common reed,
tidal marsh, and Exit 15W of the Turnpike. Regarding a concurrently studied
landfill on Staten Island, Yurlina (1998) stated that pre-existing cavities in
stems of plants such as sumacs (*Rhus*), tree-of-heaven (*Ailanthus altissima*),
sunflower (*Helianthus*), and common reed offered nest sites for several types
of native bees. Smaller stems (cavity generally less than 1 cm diameter) are
suitable for yellow-faced bees (*Hylaeus*), small carpenter bees (*Ceratina*),
and many types of leafcutter bees (several genera in the Megachilidae). The
megachilids may also use cavities left by wood-boring insects such as beetles,
or smaller cavities in the soil. Larger pre-existing cavities are used by bumble
bee (*Bombus*) colonies; these may be found in live trees, but are more commonly underground in rodent burrows, beneath tree stumps, or in spaces
within or beneath human-generated debris. Most other native bees create
nests by excavating exposed dry soil or decaying wood.

Many bees require flowers as sources of nectar and pollen, and there has
to be a dependable sequence of bloom through the active foraging season of
that particular bee species. We infer that abundance of potential pollen and
nectar sources on upland meadows, and the abundance of potential nest sites
both in friable, dry, sparsely vegetated mineral soil used as landfill cover or
wetland fill, and plants such as sumacs, tree-of-heaven, and reed, may make

Table 10.3. Bees of the New Jersey Meadowlands.

Common Name	Scientific name	N/I	Source
	Colletidae[a]		
Eastern masked bee	*Hylaeus affinis*	N	K, Y
Masked bee	*Hylaeus illinoisensis*	N	K, Y
Slender-faced masked bee	*Hylaeus leptocephalus*	N	K, Y
Mesilla masked bee	*Hylaeus mesillae*	N	K, Y
Modest masked bee	*Hylaeus modestus*	N	K, Y
Masked bee	*Hylaeus purpurissitus*	I	Y
Masked bee	*Hylaeus schwartzi*	N	K
Masked bee	*Hylaeus cf. praenotatus*	I	K
	Andrenidae		
Two-spotted mining bee	*Andrena accepta*	N	Y
Mustard mining bee	*Andrena arabis*	N	Y
Short-tongued mining bee	*Andrena brevipalpis*	N	K
Hawthorn mining bee	*Andrena crataegi*	N	K, Y
Mining bee	*Andrena cessonii*	N	K
Fragile mining bee	*Andrena fragilis*	N	Y
Hippotes mining bee	*Andrena hippotes*	N	Y
Bumped mining bee	*Andrena nasonii*	N	K, Y
Perplexed mining bee	*Andrena perplexa*	N	K, Y
Cherry mining bee	*Andrena pruni*	N	K
Neighboring mining bee	*Andrena vicina*	N	K, Y
European legume mining bee	*Andrena wilkella*	I	K, Y
Eastern mining bee	*Calliopsis andreniformis*	N	K, Y
Cinquefoil mining bee	*Panurginus potentillae*	N	Y
	Halictidae		
Silky-striped sweat bee	*Agapostemon sericeus*	N	K
Bicolored striped sweat bee	*Agapostemon virescens*	N	K
Pure green sweat bee	*Augochlora pura*	N	K
Golden sweat bee	*Augochlorella aurata*	N	K
Metallic epaulleted-sweat bee	*Augochloropsis metallica*	N	Y
Southern bronze furrow bee	*Halictus confusus*	N	K
Ligated furrow bee	*Halictus ligatus*	N	K, Y
Yellow-legged furrow bee	*Halictus rubicundus*	N	K
Cuckoo sweat bee	*Sphecodes atlantis*	N	K
Sweat bee	*Lasioglossum coreopsis*	N	Y
Cresson's sweat bee	*Lasioglossum cressonii*	N	Y
Sweat bee	*Lasioglossum mitchelli*	N	Y
Hairy sweat bee	*Lasioglossum pilosum*	N	Y
Metallic-epauletted sweat bee	*Lasioglossum tegulare*	N	Y
Eastern green sweat bee	*Lasioglossum viridatum*	N	Y
	Megachilidae		
European wool carder bee	*Anthidium manicatum*	I	K
Oblong wool carder bee	*Anthidium oblongatum*	I	K, Y
Small wool carder bee	*Pseudoanthidium scapulare*	I	K
Short leafcutter bee	*Megachile brevis*	N	K, Y

Common Name	Scientific name	N/I	Source
Patchwork leafcutter bee	*Megachile centuncularis*	N	K, Y
Small-handed leafcutter bee	*Megachile gemula*	N	Y
Flat-tailed leafcutter bee	*Megachile mendica*	N	K, Y
Silver-tailed petalcutter bee	*Megachile montivaga*	N	K, Y
Alfalfa leafcutter bee	*Megachile rotundata*	I	K, Y
Giant resin bee	*Megachile sculpturalis*	I	K
Bufflehead mason bee	*Osmia bucephala*	N	K
Dwarf mason bee	*Osmia pumila*	N	K, Y
Resin bee	*Heriades variolosus*	N	Y
Hairy-fronted small mason bee	*Hoplitis pilosifrons*	N	K, Y
Produced small mason bee	*Hoplitis producta*	N	K
Say's cuckoo leafcutter	*Coelioxys sayi*	N	K, Y
Spot-sided dark bee	*Stelis lateralis*	N	K, Y
Apidae			
Orange-tipped wood-digger bee	*Anthophora terminalis*	N	K
Agile long-horned bee	*Melissodes agilis*	N	Y
Two-spotted long-horned bee	*Melissodes bimaculata*	N	K
Ironweed long-horned bee	*Melissodes denticulata*	N	Y
Three-knotted long-horned bee	*Melissodes trinodis*	N	K
Hibiscus bee	*Ptilothrix bombiformis*	N	K, Y
Imbricate cuckoo nomad bee	*Nomada imbricata*	N	K
Yellow nomad bee	*Nomada luteoloides*	N	K
Nomad bee	*Nomada* cf. *perplexa*	N	K
Pygmy nomad bee	*Nomada pygmaea*	N	K
Nomad bee	*Nomada subrutila*	N	Y
Bounded cuckoo nomad bee	*Nomada vincta*	N	Y
Nomad bee	*Nomada* sp.	?	K
Lunate longhorn cuckoo bee	*Triepeolus lunatus*	N	Y
Spurred small carpenter bee	*Ceratina calcarata*	N	K, Y
Doubled small carpenter bee	*Ceratina dupla*	N	K, Y
Nimble small carpenter bee	*Ceratina strenua*	N	K
Large carpenter bee	*Xylocopa virginica*	N	K, Y
Two-spotted bumble bee	*Bombus bimaculatus*	N	K
Cuckoo bumble bee	*Bombus citrinus*	N	K
Yellow bumble bee	*Bombus fervidus*	N	K, Y
Brown-belted bumble bee	*Bombus griseocollis*	N	K, Y
Common eastern bumble bee	*Bombus impatiens*	N	K, Y
European honey bee	*Apis mellifera*	I	K, Y

Note: This list was compiled from the field work of Mary Yurlina (Y) and Sarah Kornbluth (K), with taxonomic corrections and nomenclatural updates, by Parker Gambino. Many bees are small and familiar only to specialists, hence some of the common names have been coined recently and some species apparently lack common names (in which cases we have used common names from the genus).

N = native, I = introduced (nonnative) to New Jersey.

[a] Colletidae are also called *miner bees*. *Hylaeus* bees are also called *yellow-faced bees*.

certain landfills and other filled areas in the Meadowlands suitable habitat for diverse native bee faunas.

Many bees are listed as SGCN in New Jersey (NJDEP 2018a). Among the Kearny landfill bees, the following are SGCN: *Bombus fervidus* (yellow bumble bee), *Megachile centuncalaris, Melissodes agilis,* and *Hylaeus leptocephalus.* Although most bees in the Meadowlands are native species, a few are nonnative. The Meadowlands were one northeastern locality where the carder bee species *Anthidium oblongatum* was first documented in North America (Hoebeke and Wheeler 1999). Nonnative honey bees (*Apis mellifera*) are common visitors to Meadowlands wildflowers, with the hives kept by beekeepers a potential source.

Another group of Hymenoptera, the thread-waisted wasps (Sphecidae *s.l.*), can be seen digging nest burrows on dirt roads and other wetland fill, but they have not been studied in the Meadowlands. Nor have ants been studied, despite their usefulness as environmental indicators. In the city of Raleigh, North Carolina, warm and dry conditions favored ant species of arid environments (Menke et al. 2011), and we would expect arid land species to do well in the Meadowlands on dry wetland fill and in quarried or built areas.

TERRESTRIAL MOLLUSKS (TRUE LAND SNAILS AND SLUGS, GASTROPODA IN PART)

Land snails are mostly recognizable as such, even by a non-biologist: they have shells and, well, they look like snails (although many species are tiny—with shells as small as 1 mm long when adult). Terrestrial slugs are also mollusks but they do not have an external calcareous shell as do the *true land snails.* Most true land snails in the northeastern states are native, although there are several introduced species. The commonly noticed terrestrial slugs are mostly introduced species, and there are native species as well. A few of the terrestrial snails and slugs do well in urban areas, with introduced species usually predominating in the urban environment (Emerson and Jacobson 1976). Native snails (and introduced slugs) generally predominate in the northeastern countryside.

Terrestrial mollusks are usually little more than a curiosity, if noticed at all, to American biologists and naturalists. These mollusks, however, are very widespread and often abundant, occur in many habitats, and are species-rich. Because many species are associated with narrow ranges of ecological conditions, they are good indicators. In Europe, where more scientific attention has been paid to land snails, there are many intriguing studies of indicator value. For example, the shells found in archaeological sites are used to deduce ancient conditions.

In the Meadowlands we found only a handful of true snail and slug species, mostly tough nonnative or ubiquitous native species. We informally collected them wherever we found them, beneath cover objects, among the leaf litter, or when we saw empty shells on the ground. Because many true land snails are small and we are not accustomed to surveying for them, we undoubtedly missed some species. We found flamed disc (*Anguispira alternata*, native), black gloss (*Zonitoides nitidus*, native), rotund disc (*Discus rotundatus*, non-native), dark-bodied glass snail (*Oxychilus draparnaudi*, nonnative), cellar glass snail (*Oxychilus cellarius*, nonnative), grove snail (*Cepaea nemoralis*, nonnative), and amber snail (*Novisuccinea ovalis*, native). Another intro-duced snail, the garlic snail *Oxychilus alliarius*, was reported in 1929 from Branch Brook Park in Newark near the Passaic River (Alexander 1952). We found a native salt marsh snail, the eastern melampus (*Melampus bidentatus*), in smooth cordgrass (*Spartina alterniflora*) in the Meadowlands; this is an air-breathing land snail although it has an aquatic larva.

The large and distinctive leopard slug (*Limax maximus*) and smaller slugs, namely, the garden slug *Arion* cf. *hortensis* and the hedgehog arion *Arion* cf. *intermedius*, all nonnative species, occur in the Meadowlands, with at least the first species locally common. Another slug, the meadow slug *Deroceras laeve*, which has an affinity for wet habitats, is considered native. Undoubtedly there are other slugs and true snails, but probably few. Hotopp and Pearce (2007) reported that 60 terrestrial mollusk species occurred in Rockland County (New York, just north of the Greater Meadowlands), of which 8 were introduced species. A statewide New Jersey list has 76 native vs. 10 nonnative snails, and 3 native vs. 5 nonnative slugs (Hotopp et al. 2013). By comparison, the Meadowlands data to date show 4 native vs. 5 nonnative snails, and 1 native vs. 3 nonnative slugs. Dirrigl and Bogan (1996) listed 17 terrestrial mollusks from Bergen and Hudson counties, omitting 2 of the species that we found in the Meadowlands.

Terrestrial mollusks are, by and large, animals of the surface soil and plant litter, and many species are associated with forests or natural rock outcrops. These habitats are scarce or of reduced quality in the Meadowlands. Land snails may be affected adversely by soil contaminants (Hotopp et al. 2013). Land snails also have limited powers of dispersal, they do not move long distances easily, and they may have trouble crossing barriers such as high-ways. Yet mollusks can be transported passively in soil (e.g., with tree balls or potted plants) or rafted downstream (Alexander 1952). Small snails, when attached to dead leaves, can be blown in the wind (Bequaert and Miller 1973, 47). Species surviving human- or wind-aided dispersal, and finding suitable habitats once in the Meadowlands, are likely to be hardy, urban-tolerant, nonnative or native species. In the case of terrestrial mollusks, urban toler-ance probably means toleration of microhabitat drought, soils low in organic matter, various pollutants, and in some environments, salinity. One factor that should help the mollusks that can survive these harsh conditions is the abundant calcium (from cement and gypsum waste) in many urban areas.

The most dramatic-looking land mollusk of the Meadowlands is arguably the leopard slug. This large nonnative slug reaches a length of about 7 cm in the Meadowlands and is locally common beneath cover objects. The large, nonnative grove snail, about 2 cm in diameter, is also a strikingly beautiful animal. We found this colorful snail also locally common. Large individuals may be seen attached to plants 1 m or more above the ground.

FRESHWATER MOLLUSKS

We found few data on freshwater snails (Gastropoda, in part) or fingernail clams (Sphaeriidae) in the Meadowlands although some interesting potential habitats are present. The surveys of two Hackensack River tributaries (AMNET 1993–2018, see below) revealed small numbers of five snails (*Physella integra, Physella gyrina, Ferrissia rivularis, Stagnicola catascopium, Amnicola limosus*) and three fingernail clams (*Sphaerium transversum, Pisidium casertanum, Musculium*). About 10 km west of the upper Hackensack River estuary and outside our study region, in 5.5 ha Barbour's Pond in the Garrett Mountain Reservation, Paterson, a diverse freshwater mollusk fauna comprising 18 species (13 native and 1 nonnative snails, 3 fingernail clams, and 1 freshwater mussel) was documented by Chapman and Prezant (2006). Although Barbour's Pond is urban it could serve as a more-natural reference site for comparison to the Meadowlands.

Freshwater mussels (or pearly mussels, Unionoidea) are a large group of bivalve mollusks. The U.S. is a megadiversity region for this taxon, and the group is considered to contain the most species of conservation concern of any higher taxon studied in this country (Stein et al. 2000). Freshwater mussels are ecologically distinctive in their relatively large size among the bivalve mollusks, the requirement of most species for fish hosts for their larvae (*glochidia*), their sensitivity to hydrology and water quality (Van Hassel and Farris 2007), and the historic use of the shells of certain species for manufacture of buttons (Strayer and Jirka 1997, Strayer 2008). The percentage urban land cover was negatively correlated with mussel density, species richness, and proportion of young individuals in an analysis by Van Hassel and Farris (2007). Most of the native U.S. species are apparently salinity-limited and found only in fresh water, whether tidal or nontidal. In the tidal Hudson River, native freshwater mussels tolerate a maximum of only about 1–2 ppt salinity (D. Strayer, Cary Institute of Ecosystem Studies, pers. comm.). One introduced species, the basket clam (*Corbicula fluminea*), that is considered a freshwater mussel (i.e., unionoid) by some biologists, inhabits both fresh and brackish waters. Freshwater mussels are sensitive to metals, endocrine disrupting chemicals, organic or nutrient pollution, and other problems that afflict urban waters (Fuller 1974, Strayer 2008). Dams on rivers often harm freshwater mussel communities (Strayer 2008) because the dams interfere with dispersal and also alter hydrology, sediments, and water quality (e.g., reduce current speed, increase depth, and reduce dissolved oxygen upstream).

There have been no reports of native freshwater mussels in the Meadowlands. The nonnative basket clam occurs in Bergen County (Cordeiro and Bowers-Altman no date) and has been reported from the Meadowlands (Kraus et al. 1987). There are now considered to be two similar species of basket clam (*Corbicula fluminea* and *C. fluminalis*) in the U.S. (Sousa et al. 2008); we do not know which species was found in the Meadowlands. Five species of native freshwater mussels were found in recent years in the upper Hackensack River system and Ramapo River system in Rockland County, New York (Strayer and Jirka 1997; D. Strayer, pers. comm.), and these species are potentially available to colonize the Meadowlands. Five freshwater mussel species, plus the basket clam, were reported for Bergen or Hudson county by Cordeiro and Bowers-Altman (no date). Probably salinity intrusion, and perhaps water quality as well, have eliminated any freshwater mussels formerly present in the Meadowlands and prevented recolonization. Possibly freshwater mussels survive in the Little Ferry clay pit lakes if salinities are low enough.

ESTUARINE MOLLUSKS

Many species of estuarine mollusks (snails and bivalves) occur in the Meadowlands. The small snail *Hydrobia totteni* dominated the 1987 benthic samples (Bragin et al. 2009), and two other snails were found in small numbers. A single species of small (8 mm long) sea slug, the limpet nudibranch (*Doridella obscura*), occurred sparsely in 2002 estuarine samples (*ibid.*). Five bivalves were found in small-to-moderate numbers, except that the clam *Congeria leucophaeta* was extremely abundant at a single sampling station.

Bragin et al. (2009) did not sample in the marshes where a somewhat different mollusk assemblage occurs. The best known of these is the ribbed mussel (*Geukensia demissa*), a typically abundant large mollusk in more-brackish marshes. McClary (2004) found similar abundance of ribbed mussels in smooth cordgrass and common reed stands in the Saw Mill Creek area. *Melampus*, mentioned above with the land snails, can also be considered an estuarine mollusk.

Oysters were not found in the Bragin et al. studies. The eastern oyster (*Crassostrea virginica*) was undoubtedly abundant in the Hackensack River estuary in Colonial and pre-Columbian times (Marshall 2004). Pollution and overharvest are credited with the decline of this species and its fishery there. An oyster re-establishment experiment in the river at Laurel Hill resulted in oysters with abnormally thin shells, tissue abnormalities, and high metal content in soft tissues (Ravit et al. 2014).

BENTHIC INVERTEBRATES IN THE ESTUARY

The benthos, or benthic invertebrates, are the invertebrates that live on or in the bottom sediments of wetlands and waterways (and on other submerged surfaces, according to different definitions). Benthic macroinvertebrates, those forms retained by a sieve with 0.5–1.0 millimeter mesh, are commonly studied. There have been two detailed surveys of the macrobenthos in the Hackensack River estuary and a few of its major tidal tributaries (Bragin et al. 2009) with the mollusks mentioned above. These studies sampled the same stations in major channels in 1987 and 2002, from just north of the mouth of Overpeck Creek southward to just north of the Route 7 bridge. Off-river stations were in Mill Creek, Cromakill, Berry's Creek Canal, and Saw Mill Creek. Bragin et al. found 52 taxa in the 1987 study and 67 taxa in the 2002 study, for a total of 89 taxa in both studies combined. Taxon richness decreased from downriver to upriver (i.e., with decreasing salinity), with the numbers of taxa at each sampling station ranging from 43 to 10 in 2002. Density (total individuals per sample) increased from downriver to upriver with the highest densities in Mill Creek and the Cromakill, and the main-stem at river km 14.8 (40,000 to 65,000 individuals/m^2). These high-density samples were dominated by an amphipod crustacean *Apocorophium lacustre* and a polychaete worm *Hobsonia florida*. In the 2002 study, the samples overall were dominated by polychaetes, with 22 species and 45% of the total macrobenthic individuals. In the 1987 study, snails (mostly one species, 50% of individuals) and oligochaetes (27% of individuals) dominated the samples overall.

FRESHWATER STREAM MACROINVERTEBRATES

Urban streams undergo many changes that are adverse to biodiversity (Ranta et al. 2021), and the several upland tributaries of the Hackensack River estuary are no exception. The statewide Ambient Macroinvertebrate Network (AMNET) has sampled two such tributaries at five-year intervals. AMNET has macrobenthos data from a station in Overpeck Creek near Brookside Cemetery (Englewood) well upstream of the Meadowlands (elevation ca. 18 m), and from Van Saun Mill Brook (River Edge) not far above tidal influence (elevation ca. 1 m). Sampling began in 1993 with each sample consisting of 100 individuals. The Van Saun data show taxon richness (mostly at the generic or specific level) that varied from 17 to 28. Total Chironomidae individuals ranged from 28 to 56. Ephemeroptera-Plecoptera-Trichoptera (EPT, a commonly used index with higher numbers of individuals indicating better water quality) ranged from 8 to 56, increasing almost monotonically from 1993 to 2018. The Overpeck data show taxon richness that increased from 15 to 24 also almost monotonically. Total Chironomidae individuals decreased from 85 to 36 in the first four sampling periods then increased to 58. EPT ranged from 2 to 46, peaking in the fifth sampling period then declining to 13. The first three samples were taken in early July with the last two taken in late November, confounding detection of trends. Nonetheless, the quality of the macrobenthos and water may be improving in these two tributaries.

CONCLUSIONS

This brief review of Meadowlands invertebrates suggests certain character-istics of the fauna. The fauna is poorly known. There have been quantitative studies of only a few groups, the estuarine benthos and nekton, tributary stream benthos, the bees, butterflies, and odonates, and with the exception of the Bragin et al. studies of estuarine invertebrates, these studies have been very limited in scope. The species diversity of the fauna, where known, is mostly moderate to low. Many sensitive species requiring specialized habitats or high quality environments seem scarce or absent, as are species requiring such habitats as natural freshwater streams or natural upland soils. Many of the species present are known to be more-or-less urban-tolerant, and nonna-tive species are prominent in some taxa. Finally, estuarine taxa are present in reasonable abundance and diversity, but true freshwater taxa seem less common probably due to the scarcity of fresh water of good quality. There is indication of improvement in the two freshwater tributaries studied, at least until the most recent sampling period, and also some improvement in the estuarine benthos. Climate change, in concert with other stressors, may make estuarine organisms more vulnerable to toxic contaminants (Delorenzo 2015), thus there is a need to assess the effects of commonly used insecti-cides and herbicides in combination with ambient metals and organics in the Meadowlands environment.

A globally rare species, Mattox's clam shrimp, with great significance for conservation and occupying a highly threatened habitat type, may be extirpated. Many other invertebrate taxa have been completely overlooked. For example, the farthest extent of Pleistocene glaciation is only ca. 20 km west of the southern end of the Greater Meadowlands; investigation of taxa such as earthworms (terrestrial Oligochaeta), whose native species are principally found in unglaciated regions, would be worthwhile in the natural upland soils and freshwater wetland edges of the Meadowlands. Recently published field guides to flower flies (Syrphidae; Skevington et al. 2019), singing insects (Orthoptera and Cicadidae; Capinera et al. 2004, Elliott and Hershberger 2007), ants (Ellison et al. 2012), bumble bees (Williams et al. 2014), and terrestrial mollusks (Hotopp et al. 2013) should stimulate assessment of those groups. The list of odonates may be enlarged as more naturalists use the New Jersey field guide (Barlow et al. 2009). More survey work is needed to document the diversity, ecological roles, indicator value, conservation significance, trends, and educational interest of invertebrates. The Meadowlands have proven to be a valuable laboratory for research on urban biodiversity as well as for the conservation of birds, and there is every reason to believe that the same will eventually be true of the invertebrates. And none too soon, given the pressures of continuing urban development and the need for biodiversity-sensitive land use and habitat management.

Conservation

The large number of species, minimal knowledge about most of them, and their diverse lifestyles and habitat requirements, make the conservation of native invertebrates challenging. We can make a few generalized recommendations. Many invertebrates, including a large number of butterflies, moths, sawflies, beetles, true flies, true bugs, aphids, and mites require particular plant taxa for food or shelter. Many if not most invertebrates are associated with specific habitats or habitat features such as sparsely vegetated, friable soil, abundant downed wood and leaf litter, or nectar and pollen plants. Thus, the conservation and management of habitat diversity and native plant diversity, are critical, and efforts should be partly targeted to those species, habitats, and habitat components known to be important. Nonnative plants, nonetheless, can be highly important to pollinators, bird food organisms, and generalist herbivores and carnivores, especially in urban environments where alternate native plants may not survive or thrive. Remediation of contamination, and the use of less toxic pesticides or other techniques (especially for mosquito control, vegetation management, and rodent control) will help protect invertebrates. Replacing old outdoor lighting with new technology less attractive to nocturnally flying insects will benefit many insect species and the bats, swallows, and other animals that eat them. Reduction of anthropogenic noise levels would benefit invertebrates that communicate by sound.

Good surface and ground water quality and free-flowing streams are crucial habitats. In the Meadowlands, salt marshes, all other estuarine and nontidal aquatic and wetland habitats, forests, thickets, meadows, temporary pools, springs, sparsely vegetated soils, and rocky or sandy areas are important to invertebrates. Unlike larger animals and plants, many invertebrates can be conserved on small habitat patches, whether natural, inadvertent, or intentionally created. Measures that generally benefit invertebrates will also benefit cryptogams and many other organisms.

Conclusion

We have described biological diversity in an urban-industrial region, the New Jersey Meadowlands, and discussed the environmental features that support or impede urban biodiversity. At a time in the not-so-distant past, urban ecosystems, especially wetlands, were looked upon as eyesores in need of development or cleaning up; now there is a greater emphasis on protecting biodiversity in urban regions for all the ecosystem services it provides and for its own sake. The Meadowlands have been altered, polluted, and built on for hundreds of years, yet now support much rare or uncommon native wildlife and plants of conservation concern as well as common native and nonnative species. Our description of the biota of the Meadowlands, which higher taxa are diverse or successful and which are not, and what kinds of environments allow these organisms to colonize and persist, should be instructive for biodiversity conservation in the Meadowlands and in other urban and industrial areas. A future book will elaborate on the causes of biodiversity patterns and in-depth management and planning concepts to better support urban biodiversity in a near future where impacts from climate change, additional land and water development, and even the siting of renewable energy facilities are sure to bring additional challenges to ecosystems in coastal urban regions.

SUMMARY OF BIODIVERSITY PATTERNS BY TAXA

We summarize here the general patterns of occurrence and to a lesser extent abundance for the taxa as described in earlier chapters, with highlights of which groups were winners and which are apparently losers in the Meadowlands region. The numbers of species, of course, do not tell the whole story. Many native plants, fungi and animals are scarce in the Meadowlands despite being common elsewhere in surrounding regions. These scarce species either occur at very few localities, or form scattered and small stands, in the case of plants, and are rare in the Meadowlands environment. This may be due to historical activities and urban conditions in the Meadowlands, conditions happening outside the Meadowlands at regional, continental, and global scales, or to intrinsic ecological characteristics of the higher taxa and individual species being studied including narrow habitat affinities and interactions with other organisms.

Plants and Fungi

Overall, plants are moderately high in species representation in the Meadowlands. However, there is great variation among plant taxa in their patterns of richness and abundance and, in fact, some groups are entirely missing from the Meadowlands despite being present in nearby counties. Bryophytes (mosses and liverworts) are moderate to low in species richness; peat (*Sphagnum*) mosses are few; other mosses are moderately species-rich; and liverworts are scarce. Ferns are not well represented and fern allies (quillworts, spikemosses, clubmosses, horsetails) are absent relative to the surrounding region except for a single horsetail species.

Seed plants, especially the most species-rich groups in the Northeast (Asteraceae [asters], Poaceae [grasses], and Cyperaceae [sedges]) are well represented. Species richness of trees and shrubs is moderate and of vines is high. Some seed plant taxa have not fared well in the Meadowlands: there are apparently no bur-reeds (*Sparganium*), no native orchids (Orchidaceae), and few species of bedstraw (*Galium*), willow-herb (*Epilobium*), willow (*Salix*), violet (*Viola*), bur-marigold (*Bidens*), and conifers (several genera). In general, submergent aquatic plants also have very low species richness, and true sedges (*Carex*) should be more diverse and abundant given regional species distributions.

Lichens are moderately represented and include mainly common, widespread species that are evidently urban-tolerant. Little is known of non-lichenized fungi and a survey is in progress.

Invertebrates

There have been quantitative studies of only a few groups of invertebrates: the estuarine benthos and nekton, tributary stream benthos, the bees, the butterflies, and the odonates; with a few exceptions, these studies have been very limited in scope. The species diversity of the invertebrate fauna, where known, is mostly moderate to low.

The globally rare Mattox's clam shrimp is an exciting find and provides a key example of how management practices can overlook important biodiversity elements. This species persisted until recently in pools created by vehicles on unpaved roadways in the Meadowlands.

Benthic macroinvertebrates are associated with stream and river-bottom habitats. They are widely used as water quality indicators because they vary greatly in sensitivity to pollution. Crayfish are nearly absent (a single record) and other benthic macroinvertebrates surveyed in upstream tributaries of the Hackensack River within the Greater Meadowlands indicate an impaired community of organisms dominated by pollution-tolerant species. This is not surprising given the history of industrial discharges into the surface waters and continued inputs of polluted stormwater and sewage. There has been some improvement in stream benthos assemblages.

The Meadowlands, so far as known, support a moderate number of odonate species compared to the state as a whole but lack many of the species that are specialized inhabitants of unpolluted freshwater streams. Besides the conspicuous lack of unpolluted streams, the Meadowlands have few freshwater peatlands (exceptions being the nontidal Upper Penhorn Marsh and the formerly tidal Kearny Marsh West) or other specialized habitats important for odonate diversity.

We consider the butterflies as moderately species-rich but moths have not been studied. As a taxon, moths are much more species-rich overall, less well known than the butterflies, and generally harder to identify to species. Many moths are host plant specialists or habitat specialists. The diversity of habitats and potential larval host plants could support a diverse urban moth fauna.

Fishes

The Hackensack River estuary in the Core Meadowlands has a fish fauna generally typical of estuaries in the northeastern United States. The size of this estuary and the limited sampling effort to date contribute to the relatively short list of species. The fish fauna contains species in each of several categories that one expects to see in an estuary: diadromous (anadromous and catadromous) species, euryhaline residents, freshwater and saltwater strays, and marine species using the estuary as a nursery, but there are few data on the abundance of these species or the degree to which various species spawn in the estuary. The freshwater fishes in the Meadowlands region are not well represented as compared to the species documented in the Hackensack River watershed upstream of the Meadowlands. Salinization, hydrological modifications, and pollution have probably eliminated these species from most of the drainage.

Amphibians and Reptiles

The herpetofauna (amphibians and reptiles) is poorly represented and little studied in the Meadowlands. Many reported species are unverified, rare, or apparently extirpated. A few species (red-backed salamander, eastern box turtle) persist at the ecological and geographic edges of the Meadowlands region. However, one Species of Greatest Conservation Need, the diamond-backed terrapin, is doing well in the Meadowlands. The northern water snake is surprisingly rare.

Birds

Birds have been intensively studied by scientists and naturalists in the Meadowlands for many years. The bird fauna, especially species associated with wetlands and waterways, is well-represented, albeit mostly by nonbreeders. Herons, waterfowl, gulls, and shorebirds are among the most diverse and abundant groups, and densities are extremely high at certain times of the year. However, the secretive marsh birds (rails, bitterns) seem scarce relative to habitat availability. Passerine birds, including the songbirds, are fairly species-rich, although forest species are limited by habitat. Extensive non-forested areas, such as landfills, attract many migrants. Raptors, including eagles, falcons, hawks, vultures, and owls, are well-represented.

Mammals

There is a low to moderate richness of mammal species in the Meadowlands, depending on the group being considered. Small mammals, including rodents, moles, and shrews, require intensive, specialized survey methods in order to be assessed. An ecologically important marsh species, the muskrat, is declining, and the beaver is not yet re-established. Mustelids are rare, excepting the striped skunk, and only one insectivore, short-tailed shrew, seems common. Widespread, urban-tolerant, omnivorous mammals, including raccoon, red fox, and coyote, are widespread in the region. Bats stand out as depauperate in the Meadowlands, especially given they are able to persist in other urban regions. With the exception of harbor seal, the occurrence of marine mammals is very rare in the Meadowlands.

GENERAL PATTERNS OF BIODIVERSITY

Despite the past abuses of the Meadowlands ecosystems, the region supports a remarkable and interesting biota. Yet not surprisingly, many sensitive species requiring specialized habitats or high-quality environments seem scarce or absent, as are species requiring such habitats as natural freshwater streams or natural upland soils. Many of the species present are thought to be more-or-less urban-tolerant and nonnative species are prominent in some taxa. Finally, estuarine taxa are present in reasonable abundance and diversity, but true freshwater taxa seem less common probably due to the scarcity of freshwater habitats of good quality.

There were a few pleasant surprises that turned up in our review and field studies over the years including the globally rare Mattox's clam shrimp, state-rare floating marsh pennywort, several bryophytes documented for the first time in New Jersey, and an unusual assemblage of peat mosses and other bryophytes beneath hectares of common reed (Figure 11.1). Breeding choruses of a recently described species, Atlantic Coast leopard frog, were discovered, and otter and mink sign were found at a few locations. We cite many more such examples. There were also some aspects of biodiversity that were expected to be present but were surprisingly depauperate where surveyed—for example, bats, snakes, and fern allies. The most striking problem encountered was the complete lack of records and formal surveys for many taxa including most mammals, reptiles, amphibians, most invertebrates, fungi, ferns and allies, and bryophytes.

Characteristics Shaping Patterns of Diversity in the Meadowlands

We touch on a few of the major influences that shape biodiversity in the Meadowlands and a lengthier discussion is part of the subject matter of our next book. Although every urban region is unique, there are many commonalities. The Meadowlands, although extreme in certain aspects of their

Figure 11.1 Off-road vehicle trail in freshwater peat marsh. Photo by Erik Kiviat.

ecology, share many extant or historical features with other urban areas that impede and support biodiversity. Among these challenges are: water, soil, and air pollution and the presence of toxic contaminants; alteration of waterway hydrology, and the filling, diking, and draining of wetlands for multiple purposes; fragmentation of remaining wetland and upland habitats; salinization due to the Oradell Dam at the head of tide; dominance of built structures both active and abandoned; extensive existing dumps; and brownfields and other undeveloped lands that attract new dumping and other illegal activities. But among the benefits to biodiversity are reduced human presence and thus reduced direct disturbance to wildlife. With regard to biodiversity support, other commonalities include extensive wetlands, waterways, and waste grounds; habitat diversity; food resources such as weed seeds, shrub fruits, and garbage-associated rodents; cliff-like structures; probably calcium-rich soils; eutrophication of soils and waters; presence of large trees; inactive waste dumps supporting herb-dominated vegetation with or without sparse trees; the vectors and corridors that bring nonnative species (railroads, highways, seaports, airports); stands of nonnative plants and their associated species; urban microclimates including heat islands; low-lying areas affected by rising sea level and storm surges; and sociopolitical structures that tend to overlook or dismiss biodiversity in the planning and management processes.

The Meadowlands have an incredibly varied spectrum of species diversity that is somewhat directly related to habitat diversity (e.g., MacArthur and MacArthur 1961) and habitat quality (Dures and Cumming 2010). There are many different types of habitat, including some large patches that are usually rare in urban regions, thus a mixture of specialists and generalists is found.

The biogeographic context of the Meadowlands promotes richness of the biota. Species diversity in the Meadowlands is importantly related to the greater geographic context of the Meadowlands—along the Mid-Atlantic coast of North America. New Jersey itself is a very species-rich state because it straddles the migration routes of many birds and other animals (Dunne et al. 1989), and it is located at the boundary between northern and southern climes of North America. Therefore, the ranges of species from these two zones overlap here (Robichaud and Buell 1989). The climate is moderate and winters are relatively mild, which favor overwinter survival of many species. In addition, New Jersey has a wide range of physiographic provinces from diverse coastal areas to the Highlands, both glaciated and unglaciated. Geological history and human history have played a role in shaping modern habitats and species; this includes management of habitat and land use decisions.

There are many large-scale to global-scale influences shaping species diversity in the Meadowlands. Examples of global factors that might affect diversity include large-scale loss of habitat and exposure to persistent pesticides (DDT, dieldrin) for species wintering in Latin America. Also, pressure from hunting in other areas of North America could result in lower breeding and wintering densities of waterfowl or shorebirds in the Meadowlands. Climate change is affecting the timing of arrival on the breeding grounds for many migrant species, which could have a variety of negative impacts on populations. Diseases such as West Nile virus have potentially major negative effects on populations of many species, including some common songbirds that were until recently thought to be widespread and stable (LaDeau et al. 2007). Newly emerging diseases are a constant threat; as we finalize this book in the summer of 2021 songbirds in the eastern U.S. are suffering from a strange, as yet unidentified disease that is causing crusting of the eyes, blindness, neurological symptoms and in many cases death (Quinn 2021). Amphibians and bats are other well-known examples where introduced or emerging pathogens are taking a heavy toll on populations. Invasive plants, such as common reed, have greatly affected the wetlands of the Meadowlands.

Cities support species of conservation concern for a variety of reasons including that some interspecific interactions may be relaxed (the refuge effect) or that substitute habitats and resources may be present (such as built structures instead of cliffs, or supplemental feeding), thus cities can actually be important for conservation at a larger scale. Peregrine falcons, ancestrally cliff nesters, nest on bridges and large buildings in the Meadowlands and New York City. The rare black redstart, a cliff-nesting songbird, nests on cliff-like buildings and war rubble in London (Morgan and Glue 1981). However, for some species the urban environment is not hospitable. For example, the American black duck, a species of conservation concern, was replaced by a generalist species, the mallard, with urbanization at a southern New Jersey site (Sugihara et al. 1979), and this story can be told about many other species.

LESSONS FROM THE MEADOWLANDS

How can the Meadowlands serve as a model for managing biodiversity in other cities? What can the experiences of other cities teach us about the Meadowlands? Biodiversity resources that are not identified cannot be conserved and managed, and even those that are identified may be at a high risk of destruction due to competing management goals, insufficient information, incomplete planning, or poor communication.

Lack of Study

Much of the urban biota in the U.S. is overlooked or ignored due to lack of scientific and natural history studies and insufficient concern on the part of agencies and individuals involved in conservation, planning, engineering, and management. Without deeper planning consideration and better field studies, urban biotas will continue to become simplified and less valued. There is much more to urban biodiversity than birds, fishes, and a few plants, as we have shown in our chapters about other groups of organisms. Surveys and assessments need to consider a broad range of taxa and habitats including unappealing wetland fill, shrub thickets, and abandoned quarries. Different taxa give us different kinds of information about the environment and the path to conservation, and any one taxon may not accurately predict the presence of others.

Broad-spectrum taxonomic and functional studies are generally avoided because they are expensive, demand unusual skills from large groups of experts, and are perceived as unnecessary. Moreover, natural history, the basic investigation of the What, Where, and When of the biota, is denigrated by many ecologists and difficult to fund and publish. We believe such studies and analyses are necessary in particularly valuable or important regions, and where management or development may cause intense changes and irrevocable loss of resources. Likewise, a broad geographic perspective is valuable. We learned much by considering areas that adjoin the Meadowlands District, namely the upper Hackensack River estuary, Newark Bay, the Hudson River, higher tributary reaches, and peripheral uplands. Large urban areas have an abundance of experienced scientists, science students, and talented naturalists, making diverse studies feasible with proper funding and supervision. Excellent, recently published field guides, many for obscure taxa, may help underpin a resurgence of natural history studies and multi-taxa surveys (see, e.g., Hinds and Hinds [2007] on lichens, Lincoln [2008] on liverworts, Bradley [2013] on spiders, McKnight et al. [2013] on mosses, and Skevington et al. [2019] on flower flies); however, expert taxonomic assistance is still in short supply. Biological surveys sometimes need better quality control over taxonomy and documentation. Web-based taxonomic and data storage systems such as iNaturalist are promising but need more supervision and assistance from experts. We cannot overemphasize that research and surveys must involve accurate identifications of organisms, with documentation by means of specimens, photographs, microscope slides, and field notes where appropriate. Mosses, liverworts, ferns and fern allies, algae and cyanobacteria, fungi, slime molds, most invertebrates, fishes of small or shallow waters, secretive marsh birds, reptiles, amphibians, and mammals are all in need of further study in the Meadowlands, as are most of the habitats.

Habitat Quality and Quantity

Availability of high quality habitat, especially of urban greenspaces, affects many species. The main proximate factor associated with urbanization is a lack of suitable habitats containing an adequate supply of food and other resources. Many animals require multiple habitat types during a year. Generalists, because of their flexibility in diet, overwintering sites, and nesting needs, are able to overcome this problem more than specialized species or those that have large home ranges. Relative availability of preferred habitats will also affect the presence and distribution of species.

The Meadowlands have extensive brackish wetland ecosystems of variable quality defined by plant diversity and composition, contamination, and disturbance. Sea level rise will result in loss of brackish wetlands in the Meadowlands. Wetlands dominated by common reed could be managed to increase habitat diversity while retaining the important non-habitat services provided by reedbeds, which may entail creative techniques other than attempts at eradication. These wetlands provide full-time or part-time habitat for many organisms, and some are completely reliant on them. Large, shallow pools surrounding radio broadcast antennas (Figure 11.2) exemplify habitat diversified inadvertently by management. Old field, meadow and shrubland habitats on abandoned landfills and industrial areas support small mammals such as meadow voles and white-footed mice. In turn, these animals are prey for meso-carnivores and raptors.

Figure 11.2 Pool at radio broadcast towers in freshwater marsh. Photo by Erik Kiviat.

Protection and active management of remaining forests, including tree plantings and selective removal of invasive plants, may be necessary for maintaining this important habitat type. Specific rare natural habitat such as forests, riparian and swamp forests in particular, and low stature uplands (meadow and shrubland) are key to diversity of many birds and mammals. Freshwater wetlands are also rare. The value of small remnants of relatively rare habitat types to biodiversity, and their protection, should be addressed in land use planning, management, and restoration (Croci et al. 2008). Maintaining or increasing the coverage of these habitats and improving their quality and long-term sustainability should also be priorities.

Attention to Rare and Unique Elements of Biodiversity

Many elements of urban biodiversity are rare in the landscape; it is among these taxa and habitats that we should look for indicators of urbanization stress.

The Meadowlands have diverse habitats. Some habitat patches have low species diversity yet support species of conservation concern not found elsewhere in the Meadowlands; temporary pools on dirt roads are an example. Some habitats must cover large areas in order to attract and hold area-sensitive species. Extensive reedbeds potentially support breeding northern harrier; extensive high salt marshes may support breeding seaside sparrow and salt-marsh sparrow. Yet, although many state-listed birds are associated with the extensive wetlands and waterways of the Meadowlands, many others are associated with upland habitats. The Meadowlands contain only limited remnants of once extensive forests, a habitat type on which many species depend.

The heterogeneity within urban areas, even more so than in the countryside, requires that biodiversity management be site-specific. Many currently practiced management methods and projects miss important biodiversity targets. The best Meadowlands example is the globally-rare Mattox's clam shrimp, a large population of which was lost in 2010 to a mitigation banking project. Other examples include state-listed species such as northern harrier. What is good for one species or group may not be good for another, and different taxa are not necessarily correlated in their occurrence on the landscape. Therefore, surveys or management for ducks, for example, may not identify or benefit many other organisms, even other waterbirds. Loss of tidal and nontidal wetlands has been extensive throughout New Jersey (Tiner 1985), thus the conservation of wetland-associated biodiversity has become an acute need.

Many species are absent or of restricted occurrence in urban regions. Yet there are many species of conservation concern in urban areas that are either mobile or able to persist in small areas of habitat, and even species that can take advantage of temporary habitats in, for example, vacant lots or brownfields. It should be possible to manage habitats and landscapes so that those species can thrive. The peregrine falcon, Atlantic Coast leopard frog, Mattox's clam shrimp, wafer-ash, floating marsh pennywort, swamp lily, and many cryptogams exemplify these opportunities in the Meadowlands.

The Meadowlands bird fauna is relatively well-known and offers some lessons for conservation. Many breeding species are only present at one or two sites in the Meadowlands. Some uncommon birds are supported by special local features, such as the nesting ravens on the 30 m high abandoned mine face at Laurel Hill; spotted sandpiper nesting on the crushed stone path at Harrier Meadow; or Virginia rails in the small springfed wet meadow west of the I-E Landfill. Each marsh or landfill is different, hosting different species of birds and other organisms. Thus, site-specific surveys and management actions are needed.

Continue to Improve Water Quality

Evidence indicates that poor water quality, including contaminants, is negatively affecting biodiversity. While there have been many improvements in water quality over the past few decades, there are still important sources of pollutants that need to be addressed. High summer temperatures, low oxygen levels, high nutrient levels, high total suspended solids (TSS), and pathogens are important stressors to address. Some critical targets for water quality are: improved stormwater management using green infrastructure and decreased impervious cover; upgrades in sewage treatment plants to include better nutrient removal; elimination of combined sewer overflows (CSOs); prevention and prompt cleanup of oil and chemical spills using nontoxic materials.

Management for the Urban Context

There is much biodiversity (and the ecosystem services it provides) in urban areas that is supporting itself and it is important to sustain and protect that rather than trying to turn urban habitats into rural-looking habitats, or into habitats we think used to be here. Furthermore, nature management must be sustained indefinitely to be of conservation value—it has limited benefit to conserve or manage habitat for a few years then ignore it and let it degrade or revert (unless similar habitats are cycling through suitability in nearby areas). Habitats and species that are thriving now will in most cases be easier to sustain and manage than habitats and species that are doing poorly in the Meadowlands (or elsewhere) because of problems such as habitat fragmentation, polluted water, rising sea level, salinization, common reed dominance of marshes, or specific threats including saltmarsh dieback.

Nonnative plants, even those species like common reed that are highly invasive in wetlands and on landfill cover, can provide important biodiversity support and other ecosystem services in urban areas. In many places it is not feasible or sustainable to remove all the nonnative plants to re-create a habitat that has been lost. Although stands of nonnative plants may need to be reduced in biomass at certain sites to manage specific habitat functions, weeds such as common reed, mugwort, purple loosestrife, knotweed, and multiflora rose provide self-sustaining vegetation that protects soils and sequesters carbon, as well as supporting certain taxa of associated organisms.

Remediate brownfields and convert them to *greenfields* (parks and preserves; Callus 2006) instead of redeveloping for industry, residence, and so forth, especially in areas prone to flooding now or in the future. Brownfields represent important temporary or transitional habitats for invertebrates, plants, and other taxa (Kattwinkel et al. 2011).

Built habitats including residential, commercial, industrial, park, and residential areas can be greatly improved for wildlife (MacDonald-Beyers 2008, Gallo et al. 2017). Maintaining large, native trees (alive and dead), downed wood, shrubs, and herbs in landscaped areas, capturing stormwater via rain gardens and other green infrastructure, allowing unmanicured areas, thickets, and woodland strips, and minimizing use of pesticides and fertilizers are a few management practices that would benefit wildlife. Large woody debris in the southern portion of Schmidt's Woods, for example, provides habitat for invertebrates, bryoids, fungi, and slime molds.

Fragmentation and Connectivity

Landscape-scale effects are also important in shaping the biota in the Meadowlands region. These include fragmentation of habitat and loss of connectivity between remnant habitats due to roads, waterways, and the matrix of human-dominated land use. Large intact areas of forest, marsh, and grassland can support wide-ranging animals as well as area-sensitive plants and their pollinators. Even some common species may be somewhat area-sensitive. For example, the number of mammal species and abundance was shown to increase with forest patch size in a Wisconsin study (Matthiae and Stearns 1981). Urban forest islands provide daytime cover and because they provide important resources including cover from predators or escape from herbivores, some species can reach high densities there. Local habitat composition was less important than land use characteristics surrounding urban greenspaces in influencing mammal abundance (VanDruff and Rowse 1986); this balance may be expressed in different ways for different taxa.

Ability of historically extirpated species to recover in the Meadowlands or for organisms to exploit newly emerging habitats, is likely hampered by an inhospitable matrix and high density of major roadways. Some species may have difficulty colonizing isolated patches. Species with large area requirements, such as carnivores and some turtles, are inhibited by numerous road crossings and mortality from collisions. Therefore, they may exist at extremely low densities or not at all.

There is a great opportunity to assess habitat connectivity for terrestrial and aquatic wildlife and improve that connectivity where appropriate using the guidelines for the New Jersey Department of Environmental Protection program *Connecting Habitat Across New Jersey* (CHANJ) and the *North Atlantic Aquatic Connectivity Consortium* (NAACC). Green corridors should be maintained or created where absent and desirable, especially among core habitats in the Meadowlands. The Meadowlands may already have a high level of habitat connectivity for some species because of the extensive watercourses, where culverts and bridges may potentially provide safe crossings under roadways for aquatic organisms and some terrestrial mammals, reptiles, and amphibians. There are also opportunities to install wildlife crossings over and under roads, such as was done across Interstate-78 elsewhere in northeastern New Jersey or retrofit abandoned rail trestles and tunnels for this purpose. Dams and weirs upstream may impede movement of some terrestrial species and are also an important target for improving fish passage. Attention should be paid to restoring and managing vegetated buffers in these places, especially along streams and estuaries.

Sources of Mortality and Morbidity

Several additional stressors could be removed or controlled to minimize their effects on wildlife survival and reproduction in urban regions. They include addressing the negative effects of industrial pollution, use of rodenticides that cause secondary poisoning of scavengers and predators, use of insecticides (bats and many birds require insect food), predation by free-ranging and feral house cats on wildlife, and barriers to movement and road mortality hazards (e.g., Jersey barriers need to have wildlife breaks; see Forman et al. [2003]). Education will help people minimize negative human-wildlife interactions, especially with carnivores such as coyotes and skunks, because these interactions often result in persecution of animals. In addition, marine mammals deserve more attention in the Meadowlands region and New York City estuarine delta given their use of these habitats and the hazards they face. Most species are federally endangered and are protected under the Endangered Species Act, Marine Mammal Protection Act, Convention on International Trade in Endangered Species of Wild Fauna and Flora (CITES), and International Whaling Commission (IWC) laws and regulations. Seals are among the marine mammal species that the New Jersey Department of

Environmental Protection Endangered and Nongame Species Program identified as in need of monitoring, research and conservation in the state (NJDEP
2006). Identified threats to this group include fishing gear, habitat loss and
degradation, disease, approach and harassment by boats and aircraft, direct
killing by people, and PCB contamination. Identifying and protecting *haul
out* sites, educating various entities to minimize disturbance, and determining
the effects of contaminants on seals are considered among the priorities for
conservation (NJDEP 2006). Speed restrictions for vessels should be implemented when whales are present as ship strikes are a major cause of mortality
in this group.

Biodiversity Can Inform Planning and Best Management Practices (BMPs) for Land Use

Habitats and species continue to be lost to construction for residences, commerce, industry, transportation, education, recreation, landfill remediation,
and wetland mitigation. For example, walking trails should not follow shorelines because shorelines are biodiversity hot zones in urban as well as rural
areas; trails should be set back from shorelines with occasional spur trails
and observation platforms to afford walkers a view of the waterway and its
wildlife. Many of the corporate properties in the Meadowlands and elsewhere
either contain habitats with the potential to support species of conservation
concern, or have habitats that could be gently and inexpensively managed
for that purpose. Solar panels, super-insulation, pervious pavement, electric
vehicles, and innovative treatment and use of stormwater and graywater are
valuable measures for reduction of the environmental footprints of office
buildings, warehouses, and factories. But true sustainability requires explicit
and active biodiversity conservation and management. The loss of biodiversity is an externality of land use and environmental management; in many
cases, marginal additional expenditures prior to permitting and during development can help identify and avoid harm to biological resources that would
be much more costly or impossible to protect or restore after the fact.

Climate Change

The effects of sea level rise and increased river and estuary flooding due to
climate change put tidal and nontidal wetland habitats and the large number
of specialized species they support at risk. It also further jeopardizes water
quality when big storms cause inputs of raw sewage into waterways from
CSOs. Because of inevitably increasing flood hazards, it makes better sense
to have less development and more biodiversity (and certain kinds of ecotourism and recreation) in the Meadowlands and other low-lying coastal lands.

Dikes, tide gates, and pumps cannot indefinitely hold off intensifying storm surges and rising sea level, as has been learned in major coastal cities like New York City, Houston, and New Orleans. Even in the Netherlands, with their famously engineered coastal protection, there have been hundreds of dike failures that were expensive in dollars or human lives (Van Baars and Van Kempen 2009).

Planning to maintain vegetated intertidal habitats and create new ones where possible should be priorities. Human adaptation to rising sea level, by means of managed realignment of coastal wetlands, undevelopment, protection of existing low-lying greenspaces, and conversion of brownfields to greenspaces can have enormous benefits for biodiversity with a modest consideration in planning, although social, economic, and political problems need to be solved equitably for the people involved.

LESSONS FROM SIMILAR URBAN REGIONS

Other regions offer some lessons that provide context for the Meadowlands and for urban biodiversity in general.

The Great Swamp, New Jersey

The Great Swamp, in a glacial lake basin like the Meadowlands, is 20 km west of the Greater Meadowlands. The Great Swamp is surrounded by suburban development with little industrial influence and has many species that the Meadowlands lack. For example, the northern water snake and vernal pool breeding salamanders are common in the Great Swamp but absent or very rare in the Meadowlands. Being the headwaters of the Passaic River, instead of an estuary, without high salinity or concentrated contaminants, retaining extensive forests and natural soils, and having a much different land use history and current status as a National Wildlife Refuge, explain many differences between these two regions.

The San Francisco Bay and Sacramento—
San Joaquin Delta, California

This region is similar to the Meadowlands in being an extensive, sheltered, estuarine delta greatly altered by urbanization and agriculture. Only 5% of the pre-European wetlands of the Bay-Delta remain (Josselyn 1983). The region is situated on a major bird migration flyway, as are the Meadowlands, and in the Bay-Delta the marsh birds, water birds and shorebirds concentrate in the fragments of remaining wetlands during migration and winter. Rising sea level and sediment starvation are contributing to marsh loss. The upland edge of the tidal marsh is a critical zone that provides shelter to marsh animals during extreme high tides (*ibid.*). Tidal wetlands historically filled with dredge spoil or converted to salt production ponds are being restored to tidal wetlands, and wildlife habitat (including for endangered species) is one of the important goals of restoration (*ibid.*). The results of restoration projects have been mixed. In one marsh restoration site in the Delta, competing goals include flood management, waterfowl habitat, and mosquito control, and there seems little opportunity for biodiversity conservation (Kiviat, pers. obs.).

The Tinicum Marsh in Philadelphia, Pennsylvania

This urban wetland (now John Heinz National Wildlife Refuge) is about 250 km southwest of the Meadowlands. Tinicum is part of a sheltered delta region of the Delaware River estuary (McCormick 1970). The site contains open tidal marshes, old fill, and tidally restricted impoundments with dikes. Tinicum supports a diverse biota, including important breeding and migrant populations of marsh and water birds, and is intensively used for passive recreation (especially walking on the dike trails and bird-watching). Some of the diked marshes have recently been reopened to the tides for ecological restoration.

Baltimore

The City of Baltimore, Maryland, an old urban area in a sheltered estuarine delta region, is about 575 km southwest of the Meadowlands. In the Baltimore Ecosystem Study (Pickett et al. 2008), the city was found to support substantial native as well as nonnative biodiversity, including newly-described invertebrates, rare plants, and a large proportion of the regional species pool of birds. Soils were found to be highly heterogeneous (as were aboveground habitats) and included remnant natural patches as well as artificial soils and paved areas. The Baltimore Ecosystem Study was heuristic in developing new knowledge of urban ecology in part because that study examined a broad spectrum of taxa, ecological functions, and human socioeconomic structures.

Other urban-industrial regions around the world echo the constraints and opportunities for biodiversity present in the Meadowlands. Although they share many characteristics, each region is unique and requires its own multi-taxa biological surveys and biology-sensitive land use planning and conservation in order to recognize and respect the biota and habitats.

CONCLUSIONS

At one time the Meadowlands and other urban wetlands and rivers were considered undesirable wastelands where pathogen-infected, chemically contaminated water was given little consideration for its habitat functions for local biodiversity and ecosystem services to people; developers, industries, and politicians took advantage of this perception for financial gain and until recently, government agencies acceded. Now, we know better but we are left with deciding how to rectify and in some cases accept the damage that has been done due to unfettered neglect of urban greenspaces and biodiversity. Perceptions of urban nature have changed greatly toward a desire to protect biodiversity and habitats where people live, work, and play. In a literal and figurative sense the Meadowlands and other urban estuaries have gone from being considered trash to treasure.

Given multiple alarms at the apparent loss of large portions of biodiversity nationally and globally, the more common species of today and their habitats may eventually become the rare species and habitats of tomorrow and thus they warrant our attention now; this is also true of the Meadowlands where some of what is common elsewhere is not so common in the urban environment. However, urban biodiversity includes many nonnative species and many rare native species. In our consideration of biodiversity, we emphasize the need to accept some aspects of the urban context and conditions in attempting to manage biodiversity, including established and ubiquitous nonnative species such as common reed, because they are expensive financially and ecologically to treat, nearly impossible to eradicate, and they contribute to important ecosystem services. We also emphasize the need for including species and habitats of conservation concern, meaning those that are rare at some level, vulnerable, ecologically specialized, exemplify aspects of nature, urban-sensitive, or important to people in other ways. Many of these species are most likely to disappear from the Meadowlands and other urban areas without explicit and timely conservation management.

Many threats to these urban biodiversity treasures still persist. Commenting on the master planning process for the Meadowlands District, Freudenberg (2019) opined:

> We have an opportunity to showcase the Meadowlands as a living case study in climate resilience. It is one of our region's cherished environmental treasures, and how we adapt it should be a model for coastal states across the country. . . . This is not a time to be timid. . . . It was the bold vision and dedicated commitment of environmentalists and state leaders that saved the Meadowlands from destruction decades ago. We now stand on the edge of an equally, if not more perilous threat—accelerating sea level rise. The way we address it must reflect

the same level of boldness and commitment to decisive action, and we must start today.

Inasmuch as most of the Meadowlands District, and large portions of the rest of the Greater Meadowlands, are within 1–3 m of sea level, the existing level of residential, commercial, industrial, transportation, solid waste, and government land use is clearly unsustainable. It only remains to devise a way to shift much of this capital higher and farther inland, such that all people are treated fairly, services are maintained, and the environment is left cleaner and more intact ecologically. Each coastal or floodplain city is unique, and the problem of long-term adaptation to climate and hydrological change must be solved for many urban areas worldwide.

Neutral, scientific reviewers or advisory committees could help policy makers and managers avoid conflicts of interest and short-sighted decisions too influenced by politics or money, and focus on long-term investments in greenspace, biodiversity, and other ecosystem services—resources that can only become more valuable as they become more scarce. The Meadowlands are a perfect setting for experiments and demonstrations guided by neutral science, because it is a marvel of the resilience of nature that there are habitats and species of conservation concern there despite the odds—and we have a trust responsibility to foster them. We will do so by making decisions that are site-specific, goal-directed, and responsive to new scientific findings.

Acknowledgments

Spider Barbour, Kerry Barringer, Chanda Bennett, Taré Gantt, Kali Bird, Brett Bragin, Megan Callus, Hugh Carola, Steve Covacci, Dwane Decker, Terry Doss, Chris Graham, Dee Ann Ipp, Richard Kane, Sarahfaye Mahon, Eric Martindale, Drew McQuade, Aleshanee Mooney, Mike Newhouse, Beth Ravit, Bob Schmidt, Tina Schvejda, Bill Sheehan, Natalie Sherwood, Don Smith, Kyle Spendiff, Bill Standaert, Lea Stickle, and Maggie Wellins guided or assisted on field trips. We also benefited from Torrey Botanical Society field trips, a New Jersey Audubon Society bird outing, a Meadowlands Environment Center boat tour with Gabrielle Bennett-Meany, and a Hudsonia staff field excursion.

We benefited from discussions with many scientists and naturalists including Francisco Artigas, Kirk Barrett, David Bart, Gabrielle Bennett-Meany, Hugh Carola, David Ehrenfeld, Joan Ehrenfeld (deceased), Ross Feltes, Michael Firth and colleagues at ELM Inc., Linda Fisher, Joan Hansen, Jean Marie Hartman, Ray Hinkle, Margaret McBrien, John Quinn (deceased), Steve Quinn, Uta Gore, Susan Fox Rogers, David Strayer, Chris Takacs, Don Torino, Judy Weis, Pete Weis, Jim Wright, and participants in several Hudsonia workshops on urban biodiversity.

Douglas Appenzeller, Alana Buonaguro, Rudi Gohl, Ben Harris, Cathy McGlynn, Natalie Narotzky, Julia Palmer, Lea Stickle, and Aven Williams assisted with preparation of the cryptogams chapter. Chris Graham and Elizabeth Castle helped compile the seed plants list. Larissa Wohl extracted data on the species pools of Bergen and Hudson counties. Lea Stickle spent many hours sorting and formatting the references cited and assisted with other aspects of the book.

Carolyn Bentivegna (deceased), Brett Bragin, Dee Ann Ipp, Drew McQuade, and Don Smith loaned or donated specimens. Stuart Findlay (Cary Institute of Ecosystem Studies) performed water quality analyses in the laboratory. Dean Bryson (New Jersey Department of Environmental Protection) contributed data on benthic macroinvertebrates. Melissa Mitchell Thomas

(Raritan Headwaters) provided the land use—land cover analysis for the Meadowlands District.

We thank the taxonomists who identified or confirmed specimens for us: Richard Harris (New York Botanical Garden, deceased), Robert Dirig (Cornell University), and Dennis Waters (lichens); Roy Halling (NYBG; fungi); Susan A. Williams (Rowe, Massachusetts), Eric Karlin (Ramapo College), Nancy Slack (the Sage Colleges), and Alan Whittemore (bryophytes); Ken Karol (NYBG, stoneworts); Christopher Graham (Hudsonia, vascular plants), James G. (Spider) Barbour (Hudsonia, vascular plants, insects), Kerry Barringer (Brooklyn Botanic Garden retired, vascular plants, bryophytes), Gerry Moore (BBG, seed plants), Mihai Costea (Laurier University, dodders), John T. Mickel (NYBG, ferns), Robert E. Schmidt (Hudsonia, fishes, invertebrates), Kathleen A. Schmidt (Hudsonia, terrestrial mollusks), Robert Daniels (New York State Museum, crayfish), C. Barre Hellquist (Massachusetts College of Liberal Arts, submergent plants), Richard Hoebeke (Cornell University, beetles), Jim Springer (North American Butterfly Association, butterflies), Chris T. Maier (Connecticut Agricultural Experiment Station, deer flies), Julian Stark (American Museum of Natural History, flies), Ernie Schuyler (Academy of Natural Sciences, plants), Ken Soltesz (deceased; odonates), and Paula Mikkelsen (Cornell University, pearl oyster).

Steve Clemants (deceased) and Kerry Barringer provided unpublished data from the New York Metropolitan Flora project at Brooklyn Botanic Garden and other sources. David Dickey provided herpetofaunal collections data and Neil Duncan mammal data from the American Museum of Natural History. The Marine Mammal Stranding Center sent data on seals and cetaceans.

Julianna Zdunich edited photographs for an early version of the manuscript. Kristen Bell Travis drafted the maps which were edited by Lauren Bell and Lea Stickle. Kristen also prepared portions of Chapter 2.

The following persons commented on drafts of chapters or portions thereof: Susan A. Williams (bryophytes), John Mickel (New York Botanical Garden, (pteridophytes), Natalie Howe (USDA, lichens), Brett Bragin (NJSEA, fishes), Parker Gambino (invertebrates), Louis Sorkin (AMNH, invertebrates), James G. (Spider) Barbour (Hudsonia, invertebrates), Kathleen A. Schmidt (invertebrates, especially terrestrial mollusks), Robert E. Schmidt (invertebrates), Jim Springer (North American Butterfly Association, butterflies), Michael May (Rutgers University, odonates), Jason Tesauro (Jason Tesauro Consulting, herpetofauna), Brian Zarate (NJDEP, herpetofauna), Nellie Tsipoura (New Jersey Audubon, birds), Michael Britt (*New Jersey Birds*, birds), Ken Witkowski (U.S. Fish and Wildlife Service, birds), Don Torino (birds), Randall FitzGerald (Montclair State University, mammals), Kathleen LoGiudice (Union College, mammals), Chanda Bennett (Columbia University, bats), Richard Harris (NYBG, lichens), Ken Karol (NYBG, stoneworts), William Buck (NYBG, bryophytes), Kerry Barringer (Brooklyn Botanic Garden retired, seed plants, liverworts), Bill Standaert (Salisbury University, seed plants), Jay Kelly (Raritan Valley Community College, uplands and forested wetlands), Emilie Stander (RVCC, marshes and ponds), Claus Holzapfel (Rutgers University, several chapters), Alexander Gates (Rutgers, environmental setting), Julianna Zdunich (Hudsonia, several chapters). We are especially grateful to Hugh Carola (Hackensack Riverkeeper) and an anonymous reviewer for valuable comments on the entire manuscript. Given the broad range of taxa and other subjects, it has been challenging to keep everything accurate, nomenclaturally correct, and up-to-date, and we take responsibility for any errors.

Courtney Morales, Emma Ebert, Dominique McIndoe, Kasey Beduhn, Mikayla Mislak, Becca Beurer, and Michael Gibson were our editors at Lexington Books. Lisamarie Windham and William Sipple peer reviewed an earlier version of the manuscript for a different publisher. James Gifford (Fairleigh Dickinson University) steered us to Lexington Books.

We benefited from use of the facilities of the Bard College Field Station and Stevenson Library, Rutgers University, and the New Jersey Meadowlands Commission—MERI - NJSEA.

Kristi: I thank my family and friends for their patience and support over the years while this project took me away from them. I am especially grateful to Gavin Beyers, John McLaughlin, and Mary, Heather, Maeve and Stella MacDonald. Also, if not for my father, Robert MacDonald, an avid birdwatcher and nature photographer in urban wild places in and around Jersey City where I spent my early childhood, I might not have developed an appreciation for the New Jersey Meadowlands and a desire to protect them. I am indebted to Cindy Ehrenclou, Executive Director of Raritan Headwaters, for understanding my need to take chunks of time away from the Upper Raritan watershed to work on this project. Finally, I thank the many people who have worked tirelessly over the years to protect the Meadowlands from further loss and harm.

Erik: I especially thank my wife, Elaine Colandrea, for her patience during the two-decade genesis of this project. I am most grateful to all who have come before, in the Meadowlands and beyond, including my parents, mentors, colleagues, collaborators, communicants, and students.

Funding for the book project was granted by the Geraldine R. Dodge Foundation, Furthermore: a program of the J.M. Kaplan Fund, Metropolitan Conservation Alliance, Geoffrey C. Hughes Foundation, Emma Barnsley Foundation, Conserve Wildlife Foundation New Jersey, Will Nixon, and New Jersey Division of Fish & Wildlife's Conserve Wildlife Matching Grants. Funding for other work in the Meadowlands that contributed data to this effort came from the New Jersey Meadowlands Commission, Mary Jean and Frank P. Smeal Foundation, the H2O Fund (Highlands to Ocean Fund), Hackensack Meadowlands Partnership, Natural Resources Defense Council, ELM Inc., New Jersey Department of Environmental Protection, Rutgers Environmental Law Clinic, and Bergen County Audubon Society.

This book is a Hudsonia - Bard College Field Station Contribution. We hope this project benefits the animals, plants, fungi, and other organisms and their habitats in cities and industrial landscapes everywhere.

Appendix 1: Seed Plants

Seed plants of the New Jersey Meadowlands.

Scientific name	Common name	Family	Year	NYMF	BKL	BCFS	EKFN	KB	TBS	SIP	BCSA	N/I	HABIT
Abutilon theophrasti	Velvetleaf	Malvaceae	2015								x	I	h
Acalypha australis	Copperleaf $	Euphorbiaceae	1991	x								I	h
Acalypha rhomboidea	Common three-seeded mercury	Euphorbiaceae	2015	x					JL		x	N	h
Acalypha virginica	Virginia three-seeded mercury	Euphorbiaceae	1919						L	H		N	h
Acer ginnala	Ginnala maple $	Aceraceae	2010			x						I	t
Acer negundo	Box-elder	Aceraceae	2015	x					LH		x	N	t
Acer platanoides	Harlequin maple $	Aceraceae	2015	x					L		x	I	t
Acer rubrum	Red maple	Aceraceae	2015	x					LBH		x	N	t
Acer saccharinum	Silver maple	Aceraceae	2015	x					LBH	HB	x	N	t
Acer saccharum	Sugar maple	Aceraceae	2015	x					LB	H	x	N	t
Achillea millefolium var. occidentalis	Yarrow	Asteraceae	2015						H		x	N	h
Aegopodium podagraria	Bishop's goutweed	Apiaceae	1889			x				B		I	h
Aesculus	Buckeye	Hippocastanaceae	2010				x					I	t
Agalinis purpurea	Purple false-foxglove	Scrophulariaceae	1919							HB		N	h
Agalinis setacea	Thread-leaved false-foxglove	Scrophulariaceae	1889		x							N	h

Scientific name	Common name	Family	Year	NYMF	BKL	BCFS	EKFN	KB	TBS	SIP	BCSA	N/I	HABIT
Agalinis tenuifolia	Slender-leaved false-foxglove	Scrophulariaceae	1893									N	h
Agastache nepetoides	Yellow giant hyssop	Lamiaceae	1889							B		N	h
Ageratina altissima var. altissima	White snakeroot	Asteraceae	2015	x		x			LH	T	x	N	h
Agrimonia parviflora	Small-flowered agrimony	Rosaceae	1889							B		N	h
Agrostis gigantea	Redtop	Poaceae	2015						HP		x	I	h
Agrostis hyemalis	Winter bentgrass	Poaceae	1948	x								N	h
Agrostis perennans	Upland bentgrass	Poaceae	2015	x					LH		x	N	h
Agrostis scabra	Ticklegrass	Poaceae	2015								x	N	h
Agrostis stolonifera	Creeping bentgrass	Poaceae	2006	x					L			I	h
Ailanthus altissima	Tree-of-heaven	Simaroubaceae	2015	x					LBH		x	I	t
Albizia julibrissin	Mimosa	Fabaceae	2006	x					PB			I	t
Aletris farinosa	Colic root	Liliaceae	1889							B		N	h
Alisma subcordatum	American water-plantain	Alismataceae	2015	x		x						N	h
Alliaria petiolata	Garlic-mustard	Brassicaceae	2015	x					LBH		x	I	h
Allium canadense	Meadow garlic	Liliaceae	1867		x							N	h
Allium vineale	Wild garlic	Liliaceae	2006	x		x			L			I	h
Alnus glutinosa	Black alder $	Betulaceae	2017			x			B			I	t
Alnus serrulata	Smooth alder	Betulaceae	1908	x								N	s
Alopecurus aequalis	Short-awned foxtail	Poaceae	1889							B		N	h
Amaranthus albus	Prostrate pigweed	Amaranthaceae	2015	x							x	I	h

Scientific name	Common name	Family	Year								
Amaranthus blitum	Purple amaranth	Amaranthaceae	2000				P			l	h
Amaranthus cannabinus	Tidewater-hemp	Amaranthaceae	2015		x		L		x	N	h
Amaranthus cruentus	Red amaranth	Amaranthaceae	1889	x	x					l	h
Amaranthus hybridus	Slim amaranth	Amaranthaceae	1902	x	x					N	h
Amaranthus retroflexus	Red-rooted amaranth	Amaranthaceae	2003			x		H		N	h
Ambrosia artemisiifolia	Common ragweed	Asteraceae	2015			x	LBH	H	x	N	h
Ambrosia trifida	Giant ragweed	Asteraceae	2007	x				EH		N	h
Amelanchier canadensis	Canadian shadbush	Rosaceae	1949	x				H		N	t
Amelanchier stolonifera	Running shadbush	Rosaceae	2012		x			EH		N	s
Ammannia coccinea	Scarlet toothcup	Lythraceae	2006				L	B		N	h
Ammannia robusta	Sessile toothcup	Lythraceae	2006			x	L			N	h
Amorpha fruticosa	False-indigo	Fabaceae	2009			x	H	B		N	s
Ampelopsis brevipedunculata	Porcelainberry	Vitaceae	2015			x	H		x	l	wv
Amphicarpaea bracteata	Hog-peanut	Fabaceae	2000			x				N	hv
Anagallis arvensis	Scarlet pimpernel	Primulaceae	2006			x	LB			l	h
Anaphalis margaritacea	Pearly everlasting	Asteraceae	2006				L			N	h
Andromeda polifolia var. glaucophylla	Bog-rosemary	Ericaceae	1882		x					N	s
Andropogon gerardii	Big bluestem	Poaceae	2017			x	Li			N	h
Andropogon virginicus	Broomsedge bluestem	Poaceae	2006			x	L	HT		N	h
Anemone canadensis	Canadian anemone	Ranunculaceae	1889					B		N	h
Anemone quinquefolia	Wood anemone	Ranunculaceae	2012			x		B		N	h
Angelica atropurpurea	Angelica	Apiaceae	1950					T		N	h

Scientific name	Common name	Family	Year	NYMF	BKL	BCFS	EKFN	KB	TBS	SIP	BCSA	N/I	HABIT
Antennaria parlinii ssp. parlinii	Parlin pussytoes	Asteraceae	1895		x							N	h
Anthoxanthum odoratum	Sweet vernal grass	Poaceae	2015	x					L		x	I	h
Apios americana	Groundnut	Fabaceae	2001	x		x			H	E		N	hv
Apocynum ×floribundum [androsaemifolium × cannabinum]	Intermediate dogbane	Apocynaceae	2006						L			N	h
Apocynum androsaemifolium	Spreading dogbane	Apocynaceae	1819							T		N	h
Apocynum cannabinum	Hemp dogbane $	Apocynaceae	2015	x	x	x			LBH	H	x	N	h
Arabidopsis thaliana	Mouse-ear cress	Brassicaceae	2002			x				T		I	h
Arabis canadensis	Rockcress	Brassicaceae	1950							T		N	h
Aralia elata	Angelica tree $	Araliaceae	2008				x					I	t
Aralia hispida	Bristly sarsaparilla	Araliaceae	1889							B		N	h
Aralia nudicaulis	Wild sarsaparilla	Araliaceae	2005	x			x		BLo			N	h
Arctium lappa	Great burdock	Asteraceae	2009									I	h
Arctium minus	Lesser burdock	Asteraceae	2015	x		x			LH	H	x	I	h
Arenaria serpyllifolia	Thymeleaf sandwort	Caryophyllaceae	2017			x						I	h
Arethusa bulbosa	Dragon's mouth	Orchidaceae	1889							BT		N	h
Arisaema triphyllum s.l.	Jack-in-the-pulpit	Araceae	2009			x						N	h
Aristida dichotoma	Churchmouse three-awn	Poaceae	2006	x					L			N	h
Aristida oligantha	Prairie three-awn	Poaceae	2006						L			N	h
Aristolochia serpentaria	Virginia snakeroot	Aristolochiaceae	1889							BT		N	h

Scientific name	Common name	Family					E		N	s
Aronia ×prunifolia [arbutifolia × melanocarpa]	Purple chokeberry	Rosaceae	1949						N	s
Aronia arbutifolia	Red chokeberry	Rosaceae	1991			Lo	H		N	s
Aronia melanocarpa	Black chokeberry	Rosaceae	2015	x		Lo	H		N	s
Artemisia annua	Sweet Annie	Asteraceae	2015	x		L		x	I	h
Artemisia dracunculus	Tarragon	Asteraceae	2007		x				I	h
Artemisia vulgaris	Mugwort	Asteraceae	2015	x		LBH		x	I	h
Arthraxon hispidus	Small carpetgrass	Poaceae	2007		x				I	h
Asarum canadense	Wild-ginger	Aristolochiaceae	2015				T	x	N	h
Asclepias amplexicaulis	Clasping milkweed	Asclepiadaceae	1889	x			B		N	h
Asclepias exaltata	Poke milkweed	Asclepiadaceae	1893		x				N	h
Asclepias incarnata	Swamp milkweed	Asclepiadaceae	2019	x			EH		N	h
Asclepias syriaca	Common milkweed	Asclepiadaceae	2015	x		LB		x	N	h
Asclepias viridiflora	Green milkweed	Asclepiadaceae	1889			LB	B		N	h
Asparagus officinalis	Asparagus	Liliaceae	1999	x					I	h
Atriplex prostrata	Orache	Chenopodiaceae	2015	x	x	LP	H	x	N	h
Avena sativa	Common oats	Poaceae	1919						I	s
Baccharis halimifolia	Groundselbush	Asteraceae	2019	x		LB		x	N	h
Barbarea verna	Early wintercress	Brassicaceae	2012	x		L			I	h
Barbarea vulgaris	Common wintercress	Brassicaceae	2016	x				x	I	h
Bassia scoparia	Summer-cypress	Chenopodiaceae	2006			L			I	h
Berberis thunbergii $	Thunberg barberry	Berberidaceae	2006			L			I	s
Berberis vulgaris	Common barberry	Berberidaceae	1889				B		I	s
Betula alleghaniensis	Yellow birch	Betulaceae	1894	x					N	t

Scientific name	Common name	Family	Year	NYMF	BKL	BCFS	EKFN	KB	TBS	SIP	BCSA	N/I	HABIT
Betula lenta	Black birch	Betulaceae	2006	x					L	H		N	t
Betula nigra	River birch	Betulaceae	2016			x			H			N	t
Betula populifolia	Gray birch	Betulaceae	2015	x					BH	EH	x	N	t
Bidens bipinnata	Needles $	Asteraceae	1895		x							N	h
Bidens cernua	Nodding bur-marigold	Asteraceae	2008							B		N	h
Bidens connata	Purple-stemmed beggarticks	Asteraceae	2007	x	x			x				N	h
Bidens coronata	Crowned beggarticks	Asteraceae	1949	x						EHT		N	h
Bidens discoidea	Small beggarticks	Asteraceae	1991	x					Lo			N	h
Bidens frondosa	Devil's beggarticks	Asteraceae	2015	x		x			LP	H		N	h
Bidens laevis	Smooth bur-marigold	Asteraceae	1948	x								N	h
Boehmeria cylindrica	False-nettle	Urticaceae	2015	x			x			T		N	h
Bolboschoenus fluviatilis	River bulrush	Cyperaceae	1894		x							N	h
Bolboschoenus robustus	Saltmarsh bulrush	Cyperaceae	2015	x		x					x	N	h
Brachyelytrum erectum	Bearded shorthusk	Poaceae	1991	x		x						N	h
Bromus ciliatus	Fringed brome	Poaceae	2009			x						N	h
Bromus inermis	Smooth brome	Poaceae	2005						B			N	h
Bromus tectorum	Cheatgrass	Poaceae	2017	x		x			LP			I	h
Bromus arvensis	Field brome	Poaceae	2015	x		x			BLP		x	I	h
Broussonetia papyrifera	Paper-mulberry	Moraceae	2006	x					L			I	t
Buddleja davidii	Butterfly bush	Buddlejaceae	2012			x						I	h

Scientific name	Common name	Family	Year										
Bulbostylis capillaris	Hair-rush	Cyperaceae	2008		x			L	H			N	h
Cakile edentula	Sea rocket	Brassicaceae	1819						N			N	h
Calamagrostis canadensis	Bluejoint	Poaceae	2015	x	x			B	H	x		N	h
Calamagrostis coarctata	Arctic reedgrass	Poaceae	1889		x				B			N	h
Calamagrostis epigeios	Feathertop reedgrass	Poaceae	2015	x				P		x		I	h
Calla palustris	Wild calla	Araceae	1819						T			N	h
Callitriche heterophylla	Greater water starwort	Callitrichaceae	2009			x			B			N	h
Callitriche stagnalis	Pond water starwort	Callitrichaceae	2009			x						I	h
Calopogon tuberosus var. *tuberosus*	Grass pink	Orchidaceae	1919						H			N	h
Caltha palustris	Marsh-marigold	Ranunculaceae	1948	x					T			N	h
Calystegia sepium	Bindweed	Convolvulaceae	2015	x	x			LH	EH	x		NI	hv
Camelina sativa	Gold-of-pleasure	Brassicaceae	1894		x							I	h
Campanula aparinoides	Marsh bellflower	Campanulaceae	1819						T			N	h
Campanula rotundifolia	Harebell	Campanulaceae	1889						B			N	h
Cannabis sativa	Marijuana	Cannabaceae	1894			x						I	h
Capsella bursa-pastoris	Shepherd's purse	Brassicaceae	2016				x					I	h
Cardamine bulbosa	Springcress	Brassicaceae	2017			x						N	h
Cardamine concatenata	Cut-leaved toothwort	Brassicaceae	1819						T			N	h
Cardamine hirsuta	Hairy bittercress	Brassicaceae	2017			x		Lo				I	h
Cardamine impatiens	Narrow-leaved bittercress	Brassicaceae	2005					B				I	h
Cardamine pensylvanica	Pennsylvania bittercress	Brassicaceae	1948	x								N	h
Cardamine pratensis	Cuckoo flower	Brassicaceae	1889						B			N	h
Carex ?atlantica	Prickly bog sedge	Cyperaceae	200?			x			T			N	h

Scientific name	Common name	Family	Year	NYMF	BKL	BCFS	EKFN	KB	TBS	SIP	BCSA	N/I	HABIT
Carex aggregatus	Glomerate sedge	Cyperaceae			x							N	h
Carex alata	Broad-winged sedge	Cyperaceae	2008			x						N	h
Carex albicans	White-tinged sedge	Cyperaceae	2017		x	x						N	h
Carex albolutescens	Greenish-white sedge	Cyperaceae	2015			x					x	N	h
Carex annectens	Yellow-fruited sedge	Cyperaceae	1894		x							N	h
Carex blanda	Eastern woodland sedge	Cyperaceae	2006			x						N	h
Carex bromoides	Brome-like sedge	Cyperaceae	2008		x							N	h
Carex canescens	Hoary sedge	Cyperaceae	1895		x							N	h
Carex cephalophora	Oval-headed sedge	Cyperaceae	2017			x					x	N	h
Carex collinsii	Collins' sedge	Cyperaceae	1889							BT		N	h
Carex comosa	Bristly sedge	Cyperaceae	1819							T		N	h
Carex crinita	Fringed sedge	Cyperaceae	2017	x		x			Lo		x	N	h
Carex cristatella	Crested sedge	Cyperaceae	2015			x						N	h
Carex echinata	Star sedge	Cyperaceae	1895		x							N	h
Carex festucacea	Fescue sedge	Cyperaceae	2013			x		x				N	h
Carex folliculata	Northern long sedge	Cyperaceae	1991	x					Lo			N	h
Carex gracillima	Graceful sedge	Cyperaceae	1889							B		N	h
Carex hirsutella	Hirsute sedge	Cyperaceae	2015			x						N	h
Carex hirta	Hairy sedge	Cyperaceae	2015			x						I	h
Carex hormathodes	Marsh straw sedge	Cyperaceae	1895		x						x	N	h
Carex lacustris	Lake sedge	Cyperaceae	1889							B		N	h

Scientific name	Common name	Family	Year						
Carex laevivaginata	Smooth-sheathed sedge	Cyperaceae	2009	x				N	h
Carex lasiocarpa	Woolly-fruited sedge	Cyperaceae	2006	x				N	h
Carex longii	Long's sedge	Cyperaceae	2008	x				N	h
Carex lupulina	Hop sedge	Cyperaceae	2016	x		BT		N	h
Carex lurida	Sallow sedge	Cyperaceae	1991		Lo		x	N	h
Carex molesta	Troublesome sedge	Cyperaceae	2013	x				N	h
Carex muehlenbergii	Muhlenberg sedge	Cyperaceae	2017	x		B		N	h
Carex normalis	Greater straw sedge	Cyperaceae	2019	x			x	N	h
Carex pellita	Woolly sedge	Cyperaceae	1894		x			N	h
Carex pensylvanica	Pennsylvania sedge	Cyperaceae	2015			B	x	N	h
Carex platyphylla	Broad-leaved sedge	Cyperaceae	1889			B		N	h
Carex prasina	Drooping sedge	Cyperaceae	1889			B		N	h
Carex pseudocyperus	Cyperus-like sedge	Cyperaceae	1894		x			N	h
Carex radiata	Eastern star sedge	Cyperaceae	2017	x	x			N	h
Carex scoparia	Broom sedge	Cyperaceae	2015	x				N	h
Carex spicata	Prickly sedge	Cyperaceae	2009					I	h
Carex squarrosa	Squarrose sedge	Cyperaceae	1819			T		N	h
Carex sterilis	Sterile sedge	Cyperaceae	1894		x			N	h
Carex stipata	Awl-fruited sedge	Cyperaceae	2017	x				N	h
Carex stricta	Tussock sedge	Cyperaceae	2021		HLo		x	N	h
Carex swanii	Swan's sedge	Cyperaceae	2015	x			x	N	h
Carex tetanica	Rigid sedge	Cyperaceae	1889			B		N	h
Carex tribuloides	Blunt broom sedge	Cyperaceae	2017	x			x	N	h
Carex trisperma	Three-seeded sedge	Cyperaceae	1889			B		N	h

Scientific name	Common name	Family	Year	NYMF	BKL	BCFS	EKFN	KB	TBS	SIP	BCSA	N/I	HABIT
Carex utriculata	Common beaked sedge	Cyperaceae	1889							B		N	h
Carex vesicaria var. monile	Blister sedge	Cyperaceae	1889							B		N	h
Carex vulpinoidea	Fox sedge	Cyperaceae	2018			x			BP		x	N	h
Carpinus caroliniana	American hornbeam	Betulaceae	1970	x			x					N	t
Carya cordiformis	Bitternut	Juglandaceae	2006	x					PL			N	t
Carya glabra	Pignut	Juglandaceae	2016			x						N	t
Carya ovata	Shagbark	Juglandaceae	2008			x						N	t
Carya tomentosa	Mockernut	Juglandaceae	2006	x		x			L			N	t
Castanea dentata	American chestnut	Fagaceae	1991	x					Lo			N	t
Castanea pumila	Chinquapin	Fagaceae	2010			x						N	t
Catalpa bignonioides	Southern catalpa	Bignoniaceae	2009	x	x				L			I	t
Catalpa speciosa	Northern catalpa	Bignoniaceae	2008			x			P			I	t
Ceanothus americanus	New Jersey tea	Rhamnaceae	1894	x								N	s
Celastrus orbiculatus	Round-leaved bittersweet $	Celastraceae	2015	x					LBH		x	I	wv
Celastrus scandens	American bittersweet	Celastraceae	1894	x						H		N	wv
Celtis occidentalis	Hackberry	Ulmaceae	2015	x	x				LH	HB	x	N	t
Cenchrus longispinus	Sandbur	Poaceae	2012			x			H			N	h
Centaurea jacea	Brown knapweed	Asteraceae	2015						F		x	I	h

Centaurea nigra	Black knapweed	Asteraceae	2006				L			I	h
Centaurea stoebe ssp. *micranthos*	Spotted knapweed	Asteraceae	2015	x	x		LB		x	I	h
Centaurium pulchellum	Centaury	Gentianaceae	2019	x		x	L			I	h
Cephalanthus occidentalis	Buttonbush	Rubiaceae	2005			x	H			N	s
Cerastium fontanum	Mouse-ear chickweed	Caryophyllaceae	2005				B			I	h
Ceratophyllum demersum	Common coontail	Ceratophyllaceae	2009		x					N	h
Chaenomeles speciosa	Flowering quince	Rosaceae	2006				P			I	s
Chaenorhinum minus	Dwarf snapdragon	Scrophulariaceae	2006	x			L			I	h
Chaerophyllum procumbens	Spreading chervil	Apiaceae	1819					T		N	h
Chaiturus marrubiastrum	Lion's-tail	Lamiaceae	2008		x					I	h
Chamaecrista fasciculata	Partridge pea	Fabaceae	1819					T		N	h
Chamaecrista nictitans	Wild sensitive plant	Fabaceae	2015				L		x	N	h
Chamaecyparis thyoides	Atlantic white cedar	Cupressaceae	1992	x				BT		N	t
Chamaedaphne calyculata	Leatherleaf	Ericaceae	1935	x				B		N	s
Chamaesyce maculata	Spotted spurge	Euphorbiaceae	2015	x	x		LBH		x	N	h
Chamaesyce nutans	Eyebane	Euphorbiaceae	2015	x	x		LBH		x	N	h
Chamaesyce vermiculata	Sandmat	Euphorbiaceae	2002			x				N	h
Chamerion angustifolium ssp. *angustifolium*	Fireweed	Onagraceae	1919					H		N	h
Chelidonium majus	Greater celandine	Papaveraceae	2003							I	h
Chenopodium album	Pigweed	Chenopodiaceae	2015	x	x		LH	H	x	NI	h
Chenopodium bonus-henricus	Good King Henry	Chenopodiaceae	2017			x				I	h

Scientific name	Common name	Family	Year	NYMF	BKL	BCFS	EKFN	KB	TBS	SIP	BCSA	N/I	HABIT
Chenopodium glaucum	Oak-leaved goosefoot	Chenopodiaceae	1889							B		I	h
Chenopodium rubrum	Red goosefoot	Chenopodiaceae	1983	x						T		N	h
Chenopodium simplex	Maple-leaved goosefoot	Chenopodiaceae	2006	x					L			N	h
Chimaphila maculata	Pipsissewa	Ericaceae	1991	x					Lo			N	s
Cichorium intybus	Chicory	Asteraceae	2006						L	H		I	h
Cicuta bulbifera	Bulblet water-hemlock	Apiaceae	1819							T		N	h
Cicuta maculata	Spotted water-hemlock	Apiaceae	2001	x		x			H	H		N	h
Cinna arundinacea	Wood-reed	Poaceae	2001	x					H	T		N	h
Circaea lutetiana	Enchanter's nightshade	Onagraceae	2006						AHT			N	h
Cirsium arvense	Spreading thistle	Asteraceae	2015	x		x			LBH		x	I	h
Cirsium discolor	Field thistle	Asteraceae	1819							T		N	h
Cirsium muticum	Swamp thistle	Asteraceae	1819							T		N	h
Cirsium vulgare	Bull thistle	Asteraceae	2015	x					B	H	x	I	h
Cladium mariscoides	Twig-rush	Cyperaceae	1895		x							N	h
Claytonia caroliniana	Carolina spring-beauty	Portulacaceae	1866		x							N	h
Clematis terniflora	Autumn clematis	Ranunculaceae	2016			x			H		x	I	wv
Clematis virginiana	Virgin's-bower	Ranunculaceae	1819							T		N	wv
Clethra alnifolia	Sweet pepperbush	Clethraceae	2005	x					B	HB		N	s
Clinopodium vulgare	Wild basil	Lamiaceae	2015								x	I	h
Clitoria mariana	Butterfly pea	Fabaceae	1889							B		N	h
Collinsonia canadensis	Horse balm	Lamiaceae	1919							H		N	h

Scientific name	Common name	Family	Year								
Commelina communis	Dayflower $	Commelinaceae	2006	x				LH	H	I	h
Conium maculatum	Poison-hemlock	Apiaceae	2009			x				I	h
Conopholis americana	Cancer-root	Orobanchaceae	2021				x			N	h
Consolida regalis	Royal knight's-spur	Ranunculaceae	1889						B	I	h
Convolvulus arvensis	Field bindweed	Convolvulaceae	2001				x			I	hv
Conyza canadensis	Horseweed	Asteraceae	2015	x			x	LH		N	h
Coptis trifolia	Goldthread	Ranunculaceae	1819	x					BT	N	h
Coreopsis tinctoria	Golden tickseed	Asteraceae	1895		x					I	h
Cornus amomum	Silky dogwood	Cornaceae	2015	x				BH	H	N	s
Cornus canadensis	Bunchberry	Cornaceae	1866	x					BT	N	s
Cornus racemosa	Gray dogwood	Cornaceae	2015	x				LH	H	N	s
Cornus rugosa	Round-leaved dogwood	Cornaceae	1862	x						N	s
Cornus sericea ssp. *sericea*	Red-osier dogwood	Cornaceae	1819						T	N	s
Corydalis sempervirens	Pale corydalis	Fumariaceae	2008				x			N	h
Corylus cornuta	Beaked hazel	Betulaceae	2006					P		N	s
Cotinus coggygria	Smoke tree	Anacardiaceae	2009			x				I	t
Crataegus	Hawthorn	Rosaceae	2005					B		?	t
Cryptotaenia canadensis	Honewort	Apiaceae	1894			x				N	h
Cunila origanoides	Dittany	Lamiaceae	1819						T	N	h
Cuscuta cephalanthi	Buttonbush dodder	Cuscutaceae	2009	x						N	hv
Cuscuta compacta	Compact dodder	Cuscutaceae	2004	x					EHB	N	hv
Cuscuta gronovii	Swamp dodder	Cuscutaceae	2009	x					H	N	hv
Cuscuta japonica	Purple dodder $	Cuscutaceae	2015	x			x			I	hv

Scientific name	Common name	Family	Year	NYMF	BKL	BCFS	EKFN	KB	TBS	SIP	BCSA	N/I	HABIT
Cuscuta obtusiflora var. glandulosa	Glandular dodder	Cuscutaceae	2006						P			N	hv
Cuscuta pentagona	Five-angled dodder	Cuscutaceae	2007			x			L			N	hv
Cuscuta polygonorum	Smartweed dodder	Cuscutaceae	2006			x						N	hv
Cynodon dactylon	Couch grass $	Poaceae	2006						L			I	h
Cynoglossum occidentale	Western hound's-tongue	Boraginaceae	1889							B		I	h
Cynoglossum virginianum	Blue hound's-tongue	Boraginaceae	1889							B		N	h
Cyperus amuricus	Lesser ricefield flatsedge	Cyperaceae	2015	x							x	I	h
Cyperus bipartitus	Slender flatsedge	Cyperaceae	2006						L			N	h
Cyperus dentatus	Toothed flatsedge	Cyperaceae	1889							B		N	h
Cyperus diandrus	Umbrella flatsedge	Cyperaceae	2015			x				T	x	N	h
Cyperus difformis	Variable flatsedge	Cyperaceae	2013			x						I	h
Cyperus erythrorhizos	Red-rooted flatsedge	Cyperaceae	2013			x						N	h
Cyperus esculentus	Chufa	Cyperaceae	2008	x		x			L			NI	h
Cyperus filicinus	Fern flatsedge	Cyperaceae	2015	x		x			Lo	HT	x	NI	h
Cyperus iria	Ricefield flatsedge	Cyperaceae	2017			x					x	I	h
Cyperus lancastriensis	Many-flowered flatsedge	Cyperaceae	2009			x						N	h
Cyperus lupulinus ssp. macilentus	Slender sand sedge	Cyperaceae	2017			x			H	H	x	N	h
Cyperus odoratus	Fragrant flatsedge	Cyperaceae	2017	x		x				B	x	N	h
Cyperus polystachyos	Many-spiked flatsedge	Cyperaceae	2013					x				N	h
Cyperus schweinitzii	Schweinitz flatsedge	Cyperaceae	2009			x						N	h

Scientific name	Common name	Family	Year							
Cyperus squarrosus	Bearded flatsedge	Cyperaceae	2015					x	N	h
Cyperus strigosus	Straw-colored flatsedge	Cyperaceae	2015	x	x		EH	x	N	h
Cypripedium reginae	Showy lady-slipper	Orchidaceae	1889				BT		N	h
Dactylis glomerata	Orchard grass	Poaceae	2015	x		LBH	H	x	I	h
Danthonia spicata	Poverty grass	Poaceae	2015	x		L		x	N	h
Dasiphora fruticosa ssp. *floribunda*	Shrubby cinquefoil	Rosaceae	1889				BT		N	s
Datura stramonium	Jimsonweed	Solanaceae	2015		x	L	H	x	I	h
Daucus carota	Queen Anne's lace	Apiaceae	2015	x	x	LBH	H	x	I	h
Decodon verticillatus	Swamp loosestrife	Lythraceae	1819			L	T		N	h
Deschampsia flexuosa	Hair grass	Poaceae	2006	x	x				N	h
Desmodium canadense	Showy tick-trefoil	Fabaceae	2004		x	H	H		N	h
Desmodium ciliare	Hairy tick-trefoil	Fabaceae	2015						N	h
Desmodium glabellum	Smooth tick-trefoil	Fabaceae	2006			L		x	N	h
Desmodium glutinosum	Pointed-leaved tick-trefoil	Fabaceae	1889				B		N	h
Desmodium paniculatum	Panicled tick-trefoil	Fabaceae	1991	x		LLo			N	h
Desmodium perplexum	Perplexed tick-trefoil	Fabaceae	2005			B			N	h
Dianthus armeria	Deptford pink	Caryophyllaceae	2005		x	BHP			I	h
Dicentra cucullaria	Dutchman's breeches	Fumariaceae	1867						N	h
Dichanthelium acuminatum	Tapered rosette grass	Poaceae	2015	x		L	H	x	N	h
Dichanthelium boscii	Bosc's panic grass	Poaceae	1991			Lo			N	h
Dichanthelium clandestinum	Deer-tongue	Poaceae	2006	x		LF			N	h

Scientific name	Common name	Family	Year	NYMF	BKL	BCFS	EKFN	KB	TBS	SIP	BCSA	N/I	HABIT
Dichanthelium commutatum	Variable panic grass	Poaceae	1819							T		N	h
Dichanthelium dichotomum	Cypress panic grass	Poaceae	2006	x					LLo			N	h
Dichanthelium latifolium	Broadleaf rosette grass	Poaceae	2006	x								N	h
Dichanthelium oligosanthes	Heller rosette grass	Poaceae	2015		x						x	N	h
Dichanthelium sphaerocarpon	Round-seeded panic grass	Poaceae	1893		x							N	h
Digitaria arenicola	Sand crabgrass	Poaceae	1991	x								I	h
Digitaria cognata	Fall witchgrass	Poaceae	2006						L			N	h
Digitaria ischaemum	Smooth crabgrass	Poaceae	2015						L		x	I	h
Digitaria sanguinalis	Hairy crabgrass	Poaceae	2015						HLP	H	x	I	h
Dioscorea villosa	Wild yam	Dioscoreaceae	2005	x			x		BLo			N	hv
Diospyros virginiana	Persimmon	Ebenaceae	1819							BT		N	t
Dipsacus fullonum	Common teasel	Dipsacaceae	2006				x					I	h
Distichlis spicata	Salt grass	Poaceae	2015	x							x	N	h
Dittrichia graveolens	Stinkwort	Asteraceae	2015	x					L		x	I	h
Doellingeria umbellata var. umbellata	Tall flat-topped white aster	Asteraceae	1991	x								N	h
Draba verna	Spring whitlow-grass	Brassicaceae	2016			x				T		I	h
Drosera intermedia	Intermediate sundew	Droseraceae	1819							T		N	h
Drosera rotundifolia	Round-leaved sundew	Droseraceae	1819							T		N	h

Species	Common name	Family	Year					A	B		Origin	Habit
Duchesnea indica	Mock-strawberry $	Rosaceae	2021				x				I	h
Dulichium arundinaceum	Threeway sedge	Cyperaceae	1819						T		N	h
Dysphania ambrosioides	Mexican tea	Chenopodiaceae	2008	x				LH	H	x	I	h
Dysphania botrys	Jerusalem-oak	Chenopodiaceae	2015	x	x						I	h
Dysphania pumilio	Clammy goosefoot	Chenopodiaceae	2015	x				LH		x	I	h
Echinochloa crus-galli	Barnyard grass	Poaceae	2015	x				BLP		x	I	h
Echinochloa walteri	Walter millet	Poaceae	1948	x					HT		N	h
Echinocystis lobata	Bur-cucumber	Cucurbitaceae	2005		x						N	hv
Echium vulgare	Viper's bugloss	Boraginaceae	2019						I		I	h
Eclipta prostrata	Yerba-de-tajo	Asteraceae	2015	x	x		x	Br			?	h
Elaeagnus umbellata	Autumn-olive	Elaeagnaceae	2015	x	x			LB		x	I	s
Eleocharis acicularis	Needle spike-rush	Cyperaceae	2013			x	x	Br			N	h
Eleocharis engelmannii	Engelmann spike-rush	Cyperaceae	2004		x						N	h
Eleocharis erythropoda	Bald spike-rush	Cyperaceae	2004		x						N	h
Eleocharis flavescens	Yellow spike-rush	Cyperaceae	2015		x		x				N	h
Eleocharis obtusa	Blunt spike-rush	Cyperaceae	2015	x	x			L			N	h
Eleocharis olivacea	Olivaceous spike-rush	Cyperaceae	1889						B		N	h
Eleocharis ovata	Ovate spike-rush	Cyperaceae	2013				x	Lo			N	h
Eleocharis palustris	Marsh spike-rush	Cyperaceae	1948	x							N	h
Eleocharis parvula	Dwarf spike-rush	Cyperaceae	2015		x					x	N	h
Eleocharis rostellata	Beaked spike-rush	Cyperaceae	1889		x			B	B		N	h
Eleocharis tenuis	Slender spike-rush	Cyperaceae	2015		x						N	h
Eleocharis tuberculosa	Tubercled spike-rush	Cyperaceae	2013				x			x	N	h
Eleusine indica	Goosegrass $	Poaceae	2015	x			x	ALH	H		I	h

Scientific name	Common name	Family	Year	NYMF	BKL	BCFS	EKFN	KB	TBS	SIP	BCSA	N/I	HABIT
Elodea nuttallii	Nuttall waterweed	Hydrocharitaceae	2009			x						N	h
Elymus canadensis	Canada wild-rye	Poaceae	1819							T		N	h
Elymus glaucus	Blue wild-rye	Poaceae	2009			x						N	h
Elymus hystrix var. hystrix	Bottlebrush	Poaceae	1919							H		N	h
Elymus repens	Quackgrass	Poaceae	2015		x				P		x	I	h
Elymus trachycaulus	Wheatgrass	Poaceae	2015								x	N	h
Elymus villosus	Hairy wild-rye	Poaceae	1889							BT		N	h
Elymus virginicus	Virginia wild-rye	Poaceae	2009			x				T		N	h
Epifagus virginiana	Beechdrops	Orobanchaceae	2010				x			T		N	h
Epilobium coloratum	Purple-leaved willowherb	Onagraceae	2015	x		x			BLo	T	x	N	h
Epilobium hirsutum	Great hairy willowherb	Onagraceae	1919							H		I	h
Epipactis helleborine	Helleborine	Orchidaceae	2019			x			LB			I	h
Eragrostis capillaris	Lace grass	Poaceae	2006	x					L			N	h
Eragrostis cilianensis	Stinkgrass	Poaceae	2015	x		x			L		x	I	h
Eragrostis frankii	Sandbar lovegrass	Poaceae	1894		x				P			I	h
Eragrostis pectinacea	Tufted lovegrass	Poaceae	2015	x					L		x	N	h
Eragrostis pilosa	Hairy lovegrass $	Poaceae	2015	x					L	H	x	N	h
Eragrostis spectabilis	Purple lovegrass	Poaceae	2015	x					BrLH		x	N	h
Erechtites hieraciifolius	Pilewort	Asteraceae	2015	x		x		x	H	E	x	N	h
Erica	Heather	Ericaceae	2006						L			I	s
Erigeron annuus	Daisy fleabane	Asteraceae	2015						B		x	N	h

Species	Common name	Family	Year							
Erigeron pulchellus	Robin-plantain	Asteraceae	2005		x				N	h
Erigeron strigosus	Prairie fleabane	Asteraceae	2006				L		N	h
Eriocaulon aquaticum	Common pipewort	Eriocaulaceae	1894		x				N	h
Eriophorum gracile	Slender cottongrass	Cyperaceae	1889					B	N	h
Eriophorum tenellum	Few-nerved cottongrass	Cyperaceae	1819					T	N	h
Eriophorum virginicum	Tawny cottongrass	Cyperaceae	1862		x				N	h
Erodium cicutarium	Stork's-bill	Geraniaceae	2012		x				I	h
Eryngium aquaticum	Rattlesnake master	Apiaceae	1819					T	N	h
Erythronium americanum	Trout-lily	Liliaceae	2016		x	x	Lo		N	h
Eubotrys racemosus	Swamp doghobble	Ericaceae	2004	x	x	x	Lo	BT	N	s
Euonymus alatus	Winged euonymus	Celastraceae	2006				AL	N	I	s
Euonymus americana	Strawberry bush	Celastraceae	1866	x	x				N	s
Euonymus atropurpureus	Burningbush	Celastraceae	1889		x			B	N	s
Euonymus europaeus	Spindle tree $	Celastraceae	2017		x	x			I	s
Euonymus fortunei	Winter creeper	Celastraceae	2013		x	x			I	wv
Euonymus obovata	Running strawberry bush	Celastraceae	1904	x	x				N	s
Eupatorium album	White thoroughwort	Asteraceae	2001				H		N	h
Eupatorium altissimum	Tall boneset	Asteraceae	2006	x			L		N	h
Eupatorium perfoliatum	Common boneset	Asteraceae	2005	x				EH	N	h
Eupatorium pilosum	Rough boneset	Asteraceae	1991	x			LLo	BT	N	h
Eupatorium rotundifolium	Round-leaved thoroughwort	Asteraceae	1991	x			Lo		N	h
Eupatorium serotinum	Late thoroughwort	Asteraceae	2015	x		x	L		N	h
Eupatorium sessilifolium	Upland boneset	Asteraceae	2006	x			L	HBT	N	h

Scientific name	Common name	Family	Year	NYMF	BKL	BCFS	EKFN	KB	TBS	SIP	BCSA	N/I	HABIT
Euphorbia cyparissias	Cypress spurge	Euphorbiaceae	2017			x	x		L		x	I	h
Euphorbia marginata	Snow-on-the-mountain	Euphorbiaceae	2006	x					L			N	h
Eurybia divaricata	White wood aster	Asteraceae	2017	x		x			LB			N	h
Euthamia caroliniana	Slender goldentop	Asteraceae	1889							B		N	h
Euthamia graminifolia	Grass-leaved goldenrod	Asteraceae	2015	x	x				LH	EHT	x	N	h
Eutrochium dubium	Coastal Plain Joe Pye weed	Asteraceae	2015	x	x						x	N	h
Eutrochium fistulosum	Hollow-stemmed Joe Pye weed	Asteraceae	2001						H			N	h
Eutrochium maculatum var. maculatum	Spotted Joe Pye weed	Asteraceae	2015								x	N	h
Eutrochium purpureum	Purple Joe Pye weed	Asteraceae	2015	x		x			L	EHT	x	N	h
Fagus grandifolia	American beech	Fagaceae	1894	x			x					N	t
Festuca ovina	Sheep fescue	Poaceae	2015								x	I	h
Festuca rubra	Red fescue	Poaceae	2006	x					L		x	N	h
Festuca subverticillata	Nodding fescue	Poaceae	1889							B		N	h
Fimbristylis castanea	Marsh fimbry	Cyperaceae	1819							T		N	h
Floerkea proserpinacoides	False mermaid	Limnanthaceae	1889							B		N	h
Forsythia	Forsythia	Oleaceae	2005						B			I	s
Frangula alnus	Glossy buckthorn	Rhamnaceae	2015			x					x	I	s
Frangula caroliniana	Carolina buckthorn	Rhamnaceae	1889							B		N	t
Fraxinus americana	White ash	Oleaceae	2015			x			LBH	H	x	N	t
Fraxinus nigra	Black ash	Oleaceae	2010			x				B		N	t

Scientific name	Common name	Family	Year								
Fraxinus pennsylvanica	Red ash	Oleaceae	2015				L		x	N	t
Froelichia gracilis	Cottonweed	Amaranthaceae	2012	x						I	h
Gaillardia pulchella	Blanketflower	Asteraceae	2006	x			L			I	h
Galearis spectabilis	Showy orchis	Orchidaceae	1819					T		N	h
Galinsoga parviflora	Gallant soldiers	Asteraceae	1919					H		I	h
Galium aparine	Cleavers	Rubiaceae	2015		x		L		x	N	hv
Galium asprellum	Rough bedstaw	Rubiaceae	1819					T		N	hv
Galium circaezans	Licorice bedstraw	Rubiaceae	2016		x					N	hv
Galium mollugo	Stockport weed	Rubiaceae	2015						x	I	hv
Galium triflorum	Fragrant bedstraw	Rubiaceae	1819					T		N	hv
Gaultheria hispidula	Creeping snowberry	Ericaceae	1889?	x				B		N	s
Gaultheria procumbens	Wintergreen	Ericaceae	1864	x				T		N	s
Gaylussacia baccata	Black huckleberry	Ericaceae	2006	x			Llo			N	s
Gaylussacia frondosa	Blue huckleberry	Ericaceae	1991	x			Lo			N	s
Gentiana saponaria	Soapwort gentian	Gentianaceae	1889					BT		N	h
Geranium bicknellii	Northern crane's-bill	Geraniaceae	2006		x					N	h
Geranium carolinianum	Carolina crane's-bill	Geraniaceae	2006	x	x		LP			N	h
Geranium maculatum	Wild geranium	Geraniaceae	2005			x				N	h
Geranium molle	Dove's-foot geranium	Geraniaceae	2006				L			I	h
Geranium robertianum	Herb Robert	Geraniaceae	1889					B		N	h
Geum canadense	White avens	Rosaceae	2005				BBr			N	h
Geum laciniatum	Rough avens	Rosaceae	2006	x	x		L			N	h
Geum macrophyllum	Large-leaved avens	Rosaceae	2015						x	N	h
Ginkgo biloba	Gingko	Ginkgoaceae	2015						x	I	t

Scientific name	Common name	Family	Year	NYMF	BKL	BCFS	EKFN	KB	TBS	SIP	BCSA	N/I	HABIT
Glechoma hederacea	Ground-ivy	Lamiaceae	2015	x					L		x	I	hv
Gleditsia triacanthos	Honey locust	Fabaceae	2015	x					L	H	x	I	t
Glyceria acutiflora	Creeping mannagrass	Poaceae	1889							B		N	h
Glyceria fluitans	Water mannagrass	Poaceae	1819							T		I	h
Glyceria obtusa	Atlantic mannagrass	Poaceae	1991	x					Lo			N	h
Glyceria striata	Fowl mannagrass	Poaceae	1949	x						E		N	h
Gymnocladus dioicus	Kentucky coffeetree	Fabaceae	1991	x								I	t
Hackelia virginiana	Stickseed	Boraginaceae	2006	x					L			N	h
Halesia tetraptera	Mountain silverbell	Styracaceae	1941	x								N	t
Hamamelis virginiana	Witch-hazel	Hamamelidaceae	2006	x					LLo	H		N	s
Hedeoma pulegioides	American pennyroyal	Lamiaceae	2006	x					L			N	h
Hedera helix	Common ivy $	Araliaceae	2014			x			AP			I	wv
Helenium autumnale	Sneezeweed	Asteraceae	2009				x		Br			N	h
Helianthus annuus	Common sunflower	Asteraceae	2015	x					LP	H	x	N	h
Helianthus decapetalus	Thin-leaved sunflower	Asteraceae	1889							B		N	h
Helianthus divaricatus	Woodland sunflower	Asteraceae	1919							H		N	h
Helianthus giganteus	Giant sunflower	Asteraceae	1991	x					Lo	EHB	x	N	h
Helianthus petiolaris	Prairie sunflower	Asteraceae	1999	x								N	h
Helianthus tuberosus	Sunflower-artichoke $	Asteraceae	2001						H			N	h
Hemerocallis fulva	Orange day-lily	Liliaceae	2006	x					ABrH	E		I	h
Heracleum maximum	Cow-parsnip	Apiaceae	1889							BT		N	h
Hesperis matronalis	Dame's rocket	Brassicaceae	2006							B		I	h

Scientific name	Common name	Family	Year								
Heterotheca subaxillaris	Camphorweed	Asteraceae	2006	x			L			I	h
Hibiscus moscheutos	Swamp rose mallow	Malvaceae	2015	x			LB		x	N	h
Hibiscus syriacus	Rose-of-Sharon	Malvaceae	2006				BBrP			I	t
Hibiscus trionum	Flower-of-an-hour	Malvaceae	2006				L			I	h
Hieracium pilosella	Mouse-ear hawkweed	Asteraceae	2006	x			L			I	h
Hieracium piloselloides	Tall hawkweed	Asteraceae	2009		x					I	h
Hieracium sabaudum	Autumn hawkweed	Asteraceae	2006	x			L			I	h
Hieracium scabrum	Rough hawkweed	Asteraceae	1991	x						N	h
Hierochloe odorata	Sweetgrass	Poaceae	1889			x		B		N	h
Holcus lanatus	Velvetgrass	Poaceae	2015	x			P		x	I	h
Hordeum jubatum	Foxtail-barley	Poaceae	2015		x		LB			N	h
Hosta ventricosa	Blue plantain-lily	Liliaceae	2001				H			I	h
Humulus japonicus	Wild hops $	Cannabaceae	2015	x		x	HP		x	I	hv
Humulus lupulus	Hops	Cannabaceae	2005	x		x	B	H		NI	hv
Hydrocotyle ranunculoides	Floating marsh pennywort	Apiaceae	2019		x					N	h
Hydrophyllum virginianum	Virginia waterleaf	Hydrophyllaceae	1889				B			N	h
Hypericum mutilum	Dwarf St. Johnswort	Clusiaceae	1991	x			Lo			N	h
Hypericum perforatum	Common St. Johnswort	Clusiaceae	2015	x			LB		x	I	h
Hypochaeris radicata	Hairy cat's-ear	Asteraceae	2005				B			I	h
Hyssopus officinalis	Hyssop	Lamiaceae	1889					B		I	h
Ilex ambigua	Carolina holly	Aquifoliaceae	1819					T		N	s
Ilex crenata	Box-leaved holly $	Aquifoliaceae	1991	x			Lo			I	s
Ilex glabra	Inkberry	Aquifoliaceae	2018	x				BT	x	N	s

Scientific name	Common name	Family	Year	NYMF	BKL	BCFS	EKFN	KB	TBS	SIP	BCSA	N/I	HABIT
Ilex laevigata	Smooth winterberry	Aquifoliaceae	1899	x						B		N	s
Ilex mucronata	Catberry	Aquifoliaceae	1889	x								N	s
Ilex opaca	American holly	Aquifoliaceae	2005	x			x		Lo			N	t
Ilex verticillata	Common winterberry	Aquifoliaceae	2008	x		x						N	s
Impatiens capensis	Spotted jewelweed	Balsaminaceae	2015	x					LBH	EH	x	N	h
Impatiens pallida	Pale jewelweed	Balsaminaceae	1889							B		N	h
Ipomoea pandurata	Man-of-the-earth	Convolvulaceae	1889							B		N	hv
Iris prismatica	Slender blue flag	Iridaceae	1889							B		N	h
Iris pseudacorus	Yellow iris	Iridaceae	2006						BP	B		I	h
Iris versicolor	Blue flag	Iridaceae	1991	x						T		N	h
Iva frutescens	High tide bush	Asteraceae	2015	x					L		x	N	h
Juglans cinerea	Butternut	Juglandaceae	2004	x						H		N	t
Juglans nigra	Black walnut	Juglandaceae	2006	x					A			N	t
Juncus ?arcticus ssp. littoralis	Arctic rush	Juncaceae	2006			x						N	h
Juncus biflorus	Bog rush	Juncaceae	1819							T		N	h
Juncus bufonius	Toad rush	Juncaceae	1894		x							N	h
Juncus debilis	Weak rush	Juncaceae	2019			x						N	h
Juncus dichotomus	Forked rush	Juncaceae	1819							T		N	h
Juncus dudleyi	Dudley rush	Juncaceae	2015								x	N	h
Juncus effusus	Soft rush	Juncaceae	2015	x					B		x	N	h
Juncus gerardii	Black rush	Juncaceae	2015	x					L		x	N	h

Species	Common name	Family	Year						
Juncus marginatus	Grass-leaved rush	Juncaceae	1889			B		N	h
Juncus tenuis	Path rush	Juncaceae	2016	x	LBH	H	x	N	h
Juncus torreyi	Torrey rush	Juncaceae	2019	x				N	h
Juniperus virginiana	Eastern red cedar	Cupressaceae	2015	x	LH		x	N	t
Kalmia angustifolia	Sheep laurel	Ericaceae	1895	x				N	s
Kalmia latifolia	Mountain laurel	Ericaceae	1819			T		N	s
Koelreuteria paniculata	Golden rain tree	Sapindaceae	2003	x		P		I	t
Kosteletzkya virginica	Seashore mallow	Malvaceae	1889			B		N	h
Kummerowia striata	Common bush-clover $	Fabaceae	2006					I	h
Kyllinga gracillima	Pasture spikesedge	Cyperaceae	2006		Br			N	h
Lactuca biennis	Tall blue lettuce	Asteraceae	2010	x	H	H		N	h
Lactuca canadensis	Tall lettuce	Asteraceae	2015		L	T	x	N	h
Lactuca floridana var. villosa	Woodland lettuce	Asteraceae	1889			B		N	h
Lactuca graminifolia	Grass-leaved lettuce	Asteraceae	1895	x				N	h
Lactuca hirsuta var. sanguinea	Hairy lettuce	Asteraceae	1889			B		N	h
Lactuca serriola	Prickly lettuce	Asteraceae	2015		L	H	x	I	h
Lamiastrum galeobdolon	Yellow archangel	Lamiaceae	2004		S			I	h
Lamium amplexicaule	Henbit	Lamiaceae	2016	x				I	h
Lamium purpureum	Purple dead-nettle	Lamiaceae	2017	x				I	h
Laportea canadensis	Wood-nettle	Urticaceae	1819			T		N	h
Larix laricina	Tamarack	Pinaceae	1904	x		BT		N	t
Lathyrus latifolius	Sweet pea	Fabaceae	2015					I	hv
Lathyrus palustris	Marsh pea	Fabaceae	1889			BT		N	hv

Scientific name	Common name	Family	Year	NYMF	BKL	BCFS	EKFN	KB	TBS	SIP	BCSA	N/I	HABIT
Leersia oryzoides	Rice cutgrass	Poaceae	2015		x					H	x	N	h
Leersia virginica	Whitegrass	Poaceae	2006	x					HLP	B	x	N	h
Lemna minor	Common duckweed	Lemnaceae	2015	x		x	x			B	x	N	h
Leonurus cardiaca	Motherwort	Lamiaceae	2016				x		A			I	h
Lepidium campestre	Field pepperweed	Brassicaceae	2019			x					x	I	h
Lepidium virginicum	Virginia pepperweed	Brassicaceae	2019	x		x			LBH	H	x	N	h
Leptochloa fusca ssp. fascicularis	Bearded sprangletop	Poaceae	2006						L			N	h
Lespedeza capitata	Round-headed bush-clover	Fabaceae	2015								x	N	h
Lespedeza cuneata	Sericea lespedeza	Fabaceae	2015		x						x	I	h
Lespedeza frutescens	Shrubby bush-clover	Fabaceae	1991	x						H		N	h
Lespedeza hirta	Hairy bush-clover	Fabaceae	1919							H		N	h
Lespedeza procumbens	Trailing bush-clover	Fabaceae	2015		x							N	hv
Lespedeza violacea	Violet bush-clover	Fabaceae	2006						L			N	h
Leucanthemum vulgare	Oxeye daisy	Asteraceae	2006	x					L	H		I	h
Liatris scariosa	Blazing star	Asteraceae	1889							BT		N	h
Ligustrum obtusifolium	Border privet	Oleaceae	1999	x								I	s
Ligustrum vulgare	Common privet	Oleaceae	2015								x	I	s
Lilium canadense	Canada lily	Liliaceae	1819							T		N	h
Lilium philadelphicum	Wood lily	Liliaceae	1919							HB		N	h
Lilium superbum	Swamp lily $	Liliaceae	2005	x				B		HB		N	h

Linaria vulgaris	Butter-and-eggs	Scrophulariaceae	2015	x			LH	H	x	I	h
Lindera benzoin	Spicebush	Lauraceae	2021	x	x		BLo			N	s
Lindernia dubia	False-pimpernel	Scrophulariaceae	2008	x						N	h
Linnaea borealis	Twinflower	Caprifoliaceae	1889					B		N	hv
Liparis liliifolia	Large twayblade	Orchidaceae	1819					T		N	h
Liparis loeselii	Loesel twayblade	Orchidaceae	1889					B		N	h
Liquidambar styraciflua	Sweetgum	Hamamelidaceae	2015	x	x		BH	H	x	N	t
Liriodendron tulipifera	Tuliptree	Magnoliaceae	2004	x	x		Lo			N	t
Liriope spicata	Creeping liriope	Liliaceae	2018							–	h
Listera convallarioides	Broad-lipped twayblade	Orchidaceae	1819					T		N	h
Lobelia inflata	Bladderpod lobelia $	Campanulaceae	2006				A			N	h
Lobelia siphilitica	Great blue lobelia	Campanulaceae	2008		x			H		N	h
Lobelia spicata	Spiked lobelia	Campanulaceae	1819					T		N	h
Lolium perenne	Perennial ryegrass	Poaceae	2006			x	LP			–	h
Lolium temulentum	Darnel	Poaceae	2015			x			x	I	h
Lonicera ×bella [*morrowii × tatarica*]	Bell's honeysuckle	Caprifoliaceae	2015			x			x	I	s
Lonicera dioica	Limber honeysuckle	Caprifoliaceae	1893	x				B		N	wv
Lonicera japonica	Gold-and-silver honeysuckle $	Caprifoliaceae	2015	x			LBH	B	x	I	wv
Lonicera maackii	Maack honeysuckle $	Caprifoliaceae	2016	x			H		x	I	s
Lonicera morrowii	Morrow honeysuckle $	Caprifoliaceae	2006	x			LH			I	s
Lonicera sempervirens	Coral trumpets	Caprifoliaceae	1889	x				B		N	wv
Lonicera tatarica	Bush honeysuckle $	Caprifoliaceae	2000			x	H		x	I	s
Lotus corniculatus	Bird's-foot trefoil	Fabaceae	2015				LBH			I	h

Scientific name	Common name	Family	Year	NYMF	BKL	BCFS	EKFN	KB	TBS	SIP	BCSA	N/I	HABIT
Ludwigia alternifolia	Seedbox	Onagraceae	2005						B	H		N	h
Ludwigia palustris	False-purslane	Onagraceae	2006	x			x		P			N	h
Lychnis flos-cuculi	Ragged robin	Caryophyllaceae	2016			x						I	h
Lycium barbarum	Matrimony vine	Solanaceae	2006						BrP			I	s
Lycopus americanus	American water-horehound	Lamiaceae	2015	x					BLP		x	N	h
Lycopus uniflorus	Northern water-horehound	Lamiaceae	1991	x						E		N	h
Lycopus virginicus	Virginia water-horehound	Lamiaceae	2009			x			LH			N	h
Lyonia ligustrina	Maleberry	Ericaceae	1894	x								N	s
Lysimachia ciliata	Fringed loosestrife	Primulaceae	2001						H			N	h
Lysimachia hybrida	Lowland yellow loosestrife	Primulaceae	1889							BT		N	h
Lysimachia lanceolata	Lance-leaved loosestrife	Primulaceae	1894		x							N	h
Lysimachia nummularia	Moneywort	Primulaceae	2015				x					I	h
Lysimachia quadrifolia	Four-leaved loosestrife	Primulaceae	2015	x		x			Lo		x	N	h
Lysimachia terrestris	Swamp candles	Primulaceae	2015	x		x			B			N	h
Lysimachia thyrsiflora	Tufted loosestrife	Primulaceae	1948							BT		N	h
Lythrum salicaria	Purple loosestrife	Lythraceae	2015	x		x			LBH	B	x	I	h
Magnolia virginiana	Sweetbay magnolia	Magnoliaceae	1903	x						B		N	t
Maianthemum canadense	Canada mayflower	Liliaceae	2015	x					BLLo	E	x	N	h

Maianthemum racemosum	False Solomon's-seal	Liliaceae	2017	x		L			N	h
Maianthemum stellatum	Starry Solomon's-seal	Liliaceae	1889				B		N	h
Malaxis unifolia	Green adder's-mouth	Orchidaceae	1889				BT		I	t
Malus baccata	Crab apple	Rosaceae	2006			P			I	t
Malus floribunda	Flowering crab $	Rosaceae	2015	x				x	I	t
Malus pumila	Apple	Rosaceae	2006	x		HP			I	t
Malus toringo	Toringo crab	Rosaceae	2015					x	I	h
Matricaria discoidea	Pineappleweed	Asteraceae	2005		x				I	h
Mazus pumilus	Mazus $	Scrophulariaceae	2008	x		Br			I	h
Medeola virginiana	Cucumber-root $	Liliaceae	1991	x		Lo			N	h
Medicago lupulina	Black medick	Fabaceae	2006	x		LBH			I	h
Medicago sativa	Alfalfa	Fabaceae	2006			L	H		I	h
Melilotus officinalis	Yellow sweet-clover, white sweet-clover	Fabaceae	2015	x		LH	H	x	I	h
Melissa officinalis	Common balm	Lamiaceae	2001			H	B		I	h
Menispermum canadense	Moonseed	Menispermaceae	2015	x	x	AP			N	wv
Mentha ×piperita[aquatica × spicata]	Peppermint	Lamiaceae	2001			H			I	h
Mentha spicata	Spearmint	Lamiaceae	2006			F			I	h
Menyanthes trifoliata	Buckbean	Menyanthaceae	1919				HB		N	h
Microstegium vimineum	Stiltgrass $	Poaceae	2017		x				I	h
Mikania scandens	Climbing hempweed	Asteraceae	2015	x	x	HLo	E		N	hv
Mimulus alatus	Winged monkeyflower	Scrophulariaceae	1889				B	x	N	h
Mimulus ringens	Common monkeyflower	Scrophulariaceae	2006			L			N	h

Scientific name	Common name	Family	Year	NYMF	BKL	BCFS	EKFN	KB	TBS	SIP	BCSA	N/I	HABIT
Miscanthus sinensis	Miscanthus $	Poaceae	2015			x					x	I	h
Moehringia lateriflora	Blunt-leaved sandwort	Caryophyllaceae	1950	x	x				BHL	T		N	h
Mollugo verticillata	Carpetweed	Molluginaceae	2017			x			LBH		x	I	h
Monarda didyma	Scarlet beebalm	Lamiaceae	1889							B		N	h
Monotropa uniflora	Ghost pipe $	Monotropaceae	2018	x								N	h
Morella caroliniensis	Southern bayberry	Myricaceae	2015								x	N	s
Morella pensylvanica	Northern bayberry	Myricaceae	2005	x					BH			N	s
Morus alba	White mulberry	Moraceae	2015	x		x			LBH		x	I	t
Morus rubra	Red mulberry	Moraceae	2015								x	N	t
Muhlenbergia capillaris	Hair-awned muhly	Poaceae	1889							B		N	h
Muhlenbergia frondosa	Wire-stemmed muhly	Poaceae	1991	x					Lo	T		N	h
Muhlenbergia glomerata	Spiked muhly	Poaceae	1819							T		N	h
Muhlenbergia racemosa	Marsh muhly	Poaceae	1889							B		N	h
Muhlenbergia schreberi	Nimble Will	Poaceae	2001						H			N	h
Muhlenbergia sobolifera	Rock muhly	Poaceae	1889							B		N	h
Muhlenbergia sylvatica	Woodland muhly	Poaceae	1819							T		N	h
Myosotis stricta	Strict forget-me-not	Boraginaceae	2011			x						I	h
Myosotis verna	Spring forget-me-not	Boraginaceae	1819							N		N	h
Narcissus	Daffodil	Amaryllidaceae	2006						BrP			I	h
Nasturtium officinale	Watercress	Brassicaceae	2008	x		x						I	h
Nekemias arborea	Peppervine	Vitaceae	2018			x						I	wv

Species	Common name	Family	Year							
Nelumbo lutea	American lotus	Nelumbonaceae	1819				T		N	h
Nepeta cataria	Catnip	Lamiaceae	2015	x		L	H	x	I	h
Nuttallanthus canadensis	Canada toadflax	Scrophulariaceae	2008	x		L			N	h
Nyssa sylvatica	Tupelo	Nyssaceae	2016	x		BH	E		N	t
Oclemena nemoralis	Bog aster	Asteraceae	1889				BT		N	h
Oenothera biennis	Common evening primrose	Onagraceae	2015	x		LBH	H	x	N	h
Oligoneuron rigidum var. rigidum	Stiff goldenrod	Asteraceae	1919				HB		N	h
Origanum vulgare	Oregano	Lamiaceae	1889				B		I	h
Ornithogalum umbellatum	Star of Bethlehem	Liliaceae	2017	x					I	h
Orobanche uniflora	One-flowered broomrape	Orobanchaceae	1894		x				N	h
Orontium aquaticum	Goldenclub	Araceae	1889				BT		N	h
Osmorhiza claytonii	Sweet Cicely	Apiaceae	2006			L			N	h
Oxalis dillenii	Slender yellow wood-sorrel	Oxalidaceae	2015			HP		x	N	h
Oxalis stricta	Common yellow wood-sorrel	Oxalidaceae	2015	x		HLP		x	N	h
Pachysandra terminalis	Pachysandra $	Buxaceae	2011					x	I	h
Packera aurea	Golden ragwort	Asteraceae	2016					x	N	h
Panax trifolius	Dwarf ginseng	Araliaceae	2010		x				N	h
Panicum capillare	Witchgrass	Poaceae	2015	x		L	H	x	N	h
Panicum dichotomiflorum	Fall panic grass	Poaceae	2015	x		LoP	HT	x	N	h
Panicum miliaceum	Broomcorn millet	Poaceae	1889				B		I	h

Scientific name	Common name	Family	Year	NYMF	BKL	BCFS	EKFN	KB	TBS	SIP	BCSA	N/I	HABIT
Panicum virgatum	Switchgrass	Poaceae	2015	x					LH	EH	x	N	h
Parietaria pensylvanica	Pellitory-of-the-wall	Urticaceae	2006			x						N	h
Parnassia glauca	Grass-of-Parnassus	Saxifragaceae	1919							HB		N	h
Parthenocissus quinquefolia	Virginia creeper	Vitaceae	2015	x					LH	H	x	N	wv
Parthenocissus tricuspidata	Boston ivy	Vitaceae	2015			x					x	I	wv
Parthenocissus vitacea	Woodbine	Vitaceae	2003			x						N	wv
Paspalum laeve	Field paspalum	Poaceae	2017			x				B		N	h
Pastinaca sativa	Wild parsnip	Apiaceae	2006	x					L			I	h
Paulownia tomentosa	Princess tree	Scrophulariaceae	2015	x					L		x	I	t
Peltandra virginica	Arrow arum	Araceae	2021	x			x			Li		N	h
Penstemon digitalis	Common beardtongue	Scrophulariaceae	2008			x						N	h
Penstemon hirsutus	Hairy beardtongue	Scrophulariaceae	1889							B		N	h
Phacelia purshii	Miami mist	Hydrophyllaceae	1867		x							N	h
Phalaris arundinacea	Reed canary grass	Poaceae	2017				x					NI	h
Phaseolus polystachios	Thicket bean	Fabaceae	1889							B		N	hv
Philadelphus	Mock-orange	Hydrangeaceae	2006						ABr			I	s
Phleum pratense	Timothy	Poaceae	2015								x	I	h
Phragmites australis ssp. australis	Old World common reed	Poaceae	2019	x		x			LH	ET	x	I	h
Phryma leptostachya	Lopseed	Verbenaceae	1819							T		N	h

Species	Common name	Family	Year							
Physalis longifolia var. *subglabrata*	Long-leaved ground-cherry	Solanaceae	1819				T		N	h
Physalis pubescens	Downy ground-cherry	Solanaceae	1889				B		N	h
Physostegia virginiana	Obedient plant	Lamiaceae	2009	x					N	h
Phytolacca americana	Pokeweed	Phytolaccaceae	2015	x		LBH	EH	x	N	h
Picea abies	Northwoods spruce	Pinaceae	2006			L			I	t
Picea mariana	Black spruce	Pinaceae	1906	x			BT		N	t
Pilea pumila	Clearweed	Urticaceae	2013	x	x	BrL	E		N	h
Pinus nigra	White-tipped black pine $	Pinaceae	2005			B			I	t
Pinus rigida	Pitch pine	Pinaceae	2002		x				N	t
Pinus strobus	Eastern white pine	Pinaceae	2015	x	x			x	N	t
Pinus sylvestris	Scots pine	Pinaceae	2006			A			I	t
Pinus thunbergii	Black pine $	Pinaceae	2006			L			I	t
Plantago aristata	Buckhorn	Plantaginaceae	2009	x	x		BHP		N	h
Plantago lanceolata	Ribwort plantain	Plantaginaceae	2015			LBH		x	I	h
Plantago major	Common plantain	Plantaginaceae	2015	x		LBH	H	x	I	h
Plantago virginica	Virginia plantain	Plantaginaceae	2012		x				N	h
Platanthera blephariglottis var. *blephariglottis*	White fringed orchid	Orchidaceae	1889				B		N	h
Platanthera ciliaris	Yellow fringed orchid	Orchidaceae	1919				HBT		N	h
Platanthera clavellata	Green woodland orchid	Orchidaceae	1819				T		N	h
Platanthera cristata	Crested fringed orchid	Orchidaceae	1819				T		N	h
Platanthera flava	Tubercled orchid	Orchidaceae	1889				B		N	h

Scientific name	Common name	Family	Year	NYMF	BKL	BCFS	EKFN	KB	TBS	SIP	BCSA	N/I	HABIT
Platanthera grandiflora	Greater purple fringed orchid	Orchidaceae	1819							T		N	h
Platanus ×hispanica	London plane	Platanaceae	2006	x					L			I	t
Platanus occidentalis	Sycamore	Platanaceae	2015	x			x				x	N	t
Pluchea odorata	Saltmarsh-fleabane	Asteraceae	2015	x					L		x	I	h
Poa annua	Annual bluegrass	Poaceae	2015		x	x						I	h
Poa compressa	Flattened bluegrass	Poaceae	2015		x							I	h
Poa palustris	Fowl bluegrass	Poaceae	1889							B		N	h
Poa pratensis	Kentucky bluegrass	Poaceae	2015	x					LH		x	NI	h
Poa trivialis	Rough bluegrass	Poaceae	1889							B		I	h
Podophyllum peltatum	May-apple	Berberidaceae	2019			x			Lo			N	h
Pogonia ophioglossoides	Snakemouth orchid	Orchidaceae	1919							HBT		N	h
Polygala brevifolia	Short-leaved milkwort	Polygalaceae	1889							B		N	h
Polygala cruciata	Drumheads	Polygalaceae	1889							BT		N	h
Polygala paucifolia	Gaywings	Polygalaceae	1889							BT		N	h
Polygala sanguinea	Purple milkwort	Polygalaceae	1819							T		N	h
Polygala verticillata	Whorled milkwort	Polygalaceae	2006						L			N	h
Polygonatum biflorum	Solomon's-seal	Liliaceae	2016	x		x			L			N	h
Polygonum amphibium var. emersum	Water smartweed	Polygonaceae	1949	x						E		N	h
Polygonum arenastrum	Oval-leaved knotweed	Polygonaceae	2006						JLP			I	h
Polygonum arifolium	Halberd-leaved tearthumb	Polygonaceae	2001	x						EH		N	h

Species	Common name	Family	Year									
Polygonum aviculare	Prostrate knotweed	Polygonaceae	2015	x			LLo	H	x	I	h	
Polygonum baldschuanicum	Silver lace vine $	Polygonaceae	2008		x					I	h	
Polygonum careyi	Carey smartweed	Polygonaceae	1999	x					x	N	h	
Polygonum cespitosum	Bristly lady's-thumb $	Polygonaceae	2016	x	x		LP		x	I	h	
Polygonum convolvulus	Black bindweed	Polygonaceae	1919					H		I	hv	
Polygonum cuspidatum	Knotweed $	Polygonaceae	2018	x	x		LBH		x	I	h	
Polygonum extremiorientale	Big city smartweed	Polygonaceae	2017							I	h	
Polygonum hydropiper	Water-pepper	Polygonaceae	1991	x			LoP	H		I	h	
Polygonum hydropiperoides	Mild water-pepper	Polygonaceae	2015	x				T	x	N	h	
Polygonum lapathifolium	Curlytop knotweed	Polygonaceae	2015	x	x		LP	HB	x	N	h	
Polygonum orientale	Kiss-me-over-the-garden-gate	Polygonaceae	1919	x				H		I	h	
Polygonum pensylvanicum	Pennsylvania smartweed	Polygonaceae	2015	x	x		P	H	x	N	h	
Polygonum perfoliatum	Mile-a-minute $	Polygonaceae	2015	x	x		F		x	I	hv	
Polygonum persicaria	Lady's-thumb	Polygonaceae	2015	x	x		L	H	x	I	h	
Polygonum punctatum	Dotted smartweed	Polygonaceae	2017	x	x		L	E	x	N	h	
Polygonum sachalinense	Giant knotweed	Polygonaceae	2016	x	x					I	h	
Polygonum sagittatum	Arrow-leaved tearthumb	Polygonaceae	2001	x				EH		N	hv	
Polygonum scandens	Climbing false-buckwheat	Polygonaceae	2015	x	x		LH	EH	x	NI	hv	
Polygonum virginianum	Jumpseed	Polygonaceae	2006				LH	H		N	h	
Pontederia cordata	Pickerelweed	Pontederiaceae	1893			x				N	h	

Scientific name	Common name	Family	Year	NYMF	BKL	BCFS	EKFN	KB	TBS	SIP	BCSA	N/I	HABIT
Populus ×canescens[alba × tremula]	Gray poplar	Salicaceae	1991	x					Lo			I	t
Populus ×jackii [balsamifera × deltoides]	Balm-of-Gilead	Salicaceae	1819							T		N	t
Populus alba	White poplar	Salicaceae	2017				x					I	t
Populus balsamifera	Balsam poplar	Salicaceae	1991	x								N	t
Populus deltoides	Eastern cottonwood	Salicaceae	2015	x					LBH		x	N	t
Populus grandidentata	Big-toothed aspen	Salicaceae	2006	x					L	H		N	t
Populus heterophylla	Swamp cottonwood	Salicaceae	1889	x						B		N	t
Populus nigra	Lombardy poplar	Salicaceae	2006						L	T		I	t
Populus tremuloides	Quaking aspen	Salicaceae	2015	x					LBH		x	N	t
Portulaca oleracea	Purslane	Portulacaceae	2015	x		x	x		LBH		x	NI	h
Potamogeton crispus	Curly pondweed	Potamogetonaceae	2009				x					I	h
Potamogeton foliosus	Leafy pondweed	Potamogetonaceae	2015			x					x	N	h
Potentilla argentea	Silver cinquefoil	Rosaceae	2006						AH			I	h
Potentilla norvegica	Rough cinquefoil $	Rosaceae	2006						AB	H		N	h
Potentilla recta	Sulphur cinquefoil	Rosaceae	2003						H			I	h
Potentilla simplex	Common cinquefoil	Rosaceae	2005						BHLo	H		N	h
Prenanthes alba	White rattlesnakeroot	Asteraceae	1991	x								N	h
Prenanthes trifoliolata	Gall-of-the-earth	Asteraceae	1919									N	h
Prunella vulgaris ssp. lanceolata	Selfheal	Lamiaceae	2001						H			N	h

Scientific name	Common name	Family	Year							
Prunus cerasifera	Cherry plum	Rosaceae	2006			P			l	t
Prunus cerasus	Sour cherry	Rosaceae	2015	x				x	l	t
Prunus maritima	Beach plum	Rosaceae	2004	x					N	s
Prunus pensylvanica	Pin cherry	Rosaceae	1889				B		N	t
Prunus serotina	Black cherry	Rosaceae	2015	x		LBH	EH	x	N	t
Prunus virginiana	Chokecherry	Rosaceae	2015	x		B			N	s
Pseudognaphalium obtusifolium ssp. *obtusifolium*	Low cudweed	Asteraceae	2015	x		HLLo		x	N	h
Ptelea trifoliata	Wafer-ash	Rutaceae	2008	x	x	L			N	t
Ptilimnium capillaceum	Mock bishopweed	Apiaceae	1948	x					N	h
Puccinellia distans	Alkali grass	Poaceae	2015					x	NI	h
Pycnanthemum incanum	Hoary mountain mint	Lamiaceae	1819				T		N	h
Pycnanthemum muticum	Clustered mountain mint	Lamiaceae	1889				BT		N	h
Pycnanthemum virginianum	Virginia mountain mint	Lamiaceae	2009	x	x	B	H	x	N	h
Pyrus calleryana	Bradford pear	Rosaceae	2016		x	P		x	l	t
Pyrus communis	Common pear	Rosaceae	2014		x				l	t
Quercus ×fontana [*coccinea × velutina*]	Hybrid oak	Fagaceae	2008		x				N	t
Quercus alba	White oak	Fagaceae	2008	x	x	LB	H		N	t
Quercus bicolor	Swamp white oak	Fagaceae	2016	x	x	H	EH		N	t
Quercus coccinea	Scarlet oak	Fagaceae	1991	x		Lo			N	t
Quercus ilicifolia	Scrub oak	Fagaceae	2006			x			N	s

Scientific name	Common name	Family	Year	NYMF	BKL	BCFS	EKFN	KB	TBS	SIP	BCSA	N/I	HABIT
Quercus macrocarpa var. macrocarpa	Bur oak **	Fagaceae	2015	x							x	N	t
Quercus montana	Chestnut oak	Fagaceae	2006	x					L	H		N	t
Quercus palustris	Pin oak	Fagaceae	2015	x		x			BH	E	x	N	t
Quercus rubra	Northern red oak	Fagaceae	2006	x					LBH	E		N	t
Quercus stellata	Post oak	Fagaceae	2001	x		x				B		N	t
Quercus velutina	Black oak	Fagaceae	2008	x		x			L	H		N	t
Ranunculus ficaria	Lesser-celandine	Ranunculaceae	2010			x						I	h
Ranunculus hispidus var. nitidus	Bristly buttercup	Ranunculaceae	1889							B		N	h
Ranunculus pensylvanicus	Pennsylvania buttercup	Ranunculaceae	1889							B		N	h
Ranunculus repens	Creeping buttercup	Ranunculaceae	2005		x							I	h
Ranunculus sceleratus	Cursed crowfoot	Ranunculaceae	2012	x		x			L			N	h
Rhamnus alnifolia	Alder-leaved buckthorn	Rhamnaceae	1889							B		N	s
Rhamnus cathartica	Common buckthorn $	Rhamnaceae	2015	x							x	I	t
Rhododendron maximum	Great laurel	Ericaceae	1889	x					Lo	BT		N	s
Rhododendron periclymenoides	Pink azalea	Ericaceae	1991									N	s
Rhododendron viscosum	Swamp azalea	Ericaceae	2005	x					Lo	B		N	s
Rhus aromatica	Fragrant sumac	Anacardiaceae	2002			x	x					N	s
Rhus copallinum	Winged sumac	Anacardiaceae	2015	x					Blo		x	N	s
Rhus glabra	Smooth sumac	Anacardiaceae	2015	x					BH		x	N	s

Species	Common name	Family	Year										
Rhus typhina	Staghorn sumac	Anacardiaceae	2015	x	x				HB	x	N	s
Rhynchospora alba	White beak-rush	Cyperaceae	1862	x	x						N	h
Ribes americanum	Black current	Grossulariaceae	1889						B		N	s
Ribes triste	Red currant	Grossulariaceae	2008			x					N	s
Robinia pseudoacacia	Black locust	Fabaceae	2015	x		x		BH		x	I	t
Rorippa palustris	Marsh yellowcress	Brassicaceae	2016	x		x	x	LBH	B, EB		N	h
Rorippa sylvestris	Creeping yellowcress	Brassicaceae	2008	x		x					I	h
Rosa canina	Dog rose	Rosaceae	1999		x						I	wv
Rosa carolina	Carolina rose	Rosaceae	1949	x					EH		N	wv
Rosa micrantha	Small-flowered sweet-briar	Rosaceae	1999	x							I	wv
Rosa multiflora	Multiflora rose	Rosaceae	2015	x				BH		x	I	wv
Rosa rugosa	Beach rose	Rosaceae	2006	x			x	L			I	wv
Rotala ramosior	Lowland rotala	Lythraceae	1889						BT		N	h
Rubus allegheniensis	Northern blackberry	Rosaceae	2006	x		x		L			N	wv
Rubus flagellaris	Northern dewberry	Rosaceae	2015	x		x		L		x	N	wv
Rubus hispidus	Swamp dewberry	Rosaceae	2017	x		x		Lo	T		N	wv
Rubus idaeus	Red raspberry	Rosaceae	2006	x				L			N	wv
Rubus laciniatus	Cut-leaved blackberry	Rosaceae	2006	x				L			I	wv
Rubus occidentalis	Black raspberry	Rosaceae	2006					H			N	wv
Rubus odoratus	Flowering raspberry	Rosaceae	1889						BT		N	wv
Rubus pensilvanicus	Pennsylvania blackberry	Rosaceae	2015			x				x	N	wv
Rubus phoenicolasius	Wineberry	Rosaceae	2004								I	wv

Scientific name	Common name	Family	Year	NYMF	BKL	BCFS	EKFN	KB	TBS	SIP	BCSA	N/I	HABIT
Rubus pubescens	Dwarf red blackberry	Rosaceae	2008			x				B		N	wv
Rudbeckia laciniata	Green-headed coneflower	Asteraceae	2001			x				HT		N	h
Rumex acetosella	Sheep sorrel	Polygonaceae	2015		x						x	I	h
Rumex crispus	Curly dock	Polygonaceae	2015	x					LBH		x	I	h
Rumex obtusifolius	Bitter dock	Polygonaceae	2015			x			FH		x	I	h
Rumex orbiculatus	Greater water dock	Polygonaceae	2006	x					LB	EHB		N	h
Rumex pallidus	Seaside dock	Polygonaceae	1998						J			N	h
Rumex patientia	Patience dock	Polygonaceae	2004			x						I	h
Rumex verticillatus	Swamp dock	Polygonaceae	2009			x			H	B		N	h
Ruppia maritima	Wigeon-grass	Ruppiaceae	2019			x						N	h
Sabatia dodecandra	Marsh pink	Gentianaceae	1919							HBT		N	h
Sabatia stellaris	Sea pink	Gentianaceae	1919							H		N	h
Sagina procumbens	Pearlwort	Caryophyllaceae	1889							B		I	h
Sagittaria calycina var. *spongiosa*	Spongy arrowhead	Alismataceae	1889							B		N	h
Sagittaria latifolia	Broad-leaved arrowhead	Alismataceae	2005	x					B	T		N	h
Sagittaria subulata	Awl-leaved arrowhead	Alismataceae	1889							B		N	h
Salicornia depressa	Glasswort	Chenopodiaceae	2017			x						N	h
Salix ×*pendulina* [*fragilis* × ?*sepulcralis*]	Wisconsin weeping willow	Salicaceae	2004									I	t
Salix ×*sepulcralis* [*alba* × ?*pendulina*]	Weeping willow	Salicaceae	2006						L	H		I	t

Scientific name	Common name	Family	Year								
Salix alba	White willow	Salicaceae	2015	x					x	l	t
Salix atrocinerea	Large gray willow	Salicaceae	2009	x						l	t
Salix babylonica s.l.	Weeping willow	Salicaceae	2015		x		BrP		x	l	t
Salix bebbiana	Bebb willow	Salicaceae	2017		x			B		N	t
Salix candida	Sageleaf willow	Salicaceae	1889		x					N	s
Salix caprea	Goat willow	Salicaceae	2016		x					l	t
Salix cordata	Sand dune willow	Salicaceae	1919	x				H		N	s
Salix discolor	Pussy willow	Salicaceae	2012	x						N	s
Salix eriocephala	Heart-leaved willow	Salicaceae	1895			x				N	s
Salix fragilis	Crack willow	Salicaceae	1999	x				B		l	t
Salix lucida	Shining willow	Salicaceae	1819		x			T		N	t
Salix nigra	Black willow	Salicaceae	1999	x				H		N	t
Salix petiolaris	Meadow willow	Salicaceae	1889	x				B		N	s
Salix viminalis	Basket willow	Salicaceae	1889	x				B		l	t
Sambucus nigra	Common elderberry	Caprifoliaceae	2015			x			x	N	s
Sambucus racemosa	Red elderberry	Caprifoliaceae	1871	x						N	s
Samolus valerandi ssp. *parviflorus*	Water pimpernel	Primulaceae	2009		x		BrFJ			N	h
Sanguinaria canadensis	Bloodroot	Papaveraceae	1819	x						N	h
Sanguisorba canadensis	Canadian burnet	Rosaceae	2005	x			B	T		N	h
Sanicula marilandica	Maryland black snakeroot	Apiaceae	1895		x			H		N	h
Sanicula odorata	Clustered black snakeroot	Apiaceae	1893			x				N	h
Saponaria officinalis	Bouncing Bet	Caryophyllaceae	2006	x			LBH	H		l	h

Scientific name	Common name	Family	Year	NYMF	BKL	BCFS	EKFN	KB	TBS	SIP	BCSA	N/I	HABIT
Sarracenia purpurea	Pitcher plant	Sarraceniaceae	1819							T		N	h
Sassafras albidum	Sassafras	Lauraceae	2015	x		x			LB	H		N	t
Saururus cernuus	Lizard's-tail	Saururaceae	2008	x			x			T		N	h
Saxifraga pensylvanica	Swamp saxifrage	Saxifragaceae	1819							T		N	h
Schedonorus arundinaceus	Tall fescue	Poaceae	2015								x	I	h
Schedonorus pratensis	Meadow fescue	Poaceae	2015	x		x			L			I	h
Schizachyrium scoparium	Little bluestem	Poaceae	2015	x		x	x		LH	H	x	N	h
Schoenoplectus americanus	Olney threesquare	Cyperaceae	2019	x						B	x	N	h
Schoenoplectus pungens var. pungens	Common threesquare	Cyperaceae	2015	x					BrLLoP		x	N	h
Schoenoplectus tabernaemontani	Soft-stemmed bulrush	Cyperaceae	2015	x	x			x			x	N	h
Scirpus atrovirens	Green bulrush	Cyperaceae	1991	x						B		N	h
Scirpus cyperinus	Wool-grass	Cyperaceae	2004	x		x			Lo	B		N	h
Scirpus microcarpus	Panicled bulrush	Cyperaceae	1919							H		N	h
Scirpus pendulus	Drooping bulrush	Cyperaceae	1889							B		N	h
Scleranthus annuus	Knawel	Caryophyllaceae	2015								x	I	h
Scleria triglomerata	Whip nut-rush	Cyperaceae	1889							B		N	h
Scrophularia lanceolata	Lance-leaved figwort	Scrophulariaceae	2006	x					L	H		N	h
Scrophularia marilandica	Carpenter's square	Scrophulariaceae	1819							T		N	h
Scutellaria galericulata	Marsh skullcap	Lamiaceae	2016	x		x						N	h

Scientific name	Common name	Family	Year										
Securigera varia	Crown-vetch	Fabaceae	2015	x			x	A		x		I	hv
Sedum acre	Mossy stonecrop	Crassulaceae	2009	x	x							–	h
Senecio vulgaris	Common groundsel	Asteraceae	2017			x		LP				N	h
Senna hebecarpa	Wild senna	Fabaceae	2009									N	h
Setaria faberi	Nodding foxtail $	Poaceae	2015	x				L		x		–	h
Setaria parviflora	Marsh bristlegrass	Poaceae	2015			x						N	h
Setaria pumila	Yellow foxtail	Poaceae	2006	x				LP	H			–	h
Setaria verticillata	Hooked foxtail	Poaceae	1889						B			–	h
Setaria viridis	Green foxtail	Poaceae	2015	x				LBH		x		–	h
Sicyos angulatus	Star-cucumber	Cucurbitaceae	2015	x	x	x		LH	H	x		N	hv
Silene antirrhina	Sleepy catchfly	Caryophyllaceae	2006	x	x			L				N	h
Silene caroliniana	Sticky catchfly	Caryophyllaceae	1889						B			N	h
Silene latifolia ssp. *alba*	White lychnis	Caryophyllaceae	2015	x				FL	I	x		–	h
Silene stellata	Starry campion	Caryophyllaceae	2006	x	x	x		L	H			N	h
Silene vulgaris	Bladder campion	Caryophyllaceae	2006					L	B			–	h
Sisymbrium altissimum	Tall tumble-mustard	Brassicaceae	1919						H			–	h
Sisymbrium loeselii	Small tumble-mustard	Brassicaceae	2015							x		–	h
Sisymbrium officinale	Hedge-mustard	Brassicaceae	2006		x							–	h
Sisyrinchium atlanticum	Eastern blue-eyed grass	Iridaceae	2008		x							N	h
Sium suave	Water-parsnip	Apiaceae	1948	x					T			N	h
Smallanthus uvedalius	Hairy leafcup	Asteraceae	1889						B			N	h
Smilax glauca	Sawbriar	Smilacaceae	1991	x				Lo	B			N	wv
Smilax herbacea	Smooth carrionflower	Smilacaceae	1819						T			N	wv

Scientific name	Common name	Family	Year	NYMF	BKL	BCFS	EKFN	KB	TBS	SIP	BCSA	N/I	HABIT
Smilax rotundifolia	Catbriar	Smilacaceae	2008	x		x			B			N	wv
Solanum americanum	American black nightshade	Solanaceae	2003						P			I	h
Solanum carolinense	Horse-nettle	Solanaceae	2015				x		H	B	x	N	h
Solanum dulcamara	Climbing nightshade	Solanaceae	2015	x		x			LBH	EH	x	I	wv
Solanum lycopersicum	Garden tomato	Solanaceae	2009			x						I	h
Solanum nigrum	Black nightshade	Solanaceae	2006	x					BrLH			I	h
Solanum physalifolium	Hoe nightshade	Solanaceae	2015			x					x	I	h
Solanum ptycanthum	Eastern black nightshade $	Solanaceae	2015		x	x						N	h
Solidago altissima	Late goldenrod	Asteraceae	2015	x		x				H	x	N	h
Solidago bicolor	Silverrod	Asteraceae	2006	x					L	H		N	h
Solidago caesia	Blue-stemmed goldenrod	Asteraceae	2006	x					L			N	h
Solidago canadensis	Canada goldenrod	Asteraceae	2015	x					LBH		x	N	h
Solidago flexicaulis	Zigzag goldenrod	Asteraceae	1889							B		N	h
Solidago gigantea	Giant goldenrod	Asteraceae	2015						B	E	x	N	h
Solidago juncea	Early goldenrod	Asteraceae	2015	x					LH		x	N	h
Solidago latissimifolia	Elliott goldenrod	Asteraceae	1949	x						EB		N	h
Solidago nemoralis	Gray goldenrod	Asteraceae	2006	x					L	H		N	h
Solidago odora	Sweet goldenrod	Asteraceae	1819							T		N	h
Solidago patula	Spreading goldenrod	Asteraceae	1889							B		N	h
Solidago rugosa	Rough goldenrod	Asteraceae	2017	x		x			LBH		x	N	h

Scientific name	Common name	Family	Year								
Solidago sempervirens	Seaside goldenrod	Asteraceae	2015	x	x		L		x	N	h
Solidago uliginosa	Bog goldenrod	Asteraceae	1919	x			L	HB		N	h
Solidago ulmifolia	Elm-leaved goldenrod	Asteraceae	2006	x				HB		N	h
Sonchus arvensis	Field sow-thistle	Asteraceae	2003			x	P	BT		I	h
Sonchus asper	Spiny sow-thistle	Asteraceae	2006	x			L	HB		I	h
Sonchus oleraceus	Common sow-thistle	Asteraceae	2016	x		x				I	h
Sorbus aucuparia	Rowan	Rosaceae	1991	x			Lo			I	t
Sorghastrum nutans	Wood grass $	Poaceae	1919	x				H		N	h
Spartina alterniflora	Smooth cordgrass	Poaceae	2015	x			L		x	N	h
Spartina cynosuroides	Big cordgrass	Poaceae	2015	x				HT		N	h
Spartina patens	Saltmeadow cordgrass	Poaceae	2015	x			L		x	N	h
Spartina pectinata	Freshwater cordgrass	Poaceae	1949	x				E		N	h
Spergula arvensis	Corn spurrey	Caryophyllaceae	1819			x		T		I	h
Spergularia maritima	Greater sand spurrey	Caryophyllaceae	2015	x			P		x	N	h
Spergularia salina	Lesser sand spurrey	Caryophyllaceae	2016	x			L			N	h
Sphenopholis obtusata	Prarie wedgescale	Poaceae	1889			x		B		N	h
Spiraea alba var. latifolia	Meadowsweet	Rosaceae	2005	x			B	T		N	s
Spiraea tomentosa	Steeplebush	Rosaceae	2005	x			B	HT		N	s
Spiranthes praecox	Grass-leaved lady's tresses	Orchidaceae	1895		x					N	h
Spirodela polyrhiza	Great duckweed	Lemnaceae	1819							N	h
Sporobolus vaginiflorus	Poverty dropseed	Poaceae	2006			x	L	T		N	h
Staphylea trifolia	Bladdernut	Staphyleaceae	1889			x		B		N	s
Stellaria graminea	Lesser stitchwort	Caryophyllaceae	2004			x				I	h

Scientific name	Common name	Family	Year	NYMF	BKL	BCFS	EKFN	KB	TBS	SIP	BCSA	N/I	HABIT
Strophostyles helvola	Annual wild bean	Fabaceae	2015			x			PB	T	x	N	hv
Strophostyles umbellata	Perennial wild bean	Fabaceae	1819							T		N	hv
Styphnolobium japonicum	Pagoda tree $	Fabaceae	2015	x		x			L		x	I	t
Symphyotrichum cordifolium	Heart-leaved aster	Asteraceae	1991	x								N	h
Symphyotrichum ericoides var. ericoides	Heath aster	Asteraceae	2015								x	N	h
Symphyotrichum laeve	Smooth aster	Asteraceae	1889		x							N	h
Symphyotrichum lanceolatum ssp. lanceolatum var. lanceolatum	White-panicled aster	Asteraceae	1991	x	x					H		N	h
Symphyotrichum novi-belgii var. novi-belgii	New York aster	Asteraceae	2007	x	x			x				N	h
Symphyotrichum patens	Late purple aster	Asteraceae	1889		x							N	h
Symphyotrichum phlogifolium	Thin-leaved late purple aster	Asteraceae	1889							B		N	h
Symphyotrichum pilosum	Hairy oldfield aster	Asteraceae	2002		x							N	h
Symphyotrichum puniceum var. puniceum	Purple-stemmed aster	Asteraceae	1819							T		N	h
Symphyotrichum subulatum	Annual saltmarsh aster	Asteraceae	2015	x		x		x	LLo		x	N	h

Symphyotrichum tradescantii	Shore aster	Asteraceae	1919				H		N	h
Symphyotrichum undulatum	Wavy-leaved aster	Asteraceae	1819				T		N	h
Symplocarpus foetidus	Skunk-cabbage	Araceae	2017	x					N	h
Taenidia integerrima	Yellow pimpernel	Apiaceae	1891		x				N	h
Tanacetum vulgare	Tansy	Asteraceae	1999		x				I	h
Taraxacum laevigatum	Rec-seeded dandelion	Asteraceae	2017		x				I	h
Taraxacum officinale	Common dandelion	Asteraceae	2015		x	LBH	H	x	I	h
Taxodium distichum	Bald cypress	Pinaceae	2001			H			I	t
Taxus baccata	Common yew $	Taxaceae	2006			Br			-	s
Taxus canadensis	American yew	Taxaceae	2015		x				N	s
Teucrium canadense	American germander	Lamiaceae	2015		x	L	HT	x	N	h
Thalictrum dioicum	Early meadow-rue	Ranunculaceae	1889		x		B		N	h
Thalictrum pubescens	Tall meadow-rue	Ranunculaceae	2015		x	BLo	EH	x	N	h
Thlaspi arvense	Field pennycress	Brassicaceae	2016		x	P		x	I	t
Tilia ×europaea [cordata × platyphyllos]	Common lime $	Tiliaceae	2006		x				-	t
Tilia americana	Basswood	Tiliaceae	2015		x	L	T	x	N	t
Tilia cordata	Small-leaved lime $	Tiliaceae	2016		x	B			-	t
Tipularia discolor	Cranefly orchid	Orchidaceae	1889				B		N	h
Toxicodendron radicans	Poison-ivy	Anacardiaceae	2015		x	LBH	E	x	N	wv
Toxicodendron vernix	Poison-sumac	Anacardiaceae	1877		x				N	s
Tradescantia virginiana	Virginia spiderwort	Commelinaceae	2003		x	P			-	h

Scientific name	Common name	Family	Year	NYMF	BKL	BCFS	EKFN	KB	TBS	SIP	BCSA	N/I	HABIT
Tragopogon porrifolius	Goat's-beard	Asteraceae	1889							B		I	h
Trapa natans	Water-chestnut	Trapaceae	2004						K			I	h
Triadenum virginicum	Marsh St. Johnswort	Clusiaceae	2015	x						T		N	h
Trichostema dichotomum	Blue curls	Lamiaceae	2001				x		H			N	h
Tridens flavus	Purpletop	Poaceae	2015						LH		x	N	h
Trientalis borealis	Starflower	Primulaceae	2006	x			x		Lo	T		N	h
Trifolium arvense	Rabbit's-foot clover	Fabaceae	2006						L	H		I	h
Trifolium aureum	Golden clover	Fabaceae	2002		x							I	h
Trifolium campestre	Field clover	Fabaceae	2015							HB	x	I	h
Trifolium dubium	Little hop clover	Fabaceae	2016			x						I	h
Trifolium hybridum	Alsike clover	Fabaceae	2004		x							I	h
Trifolium pratense	Red clover	Fabaceae	2015						L	H	x	I	h
Trifolium repens	White clover	Fabaceae	2006	x					LB	H		I	h
Triglochin maritima	Seaside arrow-grass	Juncaginaceae	1819							T		N	h
Trillium undulatum	Painted trillium	Liliaceae	1889							BT		N	h
Triodanis perfoliata	Venus' looking-glass	Campanulaceae	2004			x						N	h
Triosteum aurantiacum	Early horse-gentian	Caprifoliaceae	2009		x							N	h
Triosteum perfoliatum	Wild-coffee	Caprifoliaceae	2004				x					N	h
Triplasis purpurea	Purple sandgrass	Poaceae	2015								x	N	h
Tripogon major	Five-minute grass	Poaceae	1992	x								I	h
Tripsacum dactyloides	Eastern gamagrass	Poaceae	1819							T		N	h
Trollius laxus	Globeflower	Ranunculaceae	1889							B		N	h

Scientific name	Common name	Family	Year							
Tsuga canadensis	Eastern hemlock	Pinaceae	1819				T		N	t
Tussilago farfara	Common coltsfoot	Asteraceae	2006	x	x	L	H		L	h
Typha ×glauca	Hybrid cattail	Typhaceae	2015	x				x	N	h
Typha angustifolia	Narrow-leaved cattail	Typhaceae	2015	x			B	x	N	h
Typha latifolia	Broad-leaved cattail	Typhaceae	1999	x					N	h
Ulmus americana	American elm	Ulmaceae	2016	x	x	LH	H	x	N	t
Ulmus pumila	Dwarf elm $	Ulmaceae	2015			F		x	L	t
Ulmus rubra	Slippery elm	Ulmaceae	2006			BFJ	T		N	t
Urtica dioica	Stinging nettles	Urticaceae	2015	x		P	B	x	N	h
Utricularia intermedia	Flat-leaved bladderwort	Lentibulariaceae	1919				HB		N	h
Uvularia sessilifolia	Sessile-leaved bellwort	Liliaceae	2015	x	x	LLo			_	h
Vaccaria hispanica	Cow soapwort	Caryophyllaceae	1889	x			B		N	s
Vaccinium angustifolium	Late low blueberry	Ericaceae	2006	x		L			N	s
Vaccinium corymbosum	Highbush blueberry	Ericaceae	2021	x	x	BLoTe	HB	x	N	s
Vaccinium fuscatum	Black highbush blueberry	Ericaceae	1889	x			B		N	s
Vaccinium macrocarpon	Large cranberry	Ericaceae	1889	x			B		N	s
Vaccinium oxycoccos	Small cranberry	Ericaceae	1889	x			BT		N	s
Vaccinium pallidum	Southern low blueberry	Ericaceae	2006	x		LLo	H		N	s
Valeriana officinalis	Garden valerian	Valerianaceae	2016	x	x				_	h
Veratrum viride	False-hellebore	Melanthiaceae	2021	x	x				N	h
Verbascum blattaria	Moth mullein	Scrophulariaceae	2015	x		LBH	H	x	L	h
Verbascum thapsus	Common mullein	Scrophulariaceae	2015	x		LBH	H	x	L	h
Verbena bracteata	Big-bracted vervain	Verbenaceae	2003	x	x	P			N	h
Verbena hastata	Blue vervain	Verbenaceae	2015	x		L	H	x	N	h

Scientific name	Common name	Family	Year	NYMF	BKL	BCFS	EKFN	KB	TBS	SIP	BCSA	N/I	HABIT
Verbena officinalis	Herb-of-the-cross	Verbenaceae	1819							T		I	h
Verbena simplex	Narrow-leaved vervain	Verbenaceae	1819							T		N	h
Verbena urticifolia	White vervain	Verbenaceae	2015	x					LBH		x	N	h
Vernonia noveboracensis	New York ironweed	Asteraceae	2015			x			B	H		N	h
Veronica anagallis-aquatica	Water speedwell	Scrophulariaceae	2003		x				P			N	h
Veronica arvensis	Corn speedwell	Scrophulariaceae	2017			x						I	h
Veronica beccabunga	Brooklime $	Scrophulariaceae	2006			x						I	h
Veronica hederifolia	Ivy-leaved speedwell	Scrophulariaceae	1889							B		I	h
Veronica peregrina	Neckweed	Scrophulariaceae	2017		x	x						N	h
Veronica serpyllifolia	Thymeleaf speedwell	Scrophulariaceae	2017			x						NI	h
Viburnum acerifolium	Maple-leaved viburnum	Caprifoliaceae	2014			x						N	s
Viburnum dentatum	Arrowwood	Caprifoliaceae	2015	x					BH	E	x	N	s
Viburnum lantanoides	Hobblebush	Caprifoliaceae	1819							T		N	s
Viburnum nudum var. cassinoides	Withe-rod	Caprifoliaceae	1865	x								N	s
Viburnum nudum var. nudum	Possumhaw	Caprifoliaceae	1948	x								N	s
Viburnum opulus	Highbush-cranberry	Caprifoliaceae	2008	x		x						NI	s
Viburnum opulus var. americanum	Highbush cranberry	Caprifoliaceae	1991						Lo			N	s
Viburnum plicatum	Doublefile viburnum	Caprifoliaceae	2008			x			A			I	s
Viburnum prunifolium	Blackhaw	Caprifoliaceae	2017	x		x			L	EH		N	s

Scientific name	Common name	Family	Date									Origin	Status
Viburnum rafinesquianum	Downy arrowwood	Caprifoliaceae	1919							H		N	s
Vicia sativa	Garden vetch	Fabaceae	1819							T		I	h
Vicia tetrasperma	Four-seeded vetch	Fabaceae	2009				x					I	hv
Vinca minor	Periwinkle	Apocynaceae	2006						A			I	wv
Viola ×*primulifolia[lanceolata* × *macloskeyi]*	Primrose violet	Violaceae	1991	x					Lo			N	h
Viola arvensis	Field pansy $	Violaceae	2004				x					I	h
Viola bicolor	Field pansy	Violaceae	2008				x			BT		N	h
Viola lanceolata	Bog white violet	Violaceae	2016				x					N	h
Viola macloskeyi ssp. *pallens*	Small white violet	Violaceae	1991	x					Lo	E		N	h
Viola pubescens var. *pubescens*	Downy yellow violet	Violaceae	1889							B		N	h
Viola sororia	Common blue violet	Violaceae	2010				x		BrLo			N	h
Vitis aestivalis	Summer grape	Vitaceae	2015	x			x		H, L	H	x	N	wv
Vitis labrusca	Fox grape	Vitaceae	2009	x			x		BH	E		N	wv
Vitis riparia	River grape	Vitaceae	2012	x								N	wv
Vulpia myuros	Rat-tailed fescue	Poaceae	2009		x							I	h
Wolffia brasiliensis	Watermeal $	Lemnaceae	2013				x					N	h
Wolffia columbiana	Watermeal $	Lemnaceae	2012				x					N	h
Xanthium strumarium var. *canadense*	Cocklebur	Asteraceae	2015						LP	H	x	N	h
Xyris caroliniana	Carolina yellow-eyed-grass	Xyridaceae	1889							BT		N	h
Yucca filamentosa	Adam's needle	Agavaceae	2006					x	H			N	h

Scientific name	Common name	Family	Year	NYMF	BKL	BCFS	EKFN	KB	TBS	SIP	BCSA	N/I	HABIT
Zannichellia palustris	Horned-pondweed	Zannichelliaceae	2015			x					x	N	h
Zanthoxylum americanum	Prickly-ash	Rutaceae	2004		x							N	s
Zizania aquatica	Annual wild-rice	Poaceae	1948	x						BT		N	h
Zizania palustris	Northern wild-rice	Poaceae	<1980									N	h
Zizia aptera	Heart-leaved Alexanders	Apiaceae	1819							T		N	h

Note: YEAR = most recent specimen or observation; NYMF = New York Metropolitan Flora; BKL = specimen in Brooklyn Botanic Garden Herbarium now at New York Botanical Garden; BCFS = specimen in Hudsonia - Bard College Field Station; EKFN = E. Kiviat field notes; KB = Kerry Barringer personal herbarium.

TBS = Torrey Botanical Society field trips: A=Andreas Park (2006), B=Bergen County Utilities Authority (2005), Br=Brett Park (), F=Foschini Park (2006), H=Hackensack River County Park (2001), J=Johnson Park (various), K=Kearny Marsh (2004), L=Laurel Hill (2006), Li=Lincoln Park (2006), Lo=Losen Slote Creek Park (1991), P=Hackensack River Greenway Pomander Walk (2006), S=Schmidt's Woods (2004), T=Hackensack River Greenway Teaneck (2006), Te=Teterboro Airport Woods (2004) – selected observations from NY-NJ-CT (2021).

SIP=Sipple (1972) historic sources: B=Britton (1889), T=Torrey (1918), H=Harshberger (1919), E=Heusser (1949)

BCSA = Kiviat and Graham 2016 Berry's Creek area (Kiviat and Graham 2016)

NYMF, BCFS, KB records supported by herbarium specimens. EKFN, NY-NJ-CT Botany observations mostly not associated with specimens.

$ Common name chosen to avoid geographic or ethnic references for nonnative species or ethnic references for native species.

N=Native to New Jersey, I=Introduced (nonnative); from USDA Plants Database (USDA 2021) & other sources.

Habit: h = herb, hv = herbaceous vine, s = shrub, t = tree, wv = woody vine.

Christopher Graham and Elizabeth Castle helped compile the seed plant list.

Appendix 2: Birds

Birds of the New Jersey Meadowlands.

Common name	Scientific name	References	Status
Geese			
Snow goose	Anser caerulescens	3,9,10	
Brant	Branta bernicla	3,9,10	SGCN
Canada goose	Branta canadensis	2,3,6,9,10	B,SGCN
Cackling goose	Branta hutchinsii	3,9	B
Swans			
Mute swan	Cygnus olor	3,9, 10	B
Tundra swan	Cygnus columbianus	3,9	
Dabbling Ducks			
Fulvous whistling-duck	Dendrocygna bicolor	9	
Wood duck	Aix sponsa	2,3,9,10	B,SGCN
Blue-winged teal	Spatula discors	2,3,4,6,9,10	B
Cinnamon teal	Spatula cyanoptera	3	
Northern shoveler	Spatula clypeata	2,3,9,10	
Gadwall	Mareca strepera	1,2,3,4,5,6,9,10	B
Eurasian wigeon	Mareca penelope	3,9,10	
American wigeon	Mareca americana	2,3,9,10	
Mallard	Anas platyrhynchos	1,2,3,4,5,6,9,10	B
Black duck	Anas rubripes	1,2,3,4,5,6,9,10	SGCN
Northern pintail	Anas acuta	1,2,3,9,10	SGCN
Green-winged teal	Anas crecca	1,2,3,6,9,10	B
Diving Ducks			
Canvasback	Aythya valisineria	1,2,3,9,10	SGCN
Redhead	Aythya americana	3,9	
Ring-necked duck	Aythya collaris	3,9,10	
Greater scaup	Aythya marila	2,3,9	SGCN
Lesser scaup	Aythya affinis	2,3,9,10	SGCN
Long-tailed duck	Clangula hyemalis	3,9	SGCN
Surf scoter	Melanitta perspicillata	3,9	SGCN
White-winged scoter	Melanitta deglandi	3,9	SGCN
Bufflehead	Bucephala albeola	2,3,9,10	SGCN
Common goldeneye	Bucephala clangula	2,3,9,10	
Hooded merganser	Lophodytes cucullatus	2,3,9,10	SGCN
Common merganser	Mergus merganser	2,3,9,10	
Red-breasted merganser	Mergus serrator	2,3,9,10	

Common name	Scientific name	References	Status
Ruddy duck	*Oxyura jamaicensis*	1,2,3,4,9,10	B,SGCN
Wading Birds			
American bittern	*Botaurus lentiginosus*	2,3,4,6,9,10	E
Least bittern	*Ixobrychus exilis*	1,2,3,4,6,9,10	B,SC
Great blue heron	*Ardea herodias*	2,3,4,6,9,10	SC
Great egret	*Ardea alba*	2,3,4,6,9,10	
Snowy egret	*Egretta thula*	2,3,4,6,9,10	SC
Little blue heron	*Egretta caerulea*	2,3,9,10	SC
Tricolored heron	*Egretta tricolor*	3,6,9,10	SC
Cattle egret	*Bubulcus ibis*	2,3,9	T
Green heron	*Butorides virescens*	2,3,4,6,9,10	B
Black-crowned night-heron	*Nycticorax nycticorax*	1,2,3,4,6,9,10	T
Yellow-crowned night-heron	*Nyctanassa violacea*	2,3,4,9,10	B,T
White ibis	*Eudocimus albus*	3,9	
Glossy ibis	*Plegadis falcinellus*	2,3,4,9	SC
Rails and Cranes			
King rail	*Rallus elegans*	6,9	SGCN
Clapper rail	*Rallus crepitans*	2,3,4,6,9,10	B,SGCN
Virginia rail	*Rallus limicola*	2,3,4,6,9,10	B,SGCN
Sora	*Porzana carolina*	2,3,9	SGCN
Common gallinule	*Gallinula galeata*	1,2,3,4,9	B,SGCN
American coot	*Fulica americana*	2,3,9,10	B,SGCN
Yellow rail	*Coturnicops noveboracensis*	11	
Sandhill crane	*Antigone canadensis*	3	
Shorebirds			
Black-necked stilt	*Himantopus mexicanus*	3,9,10	
American avocet	*Recurvirostra americana*	3,9	
Black-bellied plover	*Pluvialis squatarola*	3,9,10	
American golden-plover	*Pluvialis dominica*	3,9,10	SGCN
Killdeer	*Charadrius vociferus*	2,3,6,9,10	B,SGCN
Semipalmated plover	*Charadrius semipalmatus*	2,3,6,9,10	
Upland sandpiper	*Bartramia longicauda*	9	E
Whimbrel	*Numenius phaeopus*	3,9	SC
Hudsonian godwit	*Limosa haemastica*	3,9,10	SGCN
Marbled godwit	*Limosa fedoa*	3,9	SGCN
Ruddy turnstone	*Arenaria interpres*	3,9	

Red knot	Calidris canutus	3,9	T
Ruff	Calidris pugnax	3,9	
Stilt sandpiper	Calidris himantopus	3,9	
Curlew sandpiper	Calidris ferruginea	3,9	
Sanderling	Calidris alba	3,9	SC
Dunlin	Calidris alpina	2,3,9,10	SGCN
Baird's sandpiper	Calidris bairdii	3,9	
Least sandpiper	Calidris minutilla	2,3,6,9,10	
White-rumped sandpiper	Calidris fuscicollis	2,3,9,10	
Buff-breasted sandpiper	Calidris subruficollis	3,9	
Pectoral sandpiper	Calidris melanotos	2,3,9,10	
Semipalmated sandpiper	Calidris pusilla	2,3,6,9,10	SC
Western sandpiper	Calidris mauri	3,9,10	
Short-billed dowitcher	Limnodromus griseus	2,3,6,9,10	SGCN
Long-billed dowitcher	Limnodromus scolopaceus	3,9	
American woodcock	Scolopax minor	2,3,9,10	B,SGCN
Wilson's snipe	Gallinago delicata	2,3,6,9,10	
Spotted sandpiper	Actitis macularius	2,3,4,6,9,10	B,SC
Solitary sandpiper	Tringa solitaria	2,3,6,9,10	SGCN
Lesser yellowlegs	Tringa flavipes	2,3,6,9,10	SGCN
Willet	Tringa semipalmata	3,9	SGCN
Greater yellowlegs	Tringa melanoleuca	2,3,6,9,10	SGCN
Wilson's phalarope	Phalaropus tricolor	3,4,9,10	SGCN
Red-necked phalarope	Phalaropus lobatus	3,9	
Red phalarope	Phalaropus fulicaria	3,9	
Gulls and Terns			
Bonaparte's gull	Chroicocephalus philadelphia	3,9,10	
Black-headed gull	Chroicocephalus ridibundus	9	
Laughing gull	Leucophaeus atricilla	2,3,6,9,10	
Franklin's gull	Leucophaeus pipixcan	3	
Ring-billed gull	Larus delawarensis	2,3,6,9,10	
Herring gull	Larus argentatus	2,3,6,9,10	
Iceland gull	Larus glaucoides	2,3,9	
Lesser black-backed gull	Larus fuscus	2,3,9,10	
Glaucous gull	Larus hyperboreus	2,3,9,10	
Great black-backed gull	Larus marinus	2,3,6,9,10	
Least tern	Sternula antillarum	1,3,2,4,9,10	B,E
Gull-billed tern	Gelochelidon nilotica	3,9	SC

Common name	Scientific name	References	Status
Caspian tern	Hydroprogne caspia	3,9,10	SC
Black tern	Chlidonias niger	3,9	
Roseate tern	Sterna dougallii	3,9	E*
Common tern	Sterna hirundo	3,9,10	SC
Forster's tern	Sterna forsteri	2,9,10	SGCN
Royal tern	Thalasseus maximus	3,9	SGCN
Black skimmer	Rynchops niger	1,3,4,9,10	E
Other Waterbirds			
Red-throated loon	Gavia stellata	3,9,10	SGCN
Common loon	Gavia immer	3,9,10	SGCN
Great cormorant	Phalacrocorax carbo	2,3,10	
Double-crested cormorant	Phalacrocorax auritus	1,2,3,6,9,10	B
American white pelican	Pelecanus erythrorhynchos	3	
Pied-billed grebe	Podilymbus podiceps	2,4,9,10	B,E
Horned grebe	Podiceps auritus	2,3,9	SGCN
Red-necked grebe	Podiceps grisegena	3,9	
Raptors			
Black vulture	Coragyps atratus	3	
Turkey vulture	Cathartes aura	2,3,9,10	
Osprey	Pandion haliaetus	2,3,4,6,9,10	B,T
Swallow-tailed kite	Elanoides forficatus	3	
Golden eagle	Aquila chrysaetos	8	SGCN
Northern harrier	Circus hudsonius	1,2,3,4,5,6,9,10	B,E
Sharp-shinned hawk	Accipiter striatus	2,3,5,6,9,10	SC
Cooper's hawk	Accipiter cooperii	2,3,6,9,10	SC
Northern goshawk	Accipiter gentilis	2,3,6,9,10	E
Bald eagle	Haliaeetus leucocephalus	3,8,9,10	E
Red-shouldered hawk	Buteo lineatus	2,3,6,9	E
Broad-winged hawk	Buteo platypterus	3,9,10	SC
Red-tailed hawk	Buteo jamaicensis	2,3,6,9,10	B
Rough-legged hawk	Buteo lagopus	2,3,6,9,10	
American kestrel	Falco sparverius	2,3,4,5,6,9,10	B,T
Merlin	Falco columbarius	2,3,6,9,10	
Peregrine falcon	Falco peregrinus	2,3,9,10	B,E
Barn owl	Tyto alba	3,9	B,SC
Eastern screech owl	Megascops asio	3,9	
Great horned owl	Bubo virginianus	3,9	B
Snowy owl	Nyctea scandiacus	3,9	

Barred owl	*Strix varia*	9	T
Long-eared owl	*Asio otus*	2,3,9	T
Short-eared owl	*Asio flammeus*	2,3,4,9,10	E
Northern saw-whet owl	*Aegolius acadicus*	3	SGCN
Galliform Birds			
Northern bobwhite	*Colinus virginianus*	3,10	SGCN
Wild turkey	*Meleagris gallopavo*	3,8,9,10	
Ring-necked pheasant	*Phasianus colchicus*	2,3,4,9,10	B
Other Birds (Nonpasserines)			
Monk parakeet	*Myiopsitta monachus*	3,8	B
Mourning dove	*Zenaida macroura*	3,6,9,10	B
Rock pigeon	*Columba livia*	3,6,9,10	B
Ruby-throated hummingbird	*Archilochus colubris*	2,3,9,10	
Common nighthawk	*Chordeiles minor*	3,9	SC
Eastern whip-poor-will	*Antrostomus vociferus*	11	SC
Belted kingfisher	*Megaceryle alcyon*	2,3,6,9,10	B,SGCN
Black-billed cuckoo	*Coccyzus erythropthalmus*	3,9,10	SGCN
Yellow-billed cuckoo	*Coccyzus americanus*	3,9	B,SGCN
Red-headed woodpecker	*Melanerpes erythrocephalus*	3,9	T
Red-bellied woodpecker	*Melanerpes carolinus*	3,10	B
Yellow-bellied sapsucker	*Sphyrapicus varius*	2,3,9,10	
Downy woodpecker	*Dryobates pubescens*	2,3,6,9,10	B
Hairy woodpecker	*Dryobates villosus*	3,9,10	B
Northern flicker	*Colaptes auratus*	2,3,9,10	B,SGCN
Pileated woodpecker	*Dryocopus pileatus*	3	
Chimney swift	*Chaetura pelagica*	2,3,8,9,10	b,SGCN
Passerines			
Great crested flycatcher	*Myiarchus crinitus*	2,3,9,10	b,SGCN
Eastern kingbird	*Tyrannus tyrannus*	2,3,4,6,9,10	B,SGCN
Western kingbird	*Tyrannus verticalis*	2,9	
Eastern wood-pewee	*Contopus virens*	2,3,9,10	b,SGCN
Yellow-bellied flycatcher	*Empidonax flaviventris*	3,9	SGCN
Acadian flycatcher	*Empidonax virescens*	3,9	SGCN
Alder flycatcher	*Empidonax alnorum*	2,3	
Willow flycatcher	*Empidonax traillii*	2,3,4,6,9,10	SGCN
Least flycatcher	*Empidonax minimus*	3,9,10	SC
Eastern phoebe	*Sayornis phoebe*	2,3,9,10	B

Common name	Scientific name	References	Status
White-eyed vireo	*Vireo griseus*	3,9	
Yellow-throated vireo	*Vireo flavifrons*	3	SGCN
Blue-headed vireo	*Vireo solitarius*	2,3,9,10	B,SC
Philadelphia vireo	*Vireo philadelphicus*	3	
Warbling vireo	*Vireo gilvus*	2,3,6,9,10	B
Red-eyed vireo	*Vireo olivaceus*	2,3,9,10	B
Loggerhead shrike	*Lanius ludovicianus*	9	E
Northern shrike	*Lanius borealis*	2,3,9	
Blue jay	*Cyanocitta cristata*	2,3,9,10	B
American crow	*Corvus brachyrhynchos*	2,3,9,10	B
Fish crow	*Corvus ossifragus*	2,3,6,9,10	b
Common raven	*Corvus corax*	3,9,10	B
Black-capped chickadee	*Poecile atricapillus*	2,3,6,9,10	B
Tufted titmouse	*Baeolophus bicolor*	2,3,9,10	B
Horned lark	*Eremophila alpestris*	2,3,4,9,10	B,T
Bank swallow	*Riparia riparia*	2,3,6,9,10	SGCN
Tree swallow	*Tachycineta bicolor*	2,3,5,6,9,10	B
Northern rough-winged swallow	*Stelgidopteryx serripennis*	2,3,6,8,9,10	B
Cliff swallow	*Petrochelidon pyrrhonota*	3,9,10	SC
Purple martin	*Progne subis*	2,3,9,10	SGCN
Barn swallow	*Hirundo rustica*	2,3,4,5,6,9,10	B
Ruby-crowned kinglet	*Regulus calendula*	2,3,9,10	
Golden-crowned kinglet	*Regulus satrapa*	2,3,9,10	
Cedar waxwing	*Bombycilla cedrorum*	2,3,9,10	
Red-breasted nuthatch	*Sitta canadensis*	3,9	
White-breasted nuthatch	*Sitta carolinensis*	3,9,10	B
Brown creeper	*Certhia americana*	2,3,9,10	
Blue-gray gnatcatcher	*Polioptila caerulea*	2,3,9,10	
House wren	*Troglodytes aedon*	2,3,9,10	B
Winter wren	*Troglodytes hiemalis*	2,3,8.9,10	SC
Sedge wren	*Cistothorus platensis*	3,6,7,9	E
Marsh wren	*Cistothorus palustris*	1,2,3,4,5,6,9,10	B,SGCN
Carolina wren	*Thryothorus ludovicianus*	3,9,10	B
Gray catbird	*Dumetella carolinensis*	2,3,4,6,9,10	B,SGCN
Brown thrasher	*Toxostoma rufum*	2,3,4,6,9,10	B
Northern mockingbird	*Mimus polyglottos*	3,4,9,10	B
European starling	*Sturnus vulgaris*	2,3,6,9,10	B
Eastern bluebird	*Sialia sialis*	2,3,10	B

Veery	*Catharus fuscescens*	2,3,9,10	SC
Gray-cheeked thrush	*Catharus minimus*	3,9	SC
Swainson's thrush	*Catharus ustulatus*	2,3,9	
Hermit thrush	*Catharus guttatus*	2,3,9,10	
Wood thrush	*Hylocichla mustelina*	2,3,9,10	B,SC
American robin	*Turdus migratorius*	2,3,4,6,9,10	B
Northern wheatear	*Oenanthe oenanthe*	3	
House sparrow	*Passer domesticus*	3,9,10	B
American pipit	*Anthus rubescens*	2,3,9,10	
House finch	*Haemorhous mexicanus*	3,9,10	B
Purple finch	*Haemorhous purpureus*	2,3,9	SGCN
Common redpoll	*Acanthis flammea*	3,9	
Red crossbill	*Loxia curvirostra*	3	
Pine siskin	*Spinus pinus*	3,9,10	
American goldfinch	*Spinus tristis*	2,3,4,6,9,10	B
Lapland longspur	*Calcarius lapponicus*	2,3,9	
Snow bunting	*Plectrophenax nivalis*	2,3,9,10	
Grasshopper sparrow	*Ammodramus savannarum*	2,3,10	B,T
Lark sparrow	*Chondestes grammacus*	2,3	
Chipping sparrow	*Spizella passerina*	2,3,9,10	b
Clay-colored sparrow	*Spizella pallida*	2,3	
Field sparrow	*Spizella pusilla*	2,3,9,10	B,SGCN
Fox sparrow	*Passerella iliaca*	2,3,9,10	
American tree sparrow	*Spizelloides arborea*	2,3,6,9,10	
Dark-eyed junco	*Junco hyemalis*	2,3,9,10	
White-crowned sparrow	*Zonotrichia leucophrys*	2,3,9,10	
White-throated sparrow	*Zonotrichia albicollis*	2,3,9,10	SGCN
Vesper sparrow	*Pooecetes gramineus*	2,3,9	E
LeConte's sparrow	*Ammospiza leconteii*	2,3	
Seaside sparrow	*Ammospiza maritima*	3,4,9,10	B,SGCN
Saltmarsh sparrow	*Ammospiza caudacuta*	2,3,4,9,10	B,SC
Henslow's sparrow	*Centronyx henslowii*	2,3	E
Savannah sparrow	*Passerculus sandwichensis*	2,3,6,9,10	B,T
Song sparrow	*Melospiza melodia*	2,3,4,6,9,10	B
Lincoln's sparrow	*Melospiza lincolnii*	2,3,9,10	
Swamp sparrow	*Melospiza georgiana*	2,3,4,5,6,9,10	B
Eastern towhee	*Pipilo erythrophthalmus*	2,3,9,10	SGCN
Yellow-breasted chat	*Icteria virens*	3,9	SGCN

Common name	Scientific name	References	Status
Yellow-headed blackbird	*Xanthocephalus xanthocephalus*	9	
Bobolink	*Dolichonyx oryzivorus*	2,3,6,9,10	T
Eastern meadowlark	*Sturnella magna*	2,3,9,10	SC
Orchard oriole	*Icterus spurius*	2,3,9,10	
Baltimore oriole	*Icterus galbula*	2,3,6,9,10	B,SGCN
Red-winged blackbird	*Agelaius phoeniceus*	1,2,3,4,5,6,9,10	B
Brown-headed cowbird	*Molothrus ater*	2,3,9,10	B
Rusty blackbird	*Euphagus carolinus*	2,3,9,10	SGCN
Common grackle	*Quiscalus quiscula*	2,3,5,9,10	B
Boat-tailed grackle	*Quiscalus major*	3,9,10	
Ovenbird	*Seiurus aurocapilla*	3,9,10	
Worm-eating warbler	*Helmitheros vermivora*	3	SC
Louisiana waterthrush	*Parkesia motacilla*	3,9,10	SGCN
Northern waterthrush	*Parkesia noveboracensis*	2,3,6,9,10	
Golden-winged warbler	*Vermivora chrysoptera*	9	E
Blue-winged warbler	*Vermivora cyanoptera*	3,9,10	
Black-and-white warbler	*Mniotilta varia*	2,3,8,9,10	SGCN
Prothonotary warbler	*Protonotaria citrea*	3	SGCN
Tennessee warbler	*Leiothlypis peregrina*	2,3,9	
Orange-crowned warbler	*Leiothlypis celata*	2,3,9,10	
Nashville warbler	*Leiothlypis ruficapilla*	2,3,9,10	
Connecticut warbler	*Oporornis agilis*	3,9	
Mourning warbler	*Geothlypis philadelphia*	3,9	
Kentucky warbler	*Geothlypis formosus*	3	SC
Common yellowthroat	*Geothlypis trichas*	1,2,3,4,5,6,9,10	B
Hooded warbler	*Setophaga citrina*	3	SC
American redstart	*Setophaga ruticilla*	2,3,9,10	
Cape May warbler	*Setophaga tigrina*	3,9	
Cerulean warbler	*Setophaga cerulea*	11	SC
Northern parula	*Setophaga americana*	2,3,9,10	SGCN
Magnolia warbler	*Setophaga magnolia*	2,3,9,10	
Bay-breasted warbler	*Setophaga castanea*	2,3,9	SGCN
Blackburnian warbler	*Setophaga fusca*	3,9	SC
Yellow warbler	*Setophaga petechia*	2,3,4,5,6,9,10	B
Chestnut-sided warbler	*Setophaga pensylvanica*	2,3,9,10	B
Blackpoll warbler	*Setophaga striata*	2,3,9,10	
Black-throated blue warbler	*Setophaga caerulescens*	2,3,9,10	SC

Palm warbler	*Setophaga palmarum*	2,3,6,9,10	
Pine warbler	*Setophaga pinus*	2,3,10	
Yellow-rumped warbler	*Setophaga coronata*	2,3,6,9,10	
Yellow-throated warbler	*Setophaga dominica*	3	
Prairie warbler	*Setophaga discolor*	3,9,10	SGCN
Black-throated green warbler	*Setophaga virens*	2,3,9,10	SC
Wilson's warbler	*Cardellina pusilla*	2,3,8,9	
Canada warbler	*Cardellina canadensis*	2,3,9,10	SC
Scarlet tanager	*Piranga olivacea*	3,9,10	SGCN
Summer tanager	*Piranga rubra*	3	SGCN
Northern cardinal	*Cardinalis cardinalis*	3,9,10	B
Rose-breasted grosbeak	*Pheucticus ludovicianus*	2,3,9	SGCN
Blue grosbeak	*Passerina caerulea*	3,4,9,10	B
Indigo bunting	*Passerina cyanea*	2,3,4,6,9,10	B
Dickcissel	*Spiza americana*	9	SGCN

Note: B (breeding on record); b (highly probable species is breeding); State status, E (Endangered), T (Threatened), SC (Special Concern); SGCN (Species of Greatest Conservation Need) include E, T, SC among other criteria and are "those species that through a combination of low and/or declining populations or vulnerability to threats, particularly anthropogenic threats, are considered to be at risk of becoming extinct, extirpated, endangered, or threatened." See New Jersey's Wildlife Action Plan (NJDEP 2018a) for further explanation of criteria used to designate species as SGCN.

*Refers to federal listing as well as state

Occurrence references:

1 Day et al. 1999
2 Kane and Githens 1997
3 eBird 2021
4 New Jersey Turnpike Authority 1986
5 Wargo 1989
6 U.S. Army Corps of Engineers 2000
7 Quinn 1997
8 E. Kiviat observations, H. Carola observations; see text for eagles
9 New Jersey Meadowlands Commission Bird Checklist
10 Mizrahi et al. 2007
11 Cruickshank 1942
12 NJDEP DFW 2018

References Cited

Abbott, C.G. 1907. Summer bird-life of the Newark, New Jersey, marshes. Auk 24(1):1–12.

Able, K.W., C.W. Talbot and J.K. Shisler. 1983. The spotfin killifish, *Fundulus luciae*, is common in New Jersey salt marshes. Bulletin of the New Jersey Academy of Science 28:7–11.

Abraham, B.J. 1985. Species profiles: Life histories and environmental requirements of coastal fishes and invertebrates (Mid-Atlantic) – Mummichog and striped killifish. U.S. Fish and Wildlife Service Biological Report 82.

AECOM. 2018. Final Environmental Impact Statement for the Rebuild by Design Meadowlands Flood Protection Project. Report to the New Jersey Department of Environmental Protection. www.nj.gov/dep/floodresilience/rbd-meadowlands-feis .htm (17 April 2021).

Aggarwal, Y.P. and L.R. Sykes. 1978. Earthquakes, faults, and nuclear power plants in southern New York and northern New Jersey. Science 200(4340):425–429.

Agron, S.L. 1980. Environmental geology of the Hackensack Meadowlands. pp. 216–241 in W. Manspeizer, ed. Field Studies of New Jersey Geology and Guide to Field Trips: 52nd Annual Meeting of the New York State Geological Association. Newark College of Arts and Sciences, Newark, NJ.

Ainsworth, A.M. 2019. *Echinodontium ballouii*. The IUCN Red List of Threatened Species 2019:e.T71567420A71567447. dx.doi.org/10.2305/IUCN.UK.2019-1. RLTS.T71567420A71567447 (4 July 2021).

Albers, P.H., L. Sileo and B.M. Mulhern. 1986. Effects of environmental contaminants on snapping turtles of a tidal wetland. Archives of Environmental Contamination and Toxicology 15:39–49.

Aldrich, J.W. and R.W. Coffin. 1980. Breeding bird populations from forest to suburbia after thirty-seven years. American Birds 34:3–7.

Alexander, R.C. 1952. Introduced species of land snails in New Jersey. Nautilus 65(4):132–135.

Alisauskas, R.T. and T.W. Arnold. 1994. American coot. P. 127–143 in T.C. Tacha and C.E. Braun, eds. Migratory Shore and Upland Game Bird Management in North America. International Association of Fish and Wildlife Agencies, Allen Press, Lawrence, KS.

Allen, B. 1995. The genera *Ceratodon, Distichium, Pleuridium,* and *Saelania* in Maine. Evansia 12(3):107–116.

Allen, D.M., S.K. Service and M.V. Ogburn-Matthews. 1992. Factors influencing the collection efficiency of estuarine fishes. Transactions of the American Fisheries Society 121:234–244.

Allen, E.A., P.E. Fell, M.A. Peck, J.A. Gieg, C.R. Guthke and M.D. Newkirk. 1994. Gut contents of common mummichogs, *Fundulus heteroclitus* L., in a restored impounded marsh and in natural reference marshes. Estuaries 17:462–471.

Allen, J.L. and N.M. Howe. 2016. Landfill lichens: A checklist for Freshkills Park, Staten Island, New York. Opuscula Philolichenum 15:82–91.

Allen, J.L. and J.C. Lendemer. 2015. Fungal conservation in the USA. Endangered Species Research 28:33–42.

Alpert, P. 1990. Microtopography as habitat structure for mosses on rocks. pp. 120–140 in S.S. Bell, E.D. McCoy and H.R. Mushinsky, eds. Habitat Structure: The Physical Arrangement of Objects in Space. Chapman and Hall, London, U.K.

AMNET (Ambient Macroinvertebrate Network). 1993–2018. New Jersey Department of Environmental Protection, Bureau of Freshwater and Biological Monitoring. nj.gov/dep/wms/bfbm/ (31 May 2021).

Anderson, J.D. 1976. Amphibians and reptiles from Jekyll Island, Glynn County, Georgia. Herpetological Review 7(4):179–180.

Anderson, K. 1977. Food of long-eared owl. New Jersey Audubon 3:92–93.

Andrus, R.E. 1980. Sphagnaceae (peat moss family) of New York State. New York State Museum Bulletin 442. 89 p.

Andrus, R.E. 1990. Why rare and endangered bryophytes? P. 199–201 in R.S. Mitchell, C.J. Sheviak and D.J. Leopold, eds. Ecosystem Management: Rare Species and Significant Habitats. New York State Museum Bulletin 471.

Angalet, G.W., J.M. Tropp and A.N. Eggert. 1979. *Coccinella septempunctata* in the United States: Recolonizations and notes on its ecology. Environmental Entomology 8(5):896–901.

Anonymous. 1937. Notes of the recent forest history of the Newark Marshes. New Jersey Department of Conservation and Development, Division of Geology. Unpublished report.

Anonymous. 2006. New Jersey marine mammal and sea turtle conservation workshop proceedings. New Jersey Department of Environmental Protection Endangered and Nongame Species Program, 17–19 April 2006. Conservation Impact. http://www.njfishandwildlife.com/ensp/pdf/marinemammal_seaturtle_workshop06.pdf (31 May 2009).

Anonymous. 2007. The New Jersey Meadowlands Commission 2007 Economic Growth Plan. 15 p. rucore.libraries.rutgers.edu/rutgers-lib/27939/ (31 January 2020).

Anonymous. 2009. Al G. Bloom [*sic*] in DeKorte Park. The Meadowlands Blog. http://meadowblog.typepad.com/mblog/2009/06/about-the-dekorte-park-algae-bloom.html (11 July 2009).

Anonymous. 2010. *Orconectes virilis*. Global invasive species database. http://issg .org/database/species/ecology.asp?si=218&fr=1&sts=sss&lang=EN (22 January 2020).

Anonymous. 2019. Mosquito species distribution by county in New Jersey. New Jersey Agricultural Experiment Station Center for Vector Biology. http://vectorbio .rutgers.edu/outreach/mosquitospeciesdistributionsNJ.pdf (18 November 2019).

Antevs, E. 1928. The last glaciation, with special reference to the ice retreat in northeastern North America. American Geographical Society Research Series 17:1–292.

Anthony, R.G., L.J. Niles and J.D. Spring. 1981. Small-mammal associations in forested and old-field habitats: A quantitative comparison. Ecology 62:955–963.

Apfelbaum, S.I. and P. Seelbach. 1983. Nest tree, habitat selection and productivity of seven North American raptor species based on the Cornell University nest record card program. Raptor Research 17:97–113.

Armitage, P., P.S. Cranston and L.C.V. Pinder. 1995. The Chironomidae. Chapman and Hall, London, U.K. 572 p.

Arndt, R.G. 2004. Annotated checklist and distribution of New Jersey freshwater fishes, with comments on abundance. Bulletin of the New Jersey Academy of Sciences 49:1–33.

Arnold, M. 2008. A historical perspective on the urban wetlands of the Teaneck Creek Conservancy. Urban Habitats 5(1):153–165. http://www.urbanhabitats.org/v05n01/ history_pdf.pdf (17 November 2019).

Aronson, M.F.J., F.A. La Sorte, C.H. Nilon, M. Katti, M.A. Goddard, C.A. Lepczyk, P.S. Warren, N.S.G. Williams, S. Cilliers, B. Clarkson, C. Dobbs, R. Dolan, M. Hedblom, S. Klotz, J.L. Kooijmans, I. Kühn, I. MacGregor-Fors, M. McDonnell, U. Mörtberg, P. Pyšek, J. Siebert, J. Sushinsky, P. Werner and M. Winter. 2014. A global analysis of the impacts of urbanization on bird and plant diversity reveals key anthropogenic drivers. Proceedings of the Royal Society B Biological Sciences 281:e20133330. doi.org/10.1098/rspb.2013.3330 (18 December 2021).

Artigas, F., J. Grzyb and C. Yao. 2016. Water quality monitoring in a tidal estuary. Presentation to New Jersey Water Monitoring Council, May 19, 2016. Meadowlands Environmental Research Institute. www.state.nj.us/dep/wms/wmccmeetinginfo .html (23 January 2020).

Artigas F., J.M. Loh, J.Y Shin, J. Grzyb and Y. Yao. 2017. Baseline and distribution of organic pollutants and heavy metals in tidal creek sediments after Hurricane Sandy in the Meadowlands of New Jersey. Environmental Earth Sciences 76(7):e293.

Artigas, F., J.Y. Shin, C. Hobble, A. Marti-Donati, K.V.R. Shäfer and I. Pechmann. 2015. Long term carbon storage potential and CO_2 sink strength of a restored salt marsh in New Jersey. Agricultural and Forest Meteorology 200:313–321.

Asakawa, Y. 2012. Liverworts - Potential source of medicinal compounds. Medicinal and Aromatic Plants 1(3):1–2.

Askins, R.A. 1993. Population trends in grassland, shrubland, and forest birds in eastern North America. Current Ornithology 11:1–34.

Askins, R.A. 2000. Restoring North America's birds: Lessons from landscape ecology. Yale University Press, New Haven, CT. 320 p.

Ayres, E., R. van der Wal, M. Sommerkorn and R.D. Bardgett. 2006. Direct uptake of soil nitrogen by mosses. Biology Letters 2(2):286–288.

Baker, P.J., A.J. Bentley, R.J. Ansell and S. Harris. 2005. Impact of predation by domestic cats *Felis catus* in an urban area. Mammal Review 35:302–312.

Baker, P.J., C.V. Dowding, S.E. Molony, P.C. White and S. Harris. 2007. Activity patterns of urban red foxes (*Vulpes vulpes*) reduce the risk of traffic-induced mortality. Behavioral Ecology 18(4):716–724.

Baltzer, J.L., H.L. Hewlin, E.G. Reekie, P.D. Taylor and J.S. Boates. 2002. The impact of flower harvesting on seedling recruitment in sea lavender (*Limonium carolinianum*, Plumbaginaceae). Rhodora 104(919):280–295.

Bannor, B.K. and E. Kiviat. 2002. Common gallinule (*Gallinula galeata*), Version 1.0, Birds of the World. Cornell Lab of Ornithology, Ithaca, NY. birdsoftheworld. org/bow/species/comgal1/cur/introduction (24 January 2020).

Bansal, S., S.C. Lishawa, S. Newman, B.A. Tangen, D. Wilcox, D. Albert, M.J. Anteau, M.J. Chimney, R.L. Cressey, E. DeKeyser, K.J. Elgersma, S.A. Finkelstein, J. Freeland, R. Grosshans, P.E. Klug, D.J. Larkin, B.A. Lawrence, G.Linz, J. Marburger, G. Noe, C. Otto, N. Reo, J. Richards, C. Richardson, L. Rodgers, A.J. Schrank, D. Svedarsky, S. Travis, N. Tuchman and L. Windham-Myers. 2019. *Typha* (cattail) invasion in North American wetlands: Biology, regional problems, impacts, ecosystem services, and management. Wetlands 39(4):645–684.

Barbour, J.G. and E. Kiviat. 1997. Introduced purple loosestrife as host of native Saturniidae (Lepidoptera). Great Lakes Entomologist 30(3):115–122.

Barendregt, A., D. Whigham and A. Baldwin, eds. 2009. Tidal freshwater wetlands. Backhuys Publishers, Leiden, The Netherlands. 320 p.

Barker, J.R. and D.T. Tingey, eds. 1992. Air pollution effects on biodiversity. Van Nostrand Reinhold, New York, NY. 322 p.

Barkman, J.J. 1958. Phytosociology and ecology of cryptogamic epiphytes. Van Gorcum, Assen, The Netherlands. 628 p. + plates.

Barko, V.A., G.A. Feldhamer, M.C. Nicholson and D.K. Davie. 2003. Urban habitat: A determinant of white-footed mouse (*Peromyscus leucopus*) abundance in southern Illinois. Southeastern Naturalist 2:369–376.

Barlow, A.E., D.M. Golden and J. Bangma. 2009. Field guide to dragonflies and damselflies of New Jersey. New Jersey Department of Environmental Protection, Division of Fish and Wildlife, Flemington, NJ. 285 p.

Barrett, K.R. and M.A. Mcbrien. 2007. Chemical and biological assessment of an urban, estuarine marsh in northeastern New Jersey, USA. Environmental Monitoring and Assessment 124:63–88.

Barringer, K. 2021a. A checklist of the mosses of New Jersey. Version 2.04. 90 p. www.huttonstreet.org/floras/NJmosses.pdf (12 September 2021).

Barringer, K. 2021b. A checklist of the hornworts (Anthocerotophyta) and liverworts (Marchantiophyta) of New Jersey. Version 2.03. 23 p. www.huttonstreet.org/floras /NJliverworts.pdf (21 September 2021).

Bart, D., D. Burdick, R. Chambers and J.M. Hartman. 2006. Human facilitation of *Phragmites australis* invasions in tidal marshes: A review and synthesis. Wetlands Ecology and Management 14(1):53–65.

Basnet, B.B., H. Liu, L. Liu and Y. M. Suleimen. 2018. Diversity of anticancer and antimicrobial compounds from lichens and lichen-derived fungi: A systematic review (1985–2017). Current Organic Chemistry 22:1–14.

Batzer, D.P. and R.D. Sjogren. 1986. Larval habitat characteristics of *Coquillettidia perturbans* (Diptera: Culicidae) in Minnesota. Canadian Entomologist 118(11):1193–1198.

Baxter, G. 2002. All terrain vehicles as a cause of fire ignition in Alberta forests. Advantage 3(44):1–7.

Baxter-Gilbert J., J.L. Riley and J.D. Litzgus. 2013. *Plestiodon fasciatus* (five-lined skink). Artificial habitat use. Herpetological Review 44(4):680–681.

Bayard, T.S. and C.S. Elphick. 2011. Planning for sea-level rise: Quantifying patterns of saltmarsh sparrow (*Ammodramus caudacutus*) nest flooding under current sea-level conditions. Auk 128(2):393–403.

Brooklyn Botanic Garden. 2010. New York Metropolitan Flora Project. Brooklyn Botanic Garden, Brooklyn, NY. http://www.bbg.org/research/nymf/ (18 December 2021).

Beal, E.O. 1977. A manual of marsh and aquatic vascular plants of North Carolina with habitat data. North Carolina Agricultural Research Service Technical Bulletin 247, Raleigh, NC. 298 p.

Bean, T.H. 1903. The food and game fishes of New York. State of New York Forest, Fish, and Game Commission, Albany, NY. 209 p.

Beard, C.B., J. Occi, D.L. Bonilla et al. 2018. Multistate infestation with the exotic disease–vector tick *Haemaphysalis longicornis*—United States, August 2017–September 2018. Morbidity and Mortality Weekly Report 67(47):1310–1313.

Beissinger, S.R. and D.R. Osborne. 1982. Effects of urbanization on avian community organization. Condor 84(1):75–83.

Bekoff, M. and E.M. Gese. 2003. Coyote (*Canis latrans*). P. 467–481 in G.A. Feldhamer, B.C. Thompson and J.A. Chapman, eds. Wild Mammals of North America: Biology, Management, and Conservation. 2nd ed. Johns Hopkins University Press, Baltimore, MD.

Bennett, A.F., D.G. Nimmo and J.Q. Radford. 2014. Riparian vegetation has disproportionate benefits for landscape-scale conservation of woodland birds in highly modified environments. Journal of Applied Ecology 51:514–523.

Benoit, L.K. and R.A. Askins. 2002. Relationship between habitat area and the distribution of tidal marsh birds. Wilson Bulletin 114:314–323.

Bentley, E.L. 1994. Use of a landscape-level approach to determine the habitat requirements of the yellow-crowned night-heron, *Nycticorax violaceus*, in the lower Chesapeake Bay. MA thesis, College of William and Mary, Williamsburg, VA. 90 p.

Bequaert, J.C. and W.B. Miller. 1973. The mollusks of the arid Southwest with an Arizona check list. University of Arizona Press, Tucson. 271 p.

Berger, J. 1992. The Hackensack River Meadowlands. P. 510–518 in S. Maurizi and F. Poillon, eds. Restoration of Aquatic Ecosystems: Science, Technology, and Public Policy. National Academy Press, Washington, DC.

Berger (Louis) Group, Inc. 2001. Oritani Marsh mitigation site, East Rutherford, Bergen County, New Jersey: Baseline studies. Report to Hackensack Meadowlands Development Commission, Lyndhurst, NJ.

Bernick, A.J. 2007. New York City Audubon's Harbor Herons Project: 2007 nesting survey. New York City Audubon, New York, NY. 48 p. http://www.nycaudubon.org /pdf/2007_NYCA_HH_Report_Bernick-1.pdf (21 January 2020).

Bett, T.A. 1983. Influences of habitat composition on the breeding ecology of the American coot (*Fulica americana*). MS thesis, University of Wisconsin-Oshkosh, Oshkosh, WI.

Bibby, C.J., R.E. Green, G.R.M. Pepler and P.A. Pepler. 1976. Sedge warbler migration and reed aphids. British Birds 69:384–399.

Blair, R.B. 1996. Land use and avian species diversity along an urban gradient. Ecological Applications 6:506–519.

Blair, R.B. 1999. Birds and butterflies along an urban gradient: Surrogate taxa for assessing biodiversity? Ecological Applications 9:164–170.

Boarman, W.I. and B. Heinrich. 2020. Common raven (*Corvus corax*), Version 1.0, Birds of the World. Cornell Lab of Ornithology, Ithaca, NY. birdsoftheworld.org/ bow/species/comrav/cur/introduction (22 January 2020).

Boch, S., J. Müller, D. Prati, S. Blaser and M. Fischer. 2013. Up in the tree—The overlooked richness of bryophytes and lichens in tree crowns. PloS One 8(12):e84913.

Boerner, R.E. and R.T.T. Forman. 1975. Salt spray and coastal dune mosses. Bryologist 78(1):57–63.

Bordoni, L., C. Nasuti, D. Fedeli, R. Galeazzi, E. Laudadio, L. Massaccesi, G. López-Rodas and R. Gabbianelli. 2019. Early impairment of epigenetic pattern in neurodegeneration: Additional mechanisms behind pyrethroid toxicity. Experimental Gerontology 124:e110629.

Bornstein, S., T. Mörner and W.M. Samuel. 2001. *Sarcoptes scabiei* and sarcoptic mange. P. 107–119 in W.M. Samuel, M.J. Pybus and A.A. Kocan, eds. Parasitic Diseases of Wild Mammals. 2nd ed. Iowa State University Press, Ames, IA.

Bosakowski, T. 1982. Food habits of wintering *Asio* owls in the Hackensack Meadowlands. Records of New Jersey Birds 8:40–42.

Bosakowski, T. 1983. Density and roosting habits of northern harriers wintering in the Hackensack Meadowlands. Records of New Jersey Birds 9:50–54.

Bosakowski, T. 1986. Short-eared owl winter roosting strategies. American Birds 40(2):237–240.

Bosakowski, T. 1989. Observations on the evening departure and activity of wintering short-eared owls in New Jersey. Journal of Raptor Research 23:162–166.

Bosakowski, T. and D.G. Smith. 1997. Distribution and species richness of a forest raptor community in relation to urbanization. Journal of Raptor Research 31:26–33.

Boven, L., B. Vanschoenwinkel, E.R. De Roeck, A. Hulsmans and L. Brendonck. 2008. Diversity and distribution of large branchiopods in Kiskunság (Hungary) in relation to local habitat and spatial factors: Implications for their conservation. Marine and Freshwater Research 59:940–950.

Brackney, A.W. 1979. Population ecology of common gallinules in southwestern Lake Erie marshes. MS thesis, Ohio State University, Columbus, OH. 69 p.

Bradley, R. 2013. Common spiders of North America. University of California Press, Berkeley, CA. 271 p.

Brady, N.C. and R.R. Weil. 1996. The nature and properties of soils. 11th ed. Prentice Hall, Upper Saddle River, NJ. 740 p.

Bragin, A.B., D.P. McQuade and A.M. Iveson. 2019. Report on three fishery resource inventories of the lower Hackensack River within the Hackensack Meadowlands District. New Jersey Sports and Exposition Authority. s3.us-east-2.amazonaws.co m/njmc/pdfs/general/fisheries-report-13-15.pdf (24 January 2020).

Bragin, A.B., J. Misuik, C.A. Woolcott, K.B. Barrett and R. Jusino-Atrensino. 2005. A fishery resource inventory of the lower Hackensack River within the Hackensack Meadowlands District. New Jersey Meadowlands Commission, Meadowlands Environmental Research Institute. 8 p.

Bragin, A.B., C.A. Woolcott and J. Misuik. 2009. A study of the benthic macroinvertebrate community of an urban estuary: New Jersey's Hackensack Meadowlands. New Jersey Meadowlands Commission, Lyndhurst, NJ. Not continuously paginated.

Brand, A. 2019. The hidden world of the fox. HarperCollins Publishers, New York, NY. 224 p.

Breland, O.P. and J.W. Dobson. 1947. Specificity of mantid oothecae (Orthoptera: Mantidae). Annals of the Entomological Society of America 40(4):557–575.

Bricklin, R.B., E.M. Thomas, J.D. Lewis and J.A. Clark. 2016. Foraging birds during migratory stopovers in the New York metropolitan area: Associations with native and non-native plants. Urban Naturalist 11:1–16.

Brisbin, I.L., Jr. and T.B. Mowbray. 2002. American coot (*Fulica americana*), Version 1.0, Birds of the World. Cornell Lab of Ornithology, Ithaca, NY. birdsoftheworld.o rg/bow/species/y00475/cur/introduction (31 August 2010).

Britton, N.L. 1889. Catalogue of plants found in New Jersey. New Jersey Geological Survey, Final Report of the State Geologist. 642 p. [Cited in Sipple 1972; original not seen.]

Brodo, I.M. 1966. Lichen growth and cities: A study on Long Island, New York. Bryologist 69:427–449.

Brodo, I.M. 1968. The lichens of Long Island, New York: A vegetational and floristic analysis. New York State Museum and Science Service Bulletin 410. 330 p.

Brodo, I.M. 2017. A lichen hotspot in the city of Boulder. Book review. Bryologist 120(1):112–113.

Brodo, I.M., S.D. Sharnoff and S. Sharnoff. 2001. Lichens of North America. Yale University Press, New Haven, CT. 795 p.

Brooks, J. 1957. The Meadows. New Yorker (9 March):98–115, (16 March):108–127.

Brooks, R.T. 2011. Declines in summer bat activity in central New England 4 years following the initial detection of white-nose syndrome. Biodiversity and Conservation 20:2537–2541.

Brown, H.P. 1975. Trees of New York State: Native and naturalized. Dover Publications, New York, NY. 433 p. (Reprint of Technical Publication 15, New York State College of Forestry, 1921.)

Brown, P.M. 1997. Wild orchids of the northeastern United States. Cornell University Press, Ithaca, NY. 236 p.

Brua, R.B. 2002. Ruddy duck (*Oxyura jamaicensis*), Version 1.0, Birds of the World. Cornell Lab of Ornithology, Ithaca, NY. birdsoftheworld.org/bow/species/rudduc/c ur/introduction (31 August 2010).

Brundrett, M.C. 2002. Coevolution of roots and mycorrhizas of land plants. New Phytologist 154(2):275–304.

Buegler, R. and S. Parisio. 1982. A comparative flora of Staten Island including the 1879 and 1930 floras by Arthur Hollick and Nathaniel Lord Britton. Staten Island Institute of Arts and Sciences, Staten Island, NY. 92 p.

Bull, J. 1964. Birds of the New York area. Harper and Row, NY. 540 p.

Burger, J. and M. Gochfeld. 2009. Exotic monk parakeets (*Myiopsitta monachus*) in New Jersey: Nest site selection, rebuilding following removal, and their urban wildlife appeal. Urban Ecosystems 12:185–196.

Burger, J., J. Shisler and F.H. Lesser. 1982. Avian utilization on six salt marshes in New Jersey. Biological Conservation 23:187–212.

Burgess, R.L and D.M. Sharpe. 1981. Introduction. P. 1–5 in R.L. Burgess and D.M. Sharpe, eds. Forest Island Dynamics in Man-dominated Landscapes. Springer-Verlag, New York, NY. 310 p.

Burghardt, K.T., D.W. Tallamy, C. Philips and K.J. Shropshire. 2010. Nonnative plants reduce abundance, richness, and host specialization in lepidopteran communities. Ecosphere 1(5):1–22.

Burgio, K.R., C.B. van Rees, K.E. Block, P. Pyle, M.A. Patten, M.F. Spreyer and E.H. Bucher. 2016. Monk parakeet (*Myiopsitta monachus*), Version 1.0, Birds of the World. Cornell Lab of Ornithology, Ithaca, NY. birdsoftheworld.org/bow/species/ monpar/cur/introduction (22 January 2020).

Burke, D.J., J.S. Weis and P. Weis. 2000. Release of metals by the leaves of the salt marsh grasses *Spartina alterniflora* and *Phragmites australis*. Estuarine, Coastal and Shelf Science 51:153–159.

Burke, R.L. and G. Deichsel. 2008. Lacertid lizards introduced into North America: History and future. P. 347–353 in J.C. Mitchell, R.E.J. Brown and B. Bartholomew, eds. Urban Herpetology. Society for the Study of Amphibians and Reptiles, Salt Lake City, UT.

Burkhardt, D. 2014. Don't start a fire. Safety Services Company. www .safetyservicescompany.com/topic/uncategorized/dont-start-fire/ (24 November 2019).

Burton, T.M. and G.E. Likens. 1975. Salamander populations and biomass in the Hubbard Brook Experimental Forest, New Hampshire. Copeia (3):541–546.

Butler, J.R., E.S. Custer, Jr. and W.A. White. 1975. Potential geological natural landmarks Piedmont Region, eastern United States. Department of Geology, University of North Carolina, Chapel Hill, NC. Unpublished report.

Callaghan, D.A. and G. Farr. 2018. The unusual inter-tidal niche of the rare moss *Bryum marratii* Wilson. Journal of Bryology 40(4):371–376.

Callus, M. 2006. Brownfields to greenfields. A position paper of NY/NJ Baykeeper. http://raritan.rutgers.edu/wp-content/uploads/2015/10/NY_NJ-Baykeeper-2006 -Browndfield-to-greenfields.pdf (18 December 2021).

Campbell, K.R. and T.S. Campbell. 2002. A logical starting point for developing priorities for lizard and snake ecotoxicology: A review of available data. Environmental Toxicology and Chemistry 21:894–898.

Capinera, J.L., R.D. Scott and T.J. Walker. 2004. Field guide to grasshoppers, katydids, and crickets of the United States. Cornell University Press, Ithaca, NY. 249 p.

Caraco, N., J. Cole, S. Findlay and C. Wigand. 2006. Vascular plants as engineers of oxygen in aquatic systems. BioScience 56(3):219–225.

Carlson, D.M., R.A. Daniels and J.J. Wright. 2016. Atlas of inland fishes of New York. New York State Museum Record 7. 362 p.

Carmichael, D.P. 1980. A record of environmental change during recent millennia in the Hackensack tidal marsh, New Jersey. Bulletin of the Torrey Botanical Club 107(4):514–524.

Carola, H.M. 2006a. Hackensack watershed field notes. Hackensack Tidelines 9(2):15, 17.

Carola, H.M. 2006b. Hackensack watershed field notes. Hackensack Tidelines 9(4):15, 17.

Carola, H.M. 2006c. Hackensack watershed field notes. Hackensack Tidelines 9(1):15.

Carola, H.M. 2006d. Hackensack watershed field notes. Hackensack Tidelines 9(3):15, 17.

Carola, H.M. 2007. Hackensack watershed field notes. Hackensack Tidelines 10(2): 13, 17.

Carola, H.M. 2008a. Hackensack watershed field notes. Hackensack Tidelines 11(2):13, 16–17.

Carola, H.M. 2008b. Hackensack watershed field notes. Hackensack Tidelines 11(4):13, 16–17, 22.

Carola, H.M. 2008c. Hackensack watershed field notes. Hackensack Tidelines 11(3):13, 16–17.

Carola, H.M. 2008d. Hackensack watershed field notes. Hackensack Tidelines 11(1):13, 17.

Carola, H.M. 2011. Hackensack watershed field notes. Hackensack Tidelines 14(4):13.

Carola, H.M. 2015. Hackensack watershed field notes. Hackensack Tidelines 18(3):10–11. www.hackensackriverkeeper.org/wp-content/uploads/2015/10/news _Fall_2015.pdf (24 January 2020).

Carola, H.M. 2016a. Hackensack watershed field notes. Hackensack Tidelines 19(1):10–11. www.hackensackriverkeeper.org/wp-content/uploads/2016/08/2016 _Tidelines_Spring.pdf (24 January 2020).

Carola, H.M. 2016b. Hackensack watershed field notes. Hackensack Tidelines 19(2):15. www.hackensackriverkeeper.org/wp-content/uploads/2016/08/news _Summer_2016.pdf (24 January 2020).

Carola, H.M. 2016c. Hackensack watershed field notes. Hackensack Tidelines 14(3):10–11.

Carola, H.M. 2018a. Hackensack watershed field notes. Hackensack Tidelines 21(3):6–7. www.hackensackriverkeeper.org/wp-content/uploads/2018/11/news _Fall_18.pdf (24 January 2020).

Carola, H.M. 2018b. Hackensack watershed field notes. Hackensack Tidelines 21(1):12–13. www.hackensackriverkeeper.org/wp-content/uploads/2018/07/news _spring_18.pdf (24 January 2020).

Carola, H.M. 2019a. Hackensack watershed field notes. Hackensack Tidelines 22(3):16–18. www.hackensackriverkeeper.org/wp-content/uploads/2019/11/news _Fall_19.pdf (24 January 2020).

Carola, H.M. 2019b. Hackensack watershed field notes. Hackensack Tidelines 22(2):14–15. www.hackensackriverkeeper.org/wp-content/uploads/2019/07/news _Summer_19.pdf (24 January 2020).

Carola, H.M. 2019c. Hackensack watershed field notes. Hackensack Tidelines 22(1):11–12. www.hackensackriverkeeper.org/wp-content/uploads/2019/03/news _Spring_19.pdf (24 January 2020).

Cauley, D.L. and J.R. Schinner. 1973. The Cincinnati raccoons. Natural History 82:58–60.

Cech, R. and G. Tudor. 2005. Butterflies of the East Coast: An observer's guide. Princeton University Press, Princeton, NJ. 345 p.

Chapman, E.J., R.S. Prezant and R. Shell. 2012. Temporal variation in molluscan community structure in an urban New Jersey pond. Northeastern Naturalist 19(3):373–390.

Chapman, F.M. 1900. Bird studies with a camera. D. Appleton and Company, New York, NY.

Chen, S., Z. Yang, X. Liu, J. Sun, C. Xu, D. Xiong, W. Lin, Y. Li, J. Guo and Y. Yang. 2019. Moss regulates soil evaporation leading to decoupling of soil and near-surface air temperatures. Journal of Soils and Sediments 19(7):2903–2912.

Cheplick, G.P. and M. Aliotta. 2009. The abundance and size of annual herbs in a coastal beach community is related to their distance from seaside goldenrod (*Solidago sempervirens*). Journal of the Torrey Botanical Society 136(1):102–109.

Chidester, F. 1920. The behavior of *Fundulus heteroclitus* on the salt marshes of New Jersey. American Naturalist 54:551–557.

Chrysler, M.A. and J.L. Edwards. 1947. The ferns of New Jersey: Including the fern allies. Rutgers University Press, New Brunswick, NJ. 201 p.

Churcher, P.B. and J.H. Lawton. 1987. Predation by domestic cats in an English village. Journal of Zoology 212:439–455.

Ciotir, C. and J. Freeland. 2016. Cryptic intercontinental dispersal, commercial retailers, and the genetic diversity of native and non-native cattails (*Typha* spp.) in North America. Hydrobiologia 768(1):137–150.

Clark, J.D. and J.C. Lewis. 1983. A validity test of a habitat suitability index model for clapper rail. Proceedings, Annual Conference of the Southeast Association of Fish and Wildlife Agencies 37:95–102.

Clausnitzer, V., V.J. Kalkman, M. Ram, B. Collen, J.E.M. Baillie, M. Bedjanič, W.R.T. Darwell. K-D.B. Dijkstra, R. Dow, J. Hawking, H. Karube, E. Malikova, D. Paulson, K. Schütte, F. Suhling, R.J. Villanueva, N. von Ellenrieder and K. Wilson.

2009. Odonata enter the biodiversity crisis debate: The first global assessment of an insect group. Biological Conservation 142:1864–1869.

Cleavitt, N.L. 2005. Patterns, hypotheses and processes in the biology of rare bryophytes. Bryologist 108(4):554–566.

Clemants, S.E. and E.H. Ketchledge. 1990. Flora protection: The question of rare mosses in New York State. P. 211–216 in R.S. Mitchell, C.J. Sheviak and D.J. Leopold, eds. Ecosystem Management: Rare Species and Significant Habitats. New York State Museum Bulletin 471.

Clergeau, P., J.P.L. Savard, G. Mennechez and G. Falardeau. 1998. Bird abundance and diversity along an urban-rural gradient: A comparative study between two cities on different continents. Condor 100:413–425.

CNALH (Consortium of North American Lichen Herbaria). 2019. lichenportal.org/cnalh/index.php (1 December 2019).

Cobb, B., E. Farnsworth and C. Lowe. 2005. A field guide to ferns and their related families: Northeastern and central North America. 2nd ed. Houghton Mifflin Co., Boston, MA. 417 p.

Colbert, E.H. and P.E. Olsen. 2001. A new and unusual aquatic reptile from the Lockatong Formation of New Jersey (Late Triassic, Newark Supergroup). American Museum Novitates 3334(1):1–24.

Colburn, E.A. 2004. Vernal pools: Natural history and conservation. McDonald & Woodward Publishing Co., Newark, OH. 426 p.

Colding, J., J. Lundberg, S. Lundberg and E. Andersson. 2009. Golf courses and wetland fauna. Ecological Applications 19(6):1481–1491.

Collins, B.R. and K. Anderson. 1994. Plant communities of New Jersey: A study in landscape diversity. Rutgers University Press, New Brunswick, NJ. 308 p.

Conant, R. and J.T. Collins. 1991. A field guide to reptiles and amphibians of eastern and central North America. 2nd ed. Houghton Mifflin Co., Boston, MA.

Connor, B.F. 1978. The ecology of corticolous lichens in northern Dutchess County, New York. Senior Project, Bard College, Annandale, NY.

Connors, L.M., E. Kiviat, P.M. Groffman and R.S. Ostfeld. 2000. Muskrat (*Ondatra zibethicus*) disturbance to vegetation and potential net nitrogen mineralization and nitrification rates in a fresh-tidal marsh. American Midland Naturalist 143:53–63.

Conway, C.J. 1995. Virginia rail (*Rallus limicola*), Version 1.0, Birds of the World. Cornell Lab of Ornithology, Ithaca, NY. birdsoftheworld.org/bow/species/virrai/cur/introduction (1 September 2010).

Cook, R. 1996. Movement and ecology of eastern box and painted turtles repatriated to human-created habitat. PhD thesis, City University, New York, NY.

Cook, R.P. and C.A. Pinnock. 1987. Re-creating a herpetofaunal community at Gateway National Recreation Area. P. 151–154 in L.W. Adams and D.L. Leedy, eds. Integrating Man and Nature in the Urban Environment. National Institute for Urban Wildlife, Columbia, MD.

Cordeiro, J. and J. Bowers-Altman. No date. Freshwater mussels of the New York metropolitan region and New Jersey. American Museum of Natural History, Center for Biodiversity and Conservation. www.amnh.org/research/center-for-biodiversity

-conservation/resources-and-publications/biodiversity-guides-and-handbooks/ freshwater-mussels (29 January 2020).

Ćosić, M., M.M. Vujičić, M.S. Sabovljević and A.D. Sabovljević. 2019. What do we know about salt stress in bryophytes? Plant Biosystems 153(3):478–489.

Costello, D.F. 1936. Tussock meadows in southeastern Wisconsin. Botanical Gazette 97:610–648.

Cotinis, P.C. 2020. Species *Chrysopilus thoracicus*-golden-backed snipe fly. BugGuide. http://bugguide.net/node/view/483 (23 January 2020).

Coulter, M.W. 1957. Predation by snapping turtles upon aquatic birds in Maine marshes. Journal of Wildlife Management 21(1):17–21.

Cox, J., G. Colbert and G. Vollath. 2002. Nature's estuary: The historic tidelands of the New York New Jersey harbor estuary. Regional Plan Association. Map 46 x 61 cm. Scale approximately 1:140,000.

Crans, W.J. No date. *Culex salinarius* Coquillett. Rutgers University, New Brunswick, NJ. http://vectorbio.rutgers.edu/outreach/species/sp11a.htm (16 October 2021).

Crans, W.J. 2014. *Culex restuans* Theobald. http://vectorbio.rutgers.edu/outreach/ species/rest.htm (13 December 2019).

Crans, W.J. 2016. *Culex pipiens* Linnaeus. http://vectorbio.rutgers.edu/outreach/ species/pip2.htm (13 December 2019).

Crawford, D.W., N.L. Bonnevie, C.A. Gillis and R.J. Wenning. 1994. Historical changes in the ecological health of the Newark Bay estuary, New Jersey. Ecotoxicology and Environmental Safety 29:276–303.

Croci, S., A. Butet, A. Georges, R. Aguejdad and P. Clergeau. 2008. Small urban woodlands as biodiversity conservation hot-spot: A multi-taxon approach. Landscape Ecology 23:1171–1186.

Croteau, M.C., N. Hogan, J.C. Gibson, D. Lean and V.L. Trudeau. 2008. Toxicological threats to amphibians and reptiles in urban environments. P. 197–209 in J.C. Mitchell, R.E.J. Brown and B. Bartholomew, eds. Urban Herpetology. Society for the Study of Amphibians and Reptiles, Salt Lake City, UT.

Crowley, S.K. 1994. Habitat use and population monitoring of secretive waterbirds in Massachusetts. MS thesis, University of Massachusetts, Amherst, MA.

Cruickshank, A.D. 1942. Birds around New York City: Where and when to find them. American Museum of Natural History Handbook 13.

Crum, H. 1973. Mosses of the Great Lakes forest. Contributions from the University of Michigan Herbarium 10:1–404.

Daiber, F.C. 1982. Animals of the tidal marsh. Van Nostrand Reinhold Co., New York, NY. 422 p.

Daniels, R.A., K.E. Limburg, R.E. Schmidt, D.L. Strayer and R.C. Chambers. 2005. Changes in fish assemblages in the tidal Hudson River, New York. American Fisheries Society Symposium 45: 471–503.

Davis, D.D. and B. Allen. 2015. An updated list of bryophytes from Montour County, Pennsylvania. Evansia 32(4):189–194.

Davis, D.D. and J.J. Atwood. 2010. Mosses associated with coalmines and coal seams in western Pennsylvania. Evansia 27(1):11–17.

Davis, W.E., Jr., and J.A. Kushlan. 1994. Green heron (*Butorides virescens*), Version 1.0, Birds of the World. Cornell Lab of Ornithology, Ithaca, NY. birdsoftheworld. org/bow/species/grnher/cur/introduction (1 September 2010).

Day, C., J. Staples, R. Russell, G. Nieminen and A. Milliken. 1999. Hackensack Meadowlands National Wildlife Refuge: A presentation for a new establishment. U.S. Fish and Wildlife Service, New Jersey Field Office, Pleasantville, NJ.

DeCandido, R., A.A. Muir and M.B. Gargiullo. 2004. A first approximation of the historical and extant vascular flora of New York City: Implications for native plant species conservation. Journal of the Torrey Botanical Society 131(3):243–251.

De Frenne, P., L. Baeten, B.J. Graae, J. Brunet, M. Wulf, A. Orczewska, A. Kolb, I. Jansen, A. Jamoneau, H. Jacquemyn, M. Hermy, M. Diekmann, A. De Schrijver, M. De Sanctis, G. Decocq, S.A.O. Cousins and K. Verheyen. 2011. Interregional variation in the floristic recovery of post-agricultural forests. Journal of Ecology 99(2):600–609.

DeGraaf, R.M. and J.M. Wentworth. 1986. Avian guild structure and habitat associations in suburban bird communities. Urban Ecology 9:399–412.

Delendick, T.J. 1994. Notes on the lichens of eastern New York City: Kings and Queens counties, Long Island, New York. Bulletin of the Torrey Botanical Club 121(2):188–193.

Delorenzo, M.E. 2015. Impacts of climate change on the ecotoxicology of chemical contaminants in estuarine organisms. Current Zoology 61(4):641–652.

Devictor, V., R. Julliard, D. Couvet, A. Lee and F. Jiguet. 2007. Functional homogenization effect of urbanization on bird communities. Conservation Biology 21:741–751.

DiBona, M.T. 2007. Seasonal food availability for wintering and migrating dabbling ducks and its implications for management at the Hackensack Meadowlands of New Jersey. MS thesis, University of Delaware, Newark, DE.

Dignard, N. 1990. Bryophytes from the Old Québec City walls, Canada. Evansia 7(3):47–48.

Dillenburg, L.R., D.F. Whigham, A.H. Teramura and I.N. Forseth. 1993. Effects of vine competition on availability of light, water, and nitrogen to a tree host (*Liquidambar styraciflua*). American Journal of Botany 80(3):244–252.

Dirrigl, F.J., Jr. and A.E. Bogan. 1996. Revised checklist of the terrestrial gastropods of New Jersey (Mollusca: Gastropoda). Walkerana 8(20):127–138.

Dolan, P.G. and D.C. Carter. 1977. *Glaucomys volans*. Mammalian Species 78:1–6.

Dombroski, D.R., Jr. 2020. Earthquake risk in New Jersey. New Jersey Geological and Water Survey, Trenton, NJ. www.state.nj.us/dep/njgs/enviroed/eqrisk.htm (22 January 2020).

Doncaster, C.P., C.R. Dickman and D.W. Macdonald. 1990. Feeding ecology of red foxes (*Vulpes vulpes*) in the city of Oxford, England. Journal of Mammalogy 71(2):188–194.

Donovan, E., K. Unice, J.D. Roberts, M. Harris and B. Finley. 2008. Risk of gastrointestinal disease associated with exposure to pathogens in the water of the lower Passaic River. Applied and Environmental Microbiology 74:994–1003.

Dowling, Z., T. Hartwig, E. Kiviat and F. Keesing. 2010. Experimental management of nesting habitat for the Blanding's turtle (*Emydoidea blandingii*). Ecological Restoration 28(2):154–159.

Drake, A.A., Jr., R.A. Volkert, D.H. Monteverde, G.C. Herman, H.F. Houghton, R.A. Parker and R.F. Dalton. 2002. Bedrock geologic map of northern New Jersey. U.S. Geological Survey, Denver, CO. Large format map, scale 1:100,000.

Dressler, R.L. 1981. The orchids. Harvard University Press, Cambridge, MA. 332 p.

Drilling, N., R.D. Titman and F. McKinney. 2020. Mallard (*Anas platyrhynchos*), Version 1.0, Birds of the World. Cornell Lab of Ornithology, Ithaca, NY. birdsoftheworld.org/bow/species/mallar3/cur/introduction (22 January 2020).

Drury, W.H. and I.C. Nisbet. 1973. Succession. Journal of the Arnold Arboretum 54(3):331–368.

Duchak, T. and C. Holzapfel. 2011. The relative abundances of native and non-native emydid turtles across an urban to rural habitat gradient in central New Jersey. Bulletin of the New Jersey Academy of Science 56(2):23–28.

Duchamp, J.E., D.W. Sparks and J.O. Whitaker, Jr. 2004. Foraging-habitat selection by bats at an urban-rural interface: Comparison between a successful and a less successful species. Canadian Journal of Zoology 82:1157–1164.

du Châtelet, E.A., F. Frontalini and F. Francescangeli. 2018. Significance of replicates: Environmental and paleoenvironmental studies on benthic foraminifera and testate amoebae. Micropaleontology 63(5):257–274.

Duebbert, H.F., J.T. Lokemoen and D.E. Sharp. 1986. Nest sites of ducks in grazed mixed-grass prairie in North Dakota. Prairie Naturalist 18:99–108.

Dunne, P., R. Kane and P. Kerlinger. 1989. New Jersey at the crossroads of migration. New Jersey Audubon Society, Franklin Lakes, NJ. 74 p.

Dures, S.G. and G.S. Cumming. 2010. The confounding influence of homogenising invasive species in a globally endangered and largely urban biome: Does habitat quality dominate avian biodiversity? Biological Conservation 143:768–777.

Ebinger, J.E. 1997. Laurels in the wild. P. 29–51 in R.A. Jaynes. Kalmia: Mountain laurel and related species. Timber Press, Portland, OR.

eBird. 2021. eBird: An online database of bird distribution and abundance. Cornell Lab of Ornithology, Ithaca, NY. http://www.ebird.org (7 July 2021).

Eckel, P.M. and J. Shaw. 1991. *Bryum rubens* from Niagara Falls, new to New York State. Bryologist 94(1):80–81.

EFSA (European Food Safety Authority). 2007. Opinions of the EFSA Scientific Panel on Plant Health on request from the Commission on pest risk analysis made by EPPO on *Hydrocotyle ranunculoides* L. f. (floating pennywort). EFSA Journal 468:1–13. http://www.efsa.europa.eu/en/efsajournal/pub/468 (20 January 2020).

Egan, D. and E.A. Howell. 2005. The historical ecology handbook. Revised ed. Island Press, Washington, DC. 488 p.

Ehrenfeld, J.G., H.B. Cutway, R. Hamilton and E. Stander. 2003. Hydrologic description of forested wetlands in northeastern New Jersey, USA—an urban/suburban region. Wetlands 23(4):685–700.

Ehrenfeld, J.G., M. Palta and E. Stander. 2011. Wetlands in urban environments. P. 338–351 in I. Douglas, D. Goode, M.C. Houck and R. Wang, eds. The Routledge Handbook of Urban Ecology. Routledge, New York, NY.

Elliott, J.E., R.J. Norstrom and G.E.J. Smith. 1996. Patterns, trends, and toxicological significance of chlorinated hydrocarbon and mercury contaminants in bald eagle eggs from the Pacific Coast of Canada, 1990–1994. Archives of Environmental Contamination and Toxicology 31:354–367.

Elliott, L. and W. Hershberger. 2007. The songs of insects. Houghton Mifflin Co., Boston, MA. 228 p. + compact disk.

Ellis, C.J. and B.J. Coppins. 2006. Contrasting functional traits maintain lichen epiphyte diversity in response to climate and autogenic succession. Journal of Biogeography 33(9):1643–1656.

Ellison, A.M., N.J. Gotelli, E.J. Farnsworth and G.D. Alpert. 2012. A field guide to the ants of New England. Yale University Press, New Haven, CT. 416 p.

Emerson, W.K. and M.K. Jacobson. 1976. Guide to shells: Land, freshwater, and marine, from Nova Scotia to Florida. Alfred A. Knopf, New York, NY. 501 p.

Emery, N.J. 2006. Cognitive ornithology: The evolution of avian intelligence. Philosophical Transactions of the Royal Society of London B Biological Science 361:23–43.

Erwin, R.M. 1983. Feeding habitats of nesting wading birds: Spatial use and social influences. Auk 100:960–970.

Esslinger, T.L. 2019. A cumulative checklist for the lichen-forming, lichenicolous and allied fungi of the continental United States and Canada. Version 23. Opuscula Philolichenum 18:102–378. www.ndsu.edu/pubweb/~esslinge/chcklst/chcklst7.htm#S (26 October 2019).

Evans, A.W. and G.E. Nichols. 1908. The bryophytes of Connecticut. Connecticut Geological and Natural History Survey Bulletin 11. 203 p.

Everett, J.J. and R.G. Anthony. 1976. Heavy metal accumulation in muskrats in relation to water quality. Transactions of the Northeast Section of the Wildlife Society 33:105–118.

Ezyske, C.M. and Y. Deng. 2012. Landfill management and remediation practices in New Jersey, United States. P. 149–166 in S. Kumar, ed. Management of Organic Waste. InTechOpen. www.intechopen.com/books/management-of-organic-waste/landfill-management-and-remediation-practices-in-new-jersey-united-states (22 January 2020).

Facciolla, N.W. 1981. Minerals of Laurel Hill Secaucus, New Jersey. Published by the author, Teaneck, NJ. 46 p.

Fallon, S. 2019a. In the weeds: Overpeck Creek overrun by aquatic plant disrupting kayakers and fish alike. North Jersey Record (26 September).

Fallon, S. 2019b. Will threatened birds alter plans for industrial development on Meadowlands landfills? North Jersey Record (2 July).

Feinberg, J.A. 2004. Geographic Distribution: *Sceloporus undulatus* (eastern fence lizard). Herpetological Review 35(2):188.

Feinberg, J.A. 2015. An unexpected journey: Anuran decline research and the incidental elucidation of a new cryptic species endemic to the urban Northeast and Mid-Atlantic US. Ph.D. thesis, Rutgers University, New Brunswick, NJ.

Feinberg, J.A., C.E. Newman, G.J. Watkins-Colwell, M.D. Schlesinger, B. Zarate, B.R. Curry, H.B Shaffer and J. Burger. 2014. Cryptic diversity in metropolis: Confirmation of a new leopard frog species (Anura: Ranidae) from New York City and surrounding Atlantic Coast regions. PloS one 9(10):e108213.

Feltes, R. 2005. Meadowlands fish species collected by the Hartman Lab at Harrier Meadow and Mill Creek 1999–2004. Appendix L in J.M. Hartman. Harrier Meadows Wetlands Mitigation Site, Fifth Annual Monitoring Report. Rutgers University, New Brunswick, NJ.

Fenton, M.B. and R.M.R. Barclay. 1980. *Myotis lucifugus*. Mammalian Review 142:1–8.

Fernández-Juricic, E. 2000a. Local and regional effects of pedestrians on forest birds in a fragmented landscape. Condor 102:247–255.

Fernández-Juricic, E. 2000b. Avifaunal use of wooded streets in an urban landscape. Conservation Biology 14:513–521.

Fetzner, J.W., Jr. 2019. State of New Jersey—Crayfish species checklist. www.invertebratezoology.org/country_pages/state_pages/newjersey.htm (23 January 2020).

Feuerer, T., H. Hertel and H. Deuter. 2003. The corticolous and lignicolous lichens of Munich (Germany)—a preliminary evaluation. Bibliotheca Lichenologica 86:329–339.

Fidino, M.A., E.W. Lehrer and S.B. Magle. 2016. Habitat dynamics of the Virginia opossum in a highly urban landscape. American Midland Naturalist 175(2):155–167.

Fisk, E.J. 1978. The growing use of roofs by nesting birds. Bird-Banding 49(2):134–141.

Flinn, K.M. and M. Vellend. 2005. Recovery of forest plant communities in post-agricultural landscapes. Frontiers in Ecology and the Environment 3(5):243–250.

Folkeson, L. 1984. Deterioration of the moss and lichen vegetation in a forest polluted by heavy metals. Ambio 13:37–39.

Foote, M. 1983. The spatial and temporal distribution of suspended algae and nutrients in the upper Hackensack River estuary. PhD thesis, Rutgers University, New Brunswick, NJ. 238 p.

Foote, M. and R.E. Loveland. 1982. The development of phytoplankton populations and nutrients in a tidal river under drought conditions. Unpublished report, Rutgers University, New Brunswick, NJ.

Forbes, A.A., R.K. Bagley, M.A. Beer, A.C. Hippee and H.A. Widmayer. 2018. Quantifying the unquantifiable: Why Hymenoptera, not Coleoptera, is the most speciose animal order. BMC Ecology 18(1):article 21. doi.org/10.1186/s12898-018-0176-x.

Forman, R.T.T. 2008. Urban regions: Ecology and planning beyond the city. Cambridge University Press, Cambridge, U.K. 478 p.

Forman, R.T., D. Sperling, J.A. Bissonette, A.P. Clevenger, C.D. Cutshall, V.H. Dale, L. Fahrig, R.L. France, C.R. Goldman, K. Heanue, J. Jones, F.J. Swanson, T. Turrentine and T.C. Winter. 2003. Road ecology: Science and solutions. Island Press, Washington, DC. 504 p.

Fort, D.J., M.B. Mathis, R. Walker, L.K. Tuominen, M. Hansel, S. Hall, R. Richards, S.R. Grattan and K. Anderson. 2014. Toxicity of sulfate and chloride to early life stages of wild rice (*Zizania palustris*). Environmental Toxicology and Chemistry 33(12):2802–2809.

Fowells, H.A. 1965. Silvics of forest trees of the United States. U.S. Department of Agriculture, Agriculture Handbook 271. 762 p.

Frahm, J.-P. 2008. Diversity, dispersal and biogeography of bryophytes (mosses). Biodiversity and Conservation 17:277–284.

Francois, D.D. 1959. The crayfishes of New Jersey. Ohio Journal of Science 59:108–127.

Frankel, E. 1981. Ferns: A natural history. Stephen Greene Press, Brattleboro, VT. 264 p.

Frati, L., S. Santoni, V. Nicolardi, C. Gaggi, G. Brunialti, A. Guttova, S. Gaudino, A. Pati, S.A. Pirintsos and S. Loppi. 2007. Lichen biomonitoring of ammonia emission and nitrogen deposition around a pig stockfarm. Environmental Pollution 146(2):311–316.

French, S.S., A.C. Webb, S.B. Hudson and E.E. Virgin. 2018. Town and country reptiles: A review of reptilian responses to urbanization. Integrative and Comparative Biology 58(5):948–966.

Freudenberg, R. 2019. Our vision for the Meadowlands must address the urgency of climate change. www.northjersey.com/story/opinion/2019/11/08/nj-meadowlands -will-need-honest-planning-to-address-climate-change-opinion/4113478002/ (15 January 2020).

Frick, M.G. 1998. *Kinosternon baurii* (striped mud turtle). Herpetological Review 29(1):50.

Friedman, C.L., M.G. Cantwell and R. Lohmann. 2012. Passive sampling provides evidence for Newark Bay as a source of polychlorinated dibenzo-*p*-dioxins and furans to the New York/New Jersey, USA, atmosphere. Environmental Toxicology and Chemistry 31:253–261.

Fritz, Ö. 2009. Vertical distribution of epiphytic bryophytes and lichens emphasizes the importance of old beeches in conservation. Biodiversity and Conservation 18:289–304.

Fritz, Ö., J. Brunet and M. Caldiz. 2009. Interacting effects of tree characteristics on the occurrence of rare epiphytes in a Swedish beech forest area. Bryologist 112(3):488–505.

Fudali, E. 2005. Bryophyte species diversity and ecology in the parks and cemeteries of selected Polish cities. Agricultural University of Wrocław, Wrocław, Poland. 212 p.

Fuller, S.L.H. 1974. Clams and mussels (Mollusca: Bivalvia). P. 215–273 in C.W. Hart, Jr. and S.L.H. Fuller, eds. Pollution Ecology of Freshwater Invertebrates. Academic Press, New York, NY.

Gaddy, L.L. 1986. Twelve new ant-dispersed species from the southern Appalachians. Bulletin of the Torrey Botanical Club 113(3):247–251.

Galatowitsch, S.M., N.O. Anderson and P.D. Ascher. 1999. Invasiveness in wetland plants in temperate North America. Wetlands 19:733–755.

Galishoff, S. 1988. Newark: The nation's unhealthiest city 1832–1895. Rutgers University Press, New Brunswick, NJ. 260 p.

Gallagher, F.J., I. Pechmann, J.D. Bogden, J. Grabosky and P. Weis. 2008. Soil metal concentrations and productivity of *Betula populifolia* (gray birch) as measured by field spectrometry and incremental annual growth in an abandoned urban brownfield in New Jersey. Environmental Pollution 156:699–706.

Gallagher, F.J., I. Pechmann, B. Isaacson and J. Grabosky. 2010. Morphological variation in the seed of gray birch (*Betula populifolia*): The effects of soil-metal contamination. Urban Habitats 6. http://www.urbanhabitats.org/v06n01/graybirch _full.html (26 June 2010).

Gallo, T., M. Fidino, E.W. Lehrer and S.B. Magle. 2017. Mammal diversity and metacommunity dynamics in urban green spaces: Implications for urban wildlife conservation. Ecological Applications 27(8):2330–2341.

Gao, Y. 2009. Characterizations of atmospheric nitrogen oxides over the Meadowlands. Final report to the Meadowlands Environmental Research Institute and the New Jersey Meadowlands Commission. Rutgers University, Newark, NJ. 38 p.

Garbary, D.J., A.G. Miller, R. Scrosati, K.-Y. Kim and W.B. Schofield. 2007. Distribution and salinity tolerance of intertidal mosses from Nova Scotian salt marshes. Bryologist 111(2):282–291.

Gargiullo, M.B. 2007. A guide to native plants of the New York City region. Rutgers University Press, New Brunswick, NJ. 306 p.

Gatica-Colima, A., A. Robles-Hernández, L.A. Rivera-Hernández and A. Torres-Dura. 2016. *Phrynosoma cornutum* (Texas horned lizard). Mortality. Herpetological Review 47(2):301.

Gaul, R.W., Jr. 2008. Ecological observations of the northern brownsnake (*Storeria dekayi*) in an urban environment in North Carolina, USA. P. 361–363 in J.C. Mitchell, R.E.J. Brown and B. Bartholomew, eds. Urban Herpetology. Society for the Study of Amphibians and Reptiles, Salt Lake City, UT.

Gedzelman, S.D., S. Austin, R. Cermak, N. Stefano, S. Partridge, S. Quesenberry and D.A. Robinson. 2003. Mesoscale aspects of the urban heat island around New York City. Theoretical and Applied Climatology 75:29–42.

Gehrt, S.D. 2004. Ecology and management of striped skunks, raccoons, and coyotes in urban landscapes. P. 81–104 in N. Fascione, A. Delach and M. Smith, eds. Predators and People: From Conflict to Conservation. Island Press, Washington, DC.

Gehrt, S.D., C. Anchor and L.A. White. 2009. Home range and landscape use of coyotes in a major metropolitan landscape: Coexistence or conflict? Journal of Mammalogy 90:1045–1057.

Gehrt, S.D. and J.E. Chelsvig. 2004. Species-specific patterns of bat activity in an urban landscape. Ecological Applications 14:625–635.

Gehrt, S.D. and S.P.D. Riley. 2010. Coyotes (*Canis latrans*). P. 78–95 in S.D. Gehrt, S.P.D. Stanley and B.L. Cypher, eds. Urban Carnivores: Ecology, Conflict, and Conservation. The Johns Hopkins University Press, Baltimore, MD.

Geluso, K.N., J.S. Altenbach and D.E. Wilson. 1976. Bat mortality: Pesticide poisoning and migratory stress. Science 194:184–186.

George, S.B., J.R. Choate and H.H. Genoways. 1986. *Blarina brevicauda*. Mammalian Species 261:1–9.

George, S.L. and K.R. Crooks. 2006. Recreation and large mammal activity in an urban nature reserve. Biological Conservation 133:107–117.

Germine, M. and J.H. Puffer. 1981. Distribution of asbestos in the bedrock of the northern New Jersey area. Environmental Geology 3(6):337–351.

Getz, L.L., F.R. Cole and D.L. Gates. 1978. Interstate roadsides as dispersal routes for *Microtus pennsylvanicus*. Journal of Mammalogy 59:208–212.

Ghekiere, A., T. Verslycke, N. Fockedey and C.R. Janssen. 2006. Non-target effects of the insecticide methoprene on molting in the estuarine crustacean *Neomysis integer* (Crustacea: Mysidacea). Journal of Experimental Marine Biology and Ecology 332(2):226–234.

Gibbons, J.W., V.J. Burke, J.E. Lovich, R.D. Semlitsch, T.D. Tuberville, J.R. Bodie, J.L. Greene, P.H. Niewiarowski, H.H. Whiteman, D.E. Scott, J.H.K. Pechmann, C.R. Harrison, S.H. Bennett, J.D. Krenz, M.S. Mills, K.A. Buhlmann, J.R. Lee, R.A. Seigel, A.D. Tucker, T.M. Mills, T. Lamb, M.E. Dorcas, J.D. Congdon, M.H. Smith, D.H. Nelson, M.B. Dietsch, H.G. Hanlin, J.A. Ott and D.J. Karapatakis. 1997. Perceptions of species abundance, distribution, and diversity: Lessons from four decades of sampling on a government-managed reserve. Environmental Management 21(2):259–268.

Gibbons, J.W., D.E. Scott, T.J. Ryan, K.A. Buhlmann, T.D. Tuberville, B.S. Metts, J.L. Greene, T. Mills, Y. Leiden, S. Poppy and C.T. Winne. 2000. The global decline of reptiles, déjà vu amphibians. BioScience 50(8):653–666.

Gibbs, J.P. 1998. Distribution of woodland amphibians along a forest fragmentation gradient. Landscape Ecology 13(4):263–268.

Gibbs, J.P., A.R. Breisch, P.K. Ducey, G. Johnson, J.L. Behler and R.C. Bothner. 2007. The amphibians and reptiles of New York State. Oxford University Press, New York, NY. 422 p.

Gibbs, J.P., J.R. Longcore, D.G. McAuley and J.K. Ringelman. 1991. Use of wetland habitats by selected nongame waterbirds in Maine. U.S. Fish and Wildlife Service, Fish and Wildlife Research 9. 57 p.

Gilbert, O.L. 1989. The ecology of urban habitats. Chapman & Hall, London, U.K. 369 p.

Giordano, S., S. Sorbo, P. Adamo, A. Basile, V. Spagnuolo and R.C. Cobianchi. 2004. Biodiversity and trace element content of epiphytic bryophytes in urban and extra-urban sites of southern Italy. Plant Ecology 170:1–14.

Giudice, J.H. and J.T. Ratti. 2020. Ring-necked pheasant (*Phasianus colchicus*), Version 1.0, Birds of the World. Cornell Lab of Ornithology, Ithaca, NY. birdsoftheworld.org/bow/species/rinphe/cur/introduction (1 September 2010).

Gleason, J.A. and J.A. Fagliano. 2015. Associations of daily pediatric asthma emergency department visits with air pollution in Newark, NJ: Utilizing time-series and case-crossover study designs. Journal of Asthma 52:815–822.

Gochfeld, M. and J. Burger. 1997. Butterflies of New Jersey: A guide to their status, distribution, conservation, and appreciation. Rutgers University Press, New Brunswick, NJ. 327 p.

Godeau, C. 2019. Land reclamation by reindeer lichens: On the complexity of substrate and reindeer grazing on *Cladonia* spp. dispersal. Master's thesis, Umeå University, Sweden. 28 p.

Golden, H.E. and E.W. Boyer. 2009. Contemporary estimates of atmospheric nitrogen deposition to the watersheds of New York State, USA. Environmental Monitoring and Assessment 155:319–339.

Goldstein, E.L., M. Gross and R.M. DeGraaf. 1986. Breeding birds and vegetation: A quantitative assessment. Urban Ecology 9(3–4):377–385.

Gołdyn, B., S. Konwerski and J. Błoszyk. 2007. Large branchiopods (Anostraca, Notostraca, Spinicaudata, Laevicaudata) of small astatic waterbodies in the environs of Poznań (Wielkopolska Region, Western Poland). Oceanological and Hydrobiological Studies 36 (Suppl. 4):21–28.

Goodman, S. 1995. Soil survey of Bergen County, New Jersey. U.S. Department of Agriculture, Soil Conservation Service. 142 p. + folded maps.

Google Earth. 2020. http://maps.google.com/ (22 January 2020).

Gornitz, V., S. Couch and E.K. Hartig. 2001. Impacts of sea level rise in the New York City metropolitan area. Global and Planetary Change 32(1):61–88.

Gosner, K.L. 1978. A field guide to the Atlantic seashore from the Bay of Fundy to Cape Hatteras. Houghton Mifflin Co., Boston, MA. 329 p.

Gosselink, T.E., T.R. Van Deelen, R.E. Warner and P.C. Mankin. 2007. Survival and cause-specific mortality of red foxes in agricultural and urban areas of Illinois. Journal of Wildlife Management 71:1862–1873.

Goto, D. and W.G. Wallace. 2009. Biodiversity loss in benthic macroinfaunal communities and its consequence for organic mercury trophic availability to benthivorous predators in the lower Hudson River estuary, USA. Marine Pollution Bulletin 58:1909–1915.

Goward, T. 2008a. "Nameless little things." Evansia 25(3):54–56.

Goward, T. 2008b. Credo. Evansia 25(4):78–81.

Goward, T. 2009. Re-emergence. Evansia 26(1):1–5.

Green, D.M., ed. 1997. Amphibians in decline: Canadian studies of a global problem. Herpetological Conservation 1. Society for the Study of Amphibians and Reptiles, St. Louis, MO. 351 p.

Greenberg, R. and S. Droege. 1990. Adaptations to tidal marshes in breeding populations of swamp sparrow. Condor 92:393–404.

Greenlaw, J.S. and R.F. Miller. 1982. Breeding soras on a Long Island salt marsh. Kingbird 32:78–84.

Greenlaw, J.S., C.S. Elphick, W. Post and J.D. Rising. 2018. Saltmarsh sparrow (*Ammospiza caudacuta*), Version 1.0, Birds of the World. Cornell Lab of

Ornithology, Ithaca, NY. birdsoftheworld.org/bow/species/sstspa/cur/introduction (22 January 2020).

Greenwood, R.J. 1979. Relating residue in raccoon feces to food consumed. American Midland Naturalist 102:191–193.

Greven, H.C. 1992. Changes in the moss flora of the Netherlands. Biological Conservation 59:133–137.

Griffin, M.P.A. and I. Valiela. 2001. δ^{15}N isotope studies of the life history and trophic position of *Fundulus heteroclitus* and *Menidia menidia*. Marine Ecology Progress Series 214:299–305.

Grillas, P., P. Garcia-Murillo, O. Geertz-Hansen, N. Marbá, C. Montes, C.M. Duarte, L. Tan Ham and A. Grossmann. 1993. Submerged macrophyte seed bank in a Mediterranean temporary marsh: Abundance and relationship with established vegetation. Oecologia 94(1):1–6.

Grizzell, R.A., Jr. 1955. A study of the southern woodchuck *Marmota monax monax*. American Midland Naturalist 53:257–293.

Gronberg, J.M., A.S. Ludtke and D.L. Knifong. 2014. Estimates of inorganic nitrogen wet deposition from precipitation for the conterminous United States, 1955–84. United States Geological Survey Scientific Investigations Report 2014–5067. 18 p.

Grosse, A.M., J.D. van Dijk, K.L. Holcomb and J.C. Maerz. 2009. Diamondback terrapin mortality in crab pots in a Georgia tidal marsh. Chelonian Conservation and Biology 8(1):98–100.

Grossmueller, D. 2001. Empire Tract invertebrate survey. Draft report to Empire Ltd., Woodridge, NJ. Paulus, Sokolowski and Sartor, Inc., Warren, NJ.

Grout, A.J. 1916. The moss flora of New York City and vicinity. Published by the author, New Dorp, NY. 119 p. + plates.

Grout, A.J. 1924. Mosses with a hand-lens. 3rd ed. Published by the author, New York, NY. 339 p.

Gunnarsson, U. and H. Rydin. 2000. Nitrogen fertilization reduces *Sphagnum* production in bog communities. New Phytologist 147:527–537.

Guthrie, D.A. 1974. Suburban bird populations in southern California. American Midland Naturalist 92:461–466.

Gutiérrez-Larruga, B., B. Estébanez-Pérez and R. Ochoa-Hueso. 2020. Effects of nitrogen deposition on the abundance and metabolism of lichens: A meta-analysis. Ecosystems 23:783–797.

Haddad, N. 2019. The last butterflies. Princeton University Press, Princeton, NJ. 264 p.

Hadidan, J., S. Prange, R. Rosatte, S.P.D. Riley and S.D. Gehrt. 2010. Raccoons (*Procyon lotor*). P. 34–47 in S.D. Gehrt, S.P.D. Stanley and B.L. Cypher, eds. Urban Carnivores: Ecology, Conflict, and Conservation. Johns Hopkins University Press, Baltimore, MD.

Haines, A. 2003. The families Huperziaceae and Lycopodiaceae of New England: A taxonomic and ecological reference. V.T. Thomas Co., Bowdoin, ME. 100 p.

Hales, S., J. Staples and C. Day. 2007. The Hackensack Meadowlands Initiative: Preliminary conservation planning for the Hackensack Meadowlands Hudson and

Bergen counties, New Jersey. U.S. Fish and Wildlife Service, Pleasantville, NJ. 422 p. + appendix.

Hallingbäck, T. and N. Hodgetts, compilers. 2000. Mosses, liverworts and hornworts: Status survey and conservation action plan for bryophytes. IUCN/SSC Bryophyte Specialist Group. IUCN, Gland, Switzerland and Cambridge, U.K. 116 p. portals.i ucn.org/library/efiles/documents/2000-074.pdf (24 January 2020).

Hallman, C.A., M. Sorg, E. Jongejans, J. Siepel, N. Hofland, H. Schwan, W. Stenmans, A. Müller, H. Sumser, T. Hörren, D. Goulson and H. de Kroon. 2017. More than 75 percent decline over 27 years in total flying insect biomass in protected areas. PLoS ONE 12(10):e0185809.

Hallmon, C.F., E.T. Schreiber, T. Vo and A. Bloomquist. 2000. Field trials of three concentrations of Laginex as biological larvicide compared to Vectobac-12AS as a biocontrol agent for *Culex quinquefasciatus*. Journal of the American Mosquito Control Association 16(1):5–8.

Hamilton, W.J., Jr. 1934. Life history of the rufescent woodchuck. Annals of the Carnegie Museum 23: 85–178.

Hamilton, W.J. 1951. The food and feeding behavior of the garter snake in New York State. American Midland Naturalist 46(2):385–390.

Handel, S.N. 2013. Ecological restoration foundations to designing habitats in urban areas. P. 169–186 in J. Beardsley, ed. Designing Wildlife Habitats. Harvard University Press, Cambridge, MA.

Hansens, E. and S. Race. No date. The greenhead and you. Rutgers University, Department of Entomology, New Brunswick, NJ. esc.rutgers.edu/fact_sheet/the-greenhead-and-you/ (29 January 2020).

Harmens, H., D.A. Norris, K. Sharps, G. Mills, R. Alber, Y. Aleksiayenak, O. Blum, S.M. Cucu-Man, M. Dam, L. De Temmerman and A. Ene. 2015. Heavy metal and nitrogen concentrations in mosses are declining across Europe whilst some "hotspots" remain in 2010. Environmental Pollution 200:93–104.

Harmon, K.P. and J.C.F. Tedrow. 1969. A phytopedologic study of the Hackensack Meadowlands. New Jersey Agricultural Experiment Station, Rutgers University, New Brunswick, NJ. 105 p.

Harrison, C. and G. Davies. 2002. Conserving biodiversity that matters: Practitioners' perspectives on brownfield development and urban nature conservation in London. Journal of Environmental Management 65:95–108.

Harrison, D.J., J.A. Bissonette and J.A. Sherburne. 1989. Spatial relationships between coyotes and red foxes in eastern Maine. Journal of Wildlife Management 53:181–185.

Harshberger, J.W. and V.G. Burns. 1919. The vegetation of the Hackensack Marsh: A typical American fen. Wagner Institute of Science 9:1–35.

Hartig, E.K. and G.F. Rogers. 1984. *Phragmites* fire ecology. P. 1–3 in Proceedings of the Annual Meeting of the Association of American Geographers, Washington, DC.

Hauser, C. and S. Haulton. 2008. Draft Environmental Assessment: Increased emphasis on management and sustainability of oak-hickory communities on the Indiana State Forest system. Indiana Department of Natural Resources Division of Forestry, Indianapolis, IN. 128 p. + appendix.

Haya, K. 1989. Toxicity of pyrethroid insecticides to fish. Environmental Toxicology and Chemistry 8(5):381–391.

Headlee, T.J. 1945. The mosquitoes of New Jersey and their control. Rutgers University Press, New Brunswick, NJ. 326 p.

Heady, H.F. 1942. Annotated list of the mosses of the Huntington Forest. Roosevelt Wildlife Bulletin 8(2):41–58.

Heard, R.W. 1983. Observations on the food and food habits of clapper rails (*Rallus longirostris* Boddaert) from tidal marshes along the East and Gulf Coast of the United States. Gulf Research Reports 7:125–135.

Heaton, S.N., S.J. Bursian, J.P. Giesy, D.E. Tillitt, J.A. Render, P.D. Jones, D.A. Verbrugge, T.J. Kubiak and R.J. Aulerich. 1995. Dietary exposure of mink to carp from Saginaw Bay, Michigan. I. Effects on reproduction and survival and the potential risks to wild mink populations. Archives of Environmental Contamination and Toxicology 28:334–343.

Hedeen, S.E. and D.L. Hedeen. 1999. Railway-aided dispersal of an introduced *Podarcis muralis* population. Herpetological Review 30(1):57–58.

Heilmann-Clausen, J., E.S. Barron, L. Boddy, A. Dahlberg, G.W. Griffith, J. Nordén, O. Ovaskainen, C. Perini, B. Senn-Irlet and P. Halme. 2015. A fungal perspective on conservation biology. Conservation Biology 29(1):61–68.

Hendricks, A.J. and H.J. Behm. 1976. Cretaceous ponds: A unique area on Staten Island, N.Y. Staten Island Institute of Arts and Sciences, St. George, Staten Island, NY. 56 p.

Hepp, G.R. and F.C. Bellrose. 2013. Wood duck (*Aix sponsa*), Version 1.0, Birds of the World. Cornell Lab of Ornithology, Ithaca, NY. birdsoftheworld.org/bow/species/wooduc/cur/introduction (24 January 2020).

Hershey, A.E., A.R. Lima, G.J. Niemi and R.R. Regal. 1998. Effects of *Bacillus thuringiensis israelensis* (BTI) and methoprene on nontarget macroinvertebrates in Minnesota wetlands. Ecological Applications 8(1):41–60.

Heusser, C.J. 1949. History of an estuarine bog at Secaucus, New Jersey. Bulletin of the Torrey Botanical Club 76:385–406.

Heusser, C.J. 1963. Pollen diagrams from three former cedar bogs in the Hackensack tidal marsh, northeastern New Jersey. Bulletin of the Torrey Botanical Club 90(1):6–28.

Hill, M.O., C.D. Preston, S.D.S Bosanquet and D.B. Roy. 2007. BRYOATT: Attributes of British and Irish mosses, liverworts and hornworts with information on native status, size, life form, life history, geography and habitat. NERC Centre for Ecology and Hydrology and Countryside Council for Wales. 88 p. nora.nerc.ac.uk/id/eprint/1131/1/BRYOATT.pdf (18 December 2021).

Hinds, J.W. and P.L. Hinds. 2007. The macrolichens of New England. Memoirs of the New York Botanical Garden No. 96. New York Botanical Garden Press, Bronx, NY. 586 p.

HMDC (Hackensack Meadowlands Development Commission). 1999. 1999 annual report. HMDC, Lyndhurst, NJ. 16 p.

Hobbs, H.H., Jr. and E.T. Hall, Jr. 1974. Crayfishes (Decapoda: Astacidae). P. 195–214 in C.W. Hart, Jr. and S.L.H. Fuller, eds. Pollution Ecology of Freshwater Invertebrates. Academic Press, New York, NY.

Hoebeke, E.R. and A.G. Wheeler, Jr. 1999. *Anthidium oblongatum* (Illiger): An old world bee (Hymenoptera: Megachilidae) new to North America, and new North American records for another adventive species, *A. manicatum* (L.). University of Kansas Natural History Museum Special Publication 24:21–24.

Hoffman, J.R. 2019. *Cladonia leporina* (Cladoniaceae), a new macrolichen for New York State and northern range extension found in Brooklyn, New York City. Journal of the Torrey Botanical Society 146(2):138–141.

Hong, W.S. 2007. The hepatic flora and floristic affinity of hepatics around Takakia Lake, Queen Charlotte Islands, British Columbia. Canadian Field-Naturalist 121(1):24–28.

Hotopp, K. and T.A. Pearce. 2007. Land snails in New York: Statewide distributions and talus site faunas. Final report, Contract NYHER 041129. Report to the New York Natural Heritage Program, Albany, NY. 91 p.

Hotopp, K.P., T.A. Pearce, J.C. Nekola, J. Slapcinsky, D.C. Dourson, M. Winslow, G. Kimber and B. Watson. 2013. Land snails and slugs of the Mid-Atlantic and Northeastern United States. Carnegie Museum of Natural History, Pittsburgh, PA. http://www.carnegiemnh.org/science/mollusks/index.html (2 December 2019).

Hough, M.Y. 1983. New Jersey wild plants. Harmony Press, Harmony, NJ. 414 p.

Hughes, E.H. and E.B. Sherr. 1983. Subtidal food webs in a Georgia estuary: $\delta^{13}C$ analysis. Journal of Experimental Marine Biology and Ecology 67:227–242.

Hultengren, S., H. Gralén and H. Pleijel. 2004. Recovery of the epiphytic lichen flora following air quality improvement in south-west Sweden. Water, Air, and Soil Pollution 154(1):203–211.

Iannuzzi, T.J., D.F. Ludwig, J.C. Kinnell, J.M. Wallin, W.H. Desvousges and R.W. Dunford. 2002. A common tragedy: History of an urban river. Amherst Scientific Publishers, Amherst, MA. 200 p.

Jackson, B.J. and J.A. Jackson. 2000. Killdeer (*Charadrius vociferus*), Version 1.0, Birds of the World. Cornell Lab of Ornithology, Ithaca, NY. birdsoftheworld.org/bow/species/killde/cur/introduction (1 September 2010).

Jacobs, S., Sr. 2013. House centipedes. PennState College of Agricultural Sciences, Department of Entomology, University Park, PA. http://ento.psu.edu/extension/factsheets/house-centipedes (28 January 2020).

Jakob, C. and B. Poulin. 2016. Indirect effects of mosquito control using Bti on dragonflies and damselflies (Odonata) in the Camargue. Insect Conservation and Diversity 9(2):161–169.

Jamnback, H. 1969. Bloodsucking flies and other outdoor nuisance arthropods of New York State. New York State Museum and Science Service Memoir 19. 90 p.

Johnson, L.R., J.P. Cuda and N. Burkett-Cadena. 2017. Featured creatures: Cattail mosquito. University of Florida, Gainesville, FL. http://entnemdept.ufl.edu/creatures/aquatic/Coquillettidia_perturbans.htm (13 December 2019).

Johnston, R.F. 2001. Synanthropic birds of North America. P. 49–67 in J.M. Marzluff, R. Bowman and R. Donnelly, eds. Avian Ecology and Conservation in an Urbanizing World. Kluwer Academic Publishers, Boston, MA.

Jolls, C.L. 2003. Populations and threats to rare plants of the herb layer. P. 105–137 in F.S. Gilliam and M.R. Roberts, eds. The Herbaceous Layer in Forests of Eastern North America. Oxford University Press, Oxford, U.K.

Jones, E.L. and S.R. Leather. 2013. Invertebrates in urban areas: A review. European Journal of Entomology 109(4):463–478.

Josselyn, M. 1983. The ecology of San Francisco Bay tidal marshes: A community profile. U.S. Fish and Wildlife Service, Slidell, LA. FWS/OBS-83/23. 102 p.

Joyce, A.A. and S.B. Weisberg. 1986. The effects of predation by the mummichog *Fundulus heteroclitus* (L.) on the abundance and distribution of the saltmarsh snail, *Melampus bidentatus* (Say). Journal of Experimental Marine Biology and Ecology 100:295–306.

Jung, H.B. 2017. Nutrients and heavy metals contamination in an urban estuary of northern New Jersey. Geosciences 7:108.

Kalmbach, E.R. and I.N. Gabrielson. 1921. Economic value of the starling in the United States. U.S. Department of Agriculture Bulletin 868. 66 p.

Kaminski, R.M. and H.H. Prince. 1981. Dabbling duck and aquatic macroinvertebrate responses to manipulated wetland habitat. Journal of Wildlife Management 44:1–15.

Kane, R. 1974. Birds of the Hackensack Meadows, 1970–73. New Jersey Nature News 29:83–87.

Kane, R. 1978. Birds of the Kearny Marsh. Records of New Jersey Birds 7:36–43.

Kane, R. 1983. Fall shorebird migration in the Hackensack Meadows, 1971–1980. Records of New Jersey Birds 9:24–32.

Kane, R. 2001. *Phragmites* use by birds in New Jersey. Records of New Jersey Birds 26(4):122–124.

Kane, R. and D. Githens. 1997. Hackensack River migratory bird report: With recommendations for conservation. New Jersey Audubon Society, Bernardsville, NJ. 37 p.

Kantrud, H.A. and K.F. Higgins. 1992. Nest and nest site characteristics of some ground-nesting, non-passerine birds of northern grasslands. Prairie Naturalist 24:67–84.

Kapfer, J.M. and R.A. Paloski. 2011. On the threat to snakes of mesh deployed for erosion control and wildlife exclusion. Herpetological Conservation and Biology 6(1):1–9.

Karlin, E. No date. The mosses of New Jersey. http://phobos.ramapo.edu/~ekarlin/research/mosses.html (2 December 2019).

Karlin, E. 1990. Endangered bryophytes in New Jersey: Determination, protection and management. P. 208–210 in R.S. Mitchell, C.J. Sheviak and D.J. Leopold, eds. Ecosystem Management: Rare Species and Significant Habitats. New York State Museum Bulletin 471.

Karlin, E. and R.E. Andrus. 1988. The *Sphagnum* species of New Jersey. Bulletin of the Torrey Botanical Club 115(3):168–195.

Karlin, E. and K.A. Schaffroth. 1992. The mosses of New Jersey. Evansia 9:11–32.

Karol, K.G., R.M. McCourt, M.T. Cimino and C.F. Delwiche. 2001. The closest living relatives of land plants. Science 294(5550):2351–2353.

Karraker, N.E. 2008. Impacts of road deicing salts on amphibians and their habitats. P. 211–223 in J.C. Mitchell, R.E.J. Brown and B. Bartholomew, eds. Urban Herpetology. Society for the Study of Amphibians and Reptiles, Salt Lake City, UT.

Kattwinkel, M., R. Biedermann and M. Kleyer. 2011. Temporary conservation for urban biodiversity. Biological Conservation 144:2335–2343.

Kattwinkel, M., B. Strauss, R. Biedermann and M. Kleyer. 2009. Modelling multi-species response to landscape dynamics: Mosaic cycles support urban biodiversity. Landscape Ecology 24(7):929–941.

Kaur, S., A. Rao and S.S. Kumar. 2010. Studies on the effect of heavy metals on the growth of some bryophytes-I (mosses). International Journal of Pharmaceutical Sciences Review and Research 5(3):102–107.

Kays, R., A. Curtis and J.J. Kirchman. 2010. Rapid adaptive evolution of northeastern coyotes via hybridization with wolves. Biology Letters 6:89–93.

Kelly, J. 2019. Regional changes to forest understories since the mid-Twentieth Century: Effects of overabundant deer and other factors in northern New Jersey. Forest Ecology and Management 444:151–162.

Kelly, J.F., E.S. Bridge and M.J. Hamas. 2020. Belted kingfisher (*Megaceryle alcyon*), version 1.0. Birds of the World. Cornell Lab of Ornithology, Ithaca, NY. doi.org/10.2173/bow.belkin1.01 (18 December 2021).

Ketchledge, E.H. 1980. Revised checklist of the mosses of New York State. New York State Museum Bulletin 440. 19 p.

Kimmerer, R.W. 2003. Gathering moss: A natural and cultural history of mosses. Oregon State University Press, Corvallis, OR. 168 p.

Kimmerer, R.W. and C.C. Young. 1996. Effect of gap size and regeneration niche on species coexistance in bryophyte communities. Bulletin of the Torrey Botanical Club 123(1):16–24.

Kirwan, M.L., S. Temmerman, E.E., Skeehan, G.R. Guntenspergen and S. Fagherazzi. 2016. Overestimation of marsh vulnerability to sea level rise. Nature Climate Change 6(3):253–260.

Kiviat, E. 1978. Vertebrate use of muskrat lodges and burrows. Estuaries 1:196–200.

Kiviat, E. 1980. A Hudson River tidemarsh snapping turtle population. Transactions of the Northeast Section, the Wildlife Society 37:158–168.

Kiviat, E. 1982a. Black-capped chickadees eating giant ragweed seeds. Kingbird 32(1):25–26.

Kiviat, E. 1982b. Geographic distribution [Five locality records from Jekyll Island, Georgia]: *Rana grylio* (pig frog), *Scaphiopus holbrooki holbrooki* (eastern spadefoot), *Cnemidophorus sexlineatus sexlineatus* (six-lined racerunner), *Eumeces inexpectatus* (southeastern five-lined skink), *Opheodrys aestivus* (rough green snake). Herpetological Review 13(2):51–53.

Kiviat, E. 1989. The role of wildlife in estuarine ecosystems. P. 437–475 in J.W. Day, et al., eds. Estuarine Ecology. John Wiley and Sons, NY.

Kiviat, E. 1991. Wetland human ecology. PhD thesis, Union Institute, Cincinnati, OH. 180 p.

Kiviat, E., ed. 2007. Monitoring biological diversity in the Hackensack Meadowlands. Report to the Meadowlands Environmental Research Institute of the New Jersey Meadowlands Commission, Lyndhurst, NJ. Hudsonia Ltd., Annandale, NY. 54 p.

Kiviat, E. 2009. Invasive plants in tidal freshwater wetlands—North American East Coast. P. 106–114 in A. Barendregt, D. Whigham and A. Baldwin, eds. Tidal Freshwater Wetlands. Backhuys Publishers, Leiden, The Netherlands.

Kiviat, E. 2010. *Phragmites* management sourcebook for the tidal Hudson River and the northeastern states. Hudsonia Ltd., Annandale, NY. 74 p. hudsonia.org/wp-content/files/j-phragmites%20sourcebook%20generic%2013-June-2010.pdf (18 December 2021).

Kiviat, E. 2011. Frog call surveys in an urban wetland complex, the Hackensack Meadowlands, New Jersey, in 2006. Urban Habitats 6. http://www.urbanhabitats .org/v06n01/frogcallsurveys_full.html (23 January 2020).

Kiviat, E. 2013. Ecosystem services of *Phragmites* in North America with emphasis on habitat functions. AoB Plants. doi: 10.1093/aobpla/plt008.

Kiviat, E. 2014. Adaptation of human cultures to wetland environments. P. 404–415 in P. Gâştescu, W. Marszelewski and P. Bretcan, eds. 2nd International Conference. Water Resources and Wetlands Conference Proceedings 11–13 September, 2014 Tulcea (Romania). Romanian Limnogeographical Association.

Kiviat, E. 2019. Organisms using *Phragmites australis* are diverse and similar on three continents. Journal of Natural History 53(31–32):1975–2010.

Kiviat, E. 2020. Uses of wetlands in the urban coastal Meadowlands of New Jersey, USA. Urban Naturalist (36):1–16.

Kiviat, E. and J.G. Barbour. 1996. Wood turtles in fresh-tidal habitats of the Hudson River. Canadian Field-Naturalist 110(2):341–343.

Kiviat, E., S.E.G. Findlay and W.C. Nieder. 2006. Tidal wetlands. P. 279–295 in J.S. Levinton and J.R. Waldman, eds. The Hudson River Estuary. Cambridge University Press, New York, NY.

Kiviat, E. and C. Graham. 2016. Vascular flora survey, Berry's Creek Study Area. Report to ELM Group, Inc. Holicong, PA. P. 204–278 in BCSA—Final RI Report—Appendix I—August 2018. www.berryscreekstudyarea.com/document -library (8 May 2021).

Kiviat, E. and T. Hartwig. 1994. Marine mammals in the Hudson River. News from Hudsonia 10(2):1–5.

Kiviat, E. and K. MacDonald. 2002. Hackensack Meadowlands, New Jersey, bio-diversity: A review and synthesis. Hackensack Meadowlands Partnership. 112 p. hudsonia.org/wp-content/files/Publications/NJ%20Meadowlands/r-hm3.pdf (18 December 2021).

Kiviat, E. and K. MacDonald. 2004. Biodiversity patterns and conservation in the Hackensack Meadowlands, New Jersey. Urban Habitats 2(1):28–61.

Kiviat, E. and K. MacDonald. 2006. The Hackensack Meadowlands, a metropolitan wildlife refuge. Meadowlands Conservation Trust, Lyndhurst, NJ. 41 p.

Kiviat, E. and J. Stapleton. 1983. *Bufo americanus* (American toad): Estuarine habitat. Herpetological Review 14(2):46.

Kiviat, E. and G. Stevens. 2001. Biodiversity assessment manual for the Hudson River estuary corridor. New York State Department of Environmental Conservation, New Paltz, NY. 508 p.

Kiviat, E. and E. Talmage. 2006. Common reed (*Phragmites australis*) bird and invertebrate studies in Tivoli North Bay, New York. Report to New York State Department of Environmental Conservation. Hudson River Estuary Program, New Paltz, NY. 37 p.

Klauda, R.J., P.H. Muessig and J.A. Matousek. 1988. Fisheries data sets compiled by utility-sponsored research in the Hudson River Estuary. P. 7–85 in C.L. Smith, ed. Fisheries Research in the Hudson River. SUNY Press, Albany, NY.

Klein, C. and C.S. Hurlbut, Jr. 1985. Manual of mineralogy. Twentieth edition. John Wiley and Sons, New York, NY. 596 p.

Klemens, M.W. 1985. Survivors in Megalopolis: Reptiles of the urban Northeast. Discover (Yale Peabody Museum of Natural History) 18(1):22–25.

Klemens, M.W., E. Kiviat and R.E. Schmidt. 1987. Distribution of the northern leopard frog, *Rana pipiens*, in the lower Hudson and Housatonic river valleys. Northeastern Environmental Science 6(2):99–101.

Kneib, R.T., A.E. Stiven and E.B. Haines. 1980. Stable carbon isotope ratios in *Fundulus heteroclitus* (L.) muscle tissue and gut contents from a North Carolina *Spartina* marsh. Journal of Experimental Marine Biology and Ecology 46:89–98.

Köhler, H.-R. and R. Triebskorn. 2013. Wildlife ecotoxicology of pesticides: Can we track effects to the population level and beyond? Science 341:759–765.

Konsevick, E. 2004. Little Ferry water quality investigation. Meadowlands Environmental Research Institute, New Jersey Meadowlands Commission, Lyndhurst, NJ. 2 p.

Korsós, Z., E. Hornung, K. Szlávecz and J. Kontschán. 2002. Isopoda and Diplopoda of urban habitats: New data to the fauna of Budapest. Annales Historico-naturales Musei Nationalis Hungarici 94:193–208.

Kozicky, E.L. and F.W. Schmidt. 1949. Nesting habits of the clapper rail in New Jersey. Auk 66:355–364.

Kraus, M.L. 1989. Bioaccumulation of heavy metals in pre-fledgling tree swallows, *Tachycineta bicolor*. Bulletin of Environmental Contamination and Toxicology 43:407–414.

Kraus, M.L., A. Benda, P. Lupini and A. Smith. 1987. Species lists of organisms found in the Hackensack Meadowlands: Vascular plants—mammals. Unpublished report, Hackensack Meadowlands Development Commission, Lyndhurst, NJ. 39 p.

Kraus, M.L. and A.B. Bragin. 1989. Inventory of fishery resources of the Hackensack River within the jurisdictional boundary of the Hackensack Meadowlands Development Commission from Kearny, Hudson County, to Ridgefield, Bergen County, New Jersey. Hackensack Meadowlands Development Commission Division of Environmental Operations, Lyndhurst, NJ. 134 p. rucore.libraries.rutgers.edu/rutgers-lib/28931/ (20 January 2020).

Kraus, M.L. and A.B. Bragin. 1990. Utilization of the Hackensack River by the Atlantic tomcod (*Microgadus tomcod*). Bulletin of the New Jersey Academy of Science 35:25–27.

Krause, L.H., C. Rietsma and E. Kiviat. 1997. Terrestrial insects associated with *Phragmites australis*, *Typha angustifolia*, and *Lythrum salicaria* in a Hudson River tidal marsh. P. V-1 to V-35 in W.C. Nieder and J.R. Waldman, eds. Final Reports of the Tibor T. Polgar Fellowship Program 1996. Hudson River Foundation. http://hudsonandharbor.org/wp-content/uploads/library/Polgar_Krause_TP_05_96_final .pdf (18 December 2021).

Kroodsma, D.E. and J. Verner. 2013. Marsh wren (*Cistothorus palustris*), Version 1.0, Birds of the World. Cornell Lab of Ornithology, Ithaca, NY. birdsoftheworld.org/bow/species/marwre/cur/introduction (22 January 2020).

Kudish, M. 1992. Adirondack upland flora: An ecological perspective. Chauncy Press, Saranac, NY. 316 p.

Kurta, A. and J.A. Teramino. 1992. Bat community structure in an urban park. Ecography 15:257–261.

LaDeau, S. L., A.M. Kilpatrick and P.P. Marra. 2007. West Nile virus emergence and large-scale declines of North American bird populations. Nature 447(7145):710–714.

Laderman, A.D. 1989. The ecology of Atlantic white cedar wetlands: A community profile. U.S. Fish and Wildlife Service Biological Report 85(7.21).

Laliberte, A.S. and W.J. Ripple. 2004. Range contractions of North American carnivores and ungulates. BioScience 54:123–138.

Lam, E. 2004. Damselflies of the Northeast. Biodiversity Books, Forest Hills, NY. 96 p.

Lambert M.R., B.A. Goldfarb, G.J. Watkins-Colwell and C.M. Donihue. 2016. *Podarcis siculus* (Italian wall lizard). Habitat and suburban invasion. Herpetological Review 47(3):467–468.

Lavin, S.R., T.R. Van Deelen, P.W. Brown, R.E. Warner and S.H. Ambrose. 2003. Prey use by red foxes (*Vulpes vulpes*) in urban and rural areas of Illinois. Canadian Journal of Zoology 81:1070–1082.

Lawrey, J.D. and E.D. Rudolph. 1975. Lichen accumulation of some heavy metals from acidic surface substrates of coal mine ecosystems in southeastern Ohio. Ohio Journal of Science 75:113–117.

Learn, N.H., R.M. Feltes and B.J. Mohn. 2004. Monitoring of fish in the Mill Creek Marsh wetlands mitigation site. P. 1–11 in J.M. Hartman. Summary Report on Monitoring and Experimentation at Mill Creek Wetlands Mitigation Site. Report to the New Jersey Meadowlands Commission, Lyndhurst, NJ.

LeBlanc, F. and J. De Sloover. 1970. Relation between industrialization and the distribution and growth of epiphytic lichens and mosses in Montreal. Canadian Journal of Botany 48:1485–1496.

LeBlanc, F. and D.N. Rao. 1973. Evaluation of the pollution and drought hypotheses in relation to lichens and bryophytes in urban environments. Bryologist 76:1–19.

Lellinger, D.B. 1985. A field manual of the ferns and fern-allies of the United States and Canada. Smithsonian Institution Press, Washington, DC. 389 p.

Lemmon, E.M., A.R. Lemmon, J.T. Collins, J.A. Lee-Yaw and D.C. Cannatella. 2007. Phylogeny-based delimitation of species boundaries and contact zones in the trilling chorus frogs (*Pseudacris*). Molecular Phylogenetics and Evolution 44:1068–1082.

Len, C. 2015. The next big step in the river's recovery. Hackensack Tidelines 18(1):1, 3.

Leonardi, L. and E. Kiviat. 1990. Bryophytes of the Tivoli Bays tidal swamps. P. III-1 to 23 in J.R. Waldman and E.A. Blair, eds. Final Reports of the Tibor T. Polgar Fellowship Program 1989. Hudson River Foundation, New York, NY. www .hudsonriver.org/ls/reports/Polgar_Leonardi_TP_03_89_final.pdf (18 December 2021).

Leschack, C.R., S.K. McKnight and G.R. Hepp. 2020. Gadwall (*Mareca strepera*), Version 1.0, Birds of the World. Cornell Lab of Ornithology, Ithaca, NY. birdsofth eworld.org/bow/species/gadwal/cur/introduction (1 September 2010).

Li, J., X. Li and C. Chen. 2014. Degradation and reorganization of thylakoid protein complexes of *Bryum argenteum* in response to dehydration and rehydration. Bryologist 117(2):110–118.

Li, Y., B.A. Schichtel, J.T. Walker, D.B. Schwede, X. Chen, C.M. Lehmann, M.A. Puchalski, D.A. Gay and J.L. Collett, Jr. 2016. Increasing importance of deposition of reduced nitrogen in the United States. Proceedings of the National Academy of Sciences 113(21):5874–5879.

Liber, K., K.L. Schmude and D.M. Rau. 1998. Toxicity of *Bacillus thuringiensis* var. *israelensis* to chironomids in pond mesocosms. Ecotoxicology 7(6):343–354.

Lincoln, M.S.G. 2008. Liverworts of New England: A guide for the amateur naturalist. New York Botanical Garden Memoir No. 99. New York Botanical Garden Press, Bronx, NY. 161 p.

Linz, G.M., D.L. Bergman, D.C. Blixt and C. McMurl. 1997. Response of American coots and soras to herbicide-induced vegetation changes in wetlands. Journal of Field Ornithology 68:450–457.

Lioy, P.J. and P.G. Georgopoulos. 2011. New Jersey: A case study of the reduction in urban and suburban air pollution from the 1950s to 2010. Environmental Health Perspectives 119:1351–1355.

Lisowska, M. 2011. Lichen recolonisation in an urban-industrial area of southern Poland as a result of air quality improvement. Environmental Monitoring and Assessment 179:177–190.

Litvaitis, J.A. 2001. Importance of early successional habitats to mammals in eastern forests. Wildlife Society Bulletin 29:466–473.

Livezey, B.C. 1981. Duck nesting in retired croplands at Horicon National Wildlife Refuge, Wisconsin. Journal of Wildlife Management 45:27–37.

Lodge, J., R.L. Miller, D.J. Suszkowski, S. Litten and S. Douglas. 2015. Contamination assessment and reduction project summary report. Hudson River Foundation, New York, NY.

Löhmus, P. and A. Löhmus. 2009. The importance of representative inventories for lichen conservation assessments: The case of *Cladonia norvegica* and *C. parasitica*. Lichenologist 41(1):61–67.

Lohr, M.T. and R.A. Davis. 2018. Anticoagulant rodenticide use, non-target impacts and regulation: A case study from Australia. Science of the Total Environment 634:1372–1384.

Loppi, S., L. Giovannelli, S.A. Pirintsos, E. Putortì and A. Corsini. 1997. Lichens as bioindicators of recent changes in air quality (Montecatini Terme, Italy). Ecologia Mediterranea 23(3):53–56.

Lotrich, V.A. 1975. Summer home range and movements of *Fundulus heteroclitus* (Pisces: Cyprinodontidae) in a tidal creek. Ecology 56:191–198.

Lowther, P.E., A.F. Poole, J.P. Gibbs, S.M. Melvin and F.A. Reid. 2009. American bittern (*Botaurus lentiginosus*), Version 1.0, Birds of the World. Cornell Lab of Ornithology, Ithaca, NY. birdsoftheworld.org/bow/species/amebit/cur/introduction (22 January 2020).

MacArthur, R.H. and J.W. MacArthur. 1961. On bird species diversity. Ecology 42:594–598.

Maccarone, A.D. and J.N. Brzorad. 2002. Foraging patterns of breeding egrets at coastal and interior locations. Waterbirds 25:1–7.

Maccarone, A.D. and J.N. Brzorad. 2005. Foraging microhabitat selection by wading birds in a tidal estuary, with implications for conservation. Waterbirds 28:383–391.

Maccarone, A.D. and K.C. Parsons. 1994. Factors affecting the use of a freshwater and an estuarine foraging site by egrets and ibises during the breeding season in New York City. Colonial Waterbirds 17:60–68.

MacDonald-Beyers, K. 2008. Habitat interactions structuring songbird communities across forest-urban edges. PhD thesis, Rutgers University, New Brunswick, NJ. 142 p.

MacDonald, K. and T.K. Rudel. 2005. Sprawl and forest cover: What is the relationship? Applied Geography 25:67–79.

Macwhirter, R.B., P. Austin-Smith, Jr. and D.E. Kroodsma. 2002. Sanderling (*Calidris alba*), Version 1.0, Birds of the World. Cornell Lab of Ornithology, Ithaca, NY. birdsoftheworld.org/bow/species/sander/cur/introduction (1 September 2010).

Maeglin, R.R. and L.F. Ohmann. 1973. Boxelder (*Acer negundo*): A review and commentary. Bulletin of the Torrey Botanical Club 100(6):357–363.

Mager, K.J. and T.A. Nelson. 2001. Roost-site selection by eastern red bats (*Lasiurus borealis*). American Midland Naturalist 145:120–126.

Mangold, R.E. 1974. Clapper rail studies. 1974 Final Report, U.S. Fish and Wildlife Service Accelerated Research Program Contract No. 14–16–0008–937, Trenton, NJ.

Mansoor, N., L. Slater, F. Artigas and E. Auken. 2006. High-resolution geophysical characterization of shallow-water wetlands. Geophysics 71(4):B101–B109.

Måren, I.E., V. Vandvik and K. Ekelund. 2008. Restoration of bracken-invaded *Calluna vulgaris* heathlands: Effects on vegetation dynamics and non-target species. Biological Conservation 141:1032–1042.

MMSC (Marine Mammal Stranding Center). 2008. [Unpublished data on recoveries of marine mammals in New Jersey waters.] Marine Mammal Stranding Center, Brigantine, NJ.

Marquardt, W.C., R.S. Demaree and R.B. Grieve. 2000. Parasitology and vector biology. Harcourt Academic Press, San Diego, CA. 702 p.

Marshall, S. 2004. The Meadowlands before the commission: Three centuries of human use and alteration of the Newark and Hackensack Meadows. Urban Habitats 2(1):4–27. http://urbanhabitats.org/v02n01/3centuries_full.html (23 January 2020).

Martin, A.C., H.S. Zim and A.L. Nelson. 1951. American wildlife and plants: A guide to wildlife food habits. Dover Publications, New York, NY. 500 p. (Also published in 1951 by McGraw-Hill Book Company.)

Marzluff, J.M. 2001. Worldwide urbanization and its effects on birds. P. 19–47 in J.M. Marzluff, R. Bowman and R. Donnelly, eds. Avian Ecology and Conservation in an Urbanizing World. Kluwer Academic Publishers, Boston, MA.

Massachusetts Institute of Technology. 2014. The New Meadowlands. Prepared for Rebuild by Design. http://www.rebuildbydesign.org/data/files/672.pdf (20 July 2019).

Mathewson, R.F. 1955. Reptiles and amphibians of Staten Island. Proceedings of the Staten Island Institute of Arts and Sciences 17(2):29–48.

Matthiae, P.E. and F. Stearns. 1981. Mammals in forest islands in southeastern Wisconsin. P. 55–66 in R.L. Burgess and D.M. Sharpe, eds. Forest Island Dynamics in Man-Dominated Landscapes. Springer-Verlag, New York, NY.

Maxson, S.J. and L.W. Oring. 1980. Breeding season time and energy budgets of the polyandrous spotted sandpiper. Behaviour 74:200–263.

McAuley, D.G., D.M. Keppie and R.M. Whiting, Jr. 2013. American woodcock (*Scolopax minor*), Version 1.0, Birds of the World. Cornell Lab of Ornithology, Ithaca, NY. birdsoftheworld.org/bow/species/amewoo/cur/introduction (22 January 2020).

McClary, M., Jr. 2004. *Spartina alterniflora* and *Phragmites australis* as habitat for the ribbed mussel, *Geukensia demissa*, in Saw Mill Creek of New Jersey's Hackensack Meadowlands. Urban Habitats 2:83–90. http://www.urbanhabitats.org /v02n01/ribbedmussel_full.html (23 January 2020).

McClure, C.J.W., A.C. Korte, J.A. Heath and J.R. Barber. 2015. Pavement and riparian forest shape the bird community along an urban river corridor. Global Ecology and Conservation 4:291–310.

McCormick, J. 1970. The natural features of Tinicum Marsh, with particular emphasis on the vegetation. P. 1–104 in J. McCormick, R.R. Grant and R. Patrick, eds. Two Studies of Tinicum Marsh. Conservation Foundation, Washington, DC.

McCormick, J. & Associates, Inc. 1978. Full Environmental Impact Statement for the proposed Meadowlands Arena at the New Jersey Sports Complex, Borough of East Rutherford, County of Bergen. Prepared for the New Jersey Sports and Exposition Authority, East Rutherford, NJ. 277 p.

McCormick, M.K., K.M. Kettenring, H.M. Baron and D.F. Whigham. 2010. Spread of invasive *Phragmites australis* in estuaries with differing degrees of development: Genetic patterns, Allee effects and interpretation. Journal of Ecology 98(6):1369–1378.

McDonnell, M.J. and S.T.A. Pickett. 1990. Ecosystem structure and function along urban-rural gradients: An unexploited opportunity for ecology. Ecology 7:1232–1237.

McGlynn, C.A. and R.S. Ostfeld. 2000. A study of the effects of invasive plant species on small mammals in Hudson River freshwater marshes. P. VIII-1 to VIII-21 in W.C. Nieder and J.R. Waldman, eds. Final Reports of the Tibor T. Polgar Fellowship Program 1999. Hudson River Foundation, New York, NY. http://hudsonandharbor.org/wp-content/uploads/library/Polgar_McGlynn_TP_09_99_final.pdf (18 December 2021).

McIntyre, C. 2000. Heavy metal concentrations in sediment and diamondback terrapin (*Malaclemys terrapin terrapin*) tissues from two sites in New Jersey. Senior thesis, Hampshire College, Amherst, MA. 55 p.

McKinney, M.L. 2008. Effects of urbanization on species richness: A review of plants and animals. Urban Ecosystems 11:161–176.

McKnight, K.B., J.R. Rohrer, K.M. Ward and W.J. Perdrizet. 2013. Common mosses of the Northeast and Appalachians. Princeton University Press, Princeton, NJ. 392 p.

McMurray, J.A., D.W. Roberts, M.E. Fenn, L.H. Geiser and S. Jovan. 2013. Using epiphytic lichens to monitor nitrogen deposition near natural gas drilling operations in the Wind River Range, WY, USA. Water, Air, & Soil Pollution 224(3):e1487.

McQueen, C.B. 1990. Field guide to the peat mosses of boreal North America. University Press of New England, Hanover, NH. 138 p.

Meadowlands Regional Chamber. 2019. The Meadowlands economic development and relocation guide. Rutherford, NJ. 36 p. meadowlandsmedia.com/economic-development-relocation-guide/ (27 August 2019).

Mech, L.D. 2003. Incidence of mink, *Mustela vison*, and river otter, *Lutra canadensis*, in a highly urbanized area. Canadian Field-Naturalist 117(1):115–116.

Meierhofer, M.B., S.J. Leivers, R.R. Fern, L.K. Wolf, J.H. Young, Jr., B.L. Pierce, J.W. Evans and M.L. Morrison 2019. Structural and environmental predictors of presence and abundance of tri-colored bats in Texas culverts. Journal of Mammalogy 100:1274–1281.

Melquist, W.E. and A.E. Dronkert. 1987. River otter. P. 627–641 in M. Novak, J. Baker, M.E. Obbard, and B. Malloch, eds. Wild Furbearer Management and Conservation in North America. Ontario Trappers Association, Toronto, Canada.

Melvin, S.M. and J.P. Gibbs. 2012. Sora (*Porzana carolina*), Version 1.0, Birds of the World. Cornell Lab of Ornithology, Ithaca, NY. birdsoftheworld.org/bow/species/sora/cur/introduction (22 January 2020).

Mendyk, R.W. and J. Adragna. 2014. Notes on two introduced populations of the Italian wall lizard (*Podarcis siculus*) on Staten Island, New York. Reptiles & Amphibians 21(4):142–143.

Menke S.B., B. Guénard, J.O. Sexton, M.D. Weiser, R.R. Dunn and J. Silverman. 2011. Urban areas may serve as habitat and corridors for dry-adapted, heat tolerant species; an example from ants. Urban Ecosystems 14(2):135–63.

Mercurio, R.J. 2010. An annotated catalog of centipedes (Chilopoda) from the United States of America, Canada and Greenland (1758–2008). Xlibris, Bloomington, IN. 560 p.

MERI (Meadowlands Environmental Research Institute). 2008. Tidal datum analysis study. MERI, New Jersey Meadowlands Commission, Lyndhurst, NJ. http://meri .njmeadowlands.gov/scientific/tidaldatum/Tidal_Datum.html (19 August 2010).

MERI (Meadowlands Environmental Research Institute). 2019. Scientific data. MERI, New Jersey Meadowlands Commission, Lyndhurst, NJ. meri.njmeadowlan ds.gov/ (2 September 2019).

Merrill, E. and L. Leatherby. 2018. Here's how America uses its land. Bloomberg. www.bloomberg.com/graphics/2018-us-land-use/ (8 December 2019).

Mihocko, G., E. Kiviat, R.E. Schmidt, S.E.G., Findlay, W.C. Nieder and E. Blair. 2003. Assessing ecological functions of Hudson River fresh-tidal marshes: Reference data and a modified hydrogeomorphic (HGM) approach. Report to the New York State Department of Environmental Conservation, Hudson River Estuary Program, New Paltz, NY. 166 p.

Milford, N., G.C. McCoyd, L. Aronowitz and J.H. Scanlon. 1970. Air pollution by airports. P. 448–460 in Institute of Environmental Sciences Proceedings, 16th Annual Technical Meeting April 12–16 1970, Boston, MA.

Miller, R.W. and R.L. Donahue. 1995. Soils in our environment. 7th ed. Prentice Hall, Englewood Cliffs, NJ. 649 p.

Miller, N.A. and M.W. Klemens. 2004. Croton-to-Highlands biodiversity plan: Balancing development and the environment in the Hudson River estuary catchment. MCA Technical Paper No. 7. Metropolitan Conservation Alliance, Wildlife Conservation Society, Bronx, NY.

Miller, N.G. and S.F. McDaniel. 2004. Bryophyte dispersal inferred from colonization of an introduced substratum on Whiteface Mountain, New York. American Journal of Botany 91:1173–1182.

Milne, L. and M. Milne. 1980. National Audubon Society field guide to insects and spiders. Alfred A. Knopf, New York, NY. 989 p.

Mitchell, J.C. and R.E.J. Brown. 2008. Urban herpetology: Global overview, synthesis, and future directions. P. 1–30 in J.C. Mitchell, R.E.J. Brown and B. Bartholomew, eds. Urban Herpetology. Society for the Study of Amphibians and Reptiles, Salt Lake City, UT.

Mitchell, J.C., R.E.J. Brown and B. Bartholomew, eds. 2008. Urban herpetology. Society for the Study of Amphibians and Reptiles, Salt Lake City, UT. 586 p.

Mitov, P.G. and I.L. Stoyanov. 2004. The harvestmen fauna (Opiliones, Arachnida) of the city of Sofia (Bulgaria) and its adjacent regions. P. 319–354 in L. Penev, J. Niemelä, D.J. Kotze and N. Chipey, eds. Ecology of the City of Sofia. Species and Communities in an Urban Environment. Pensoft, Sofia, Moscow.

Mizrahi, D.S., N. Tsipoura, K. Witkowski and M. Bitignano. 2007. Avian abundance and distribution in the New Jersey Meadowlands District: The importance of habitat, landscape, and disturbance. A final report to the New Jersey Meadowlands Commission. New Jersey Audubon Society, Bernardsville, NJ.

Mohn, B.J. No date. Tetrapodal vertebrates of the Meadowlands. [No source, probably originated by the author.] 18 p.

Monosson, E. 2000. Reproductive and developmental effects of PCBs in fish: A synthesis of laboratory and field studies. Reviews in Toxicology 3:25–75.

Montgomery, J.D. and D.E. Fairbrothers. 1992. New Jersey ferns and fern-allies. Rutgers University Press, New Brunswick, NJ. 293 p.

Morgan, R.A. and D.E. Glue. 1981. Breeding survey of black redstarts in Britain, 1977. Bird Study 28:163–168.

Morgan, E. and J.A. Sperling. 2006. Changes in the bryophyte flora of Cunningham Park and Alley Pond Park, Queens County, Long Island, New York City. Evansia 23:56–60.

Morneau, F., R. Decarie, R. Pelletier, D. Lambert, J. DesGranges and J. Savard. 1999. Changes in breeding bird richness and abundance in Montreal parks over a period of 15 years. Landscape and Urban Planning 44:111–121.

Moulding, J.D. and J.J. Madenjian. 1979. New Jersey oak forest (Lepidoptera). Proceedings of the Entomological Society of Washington 81(1):135–144.

Muenscher, W.C. 1936. Storage and germination of seeds of aquatic plants. New York (Cornell University) Agricultural Experiment Station Bulletin 652. 17 p.

Muller, M.J. and R.W. Storer. 1999. Pied-billed grebe (*Podilymbus podiceps*), Version 1.0, Birds of the World. Cornell Lab of Ornithology, Ithaca, NY. birdsoftheworld.org/bow/species/pibgre/cur/introduction (1 September 2010).

Munch, S. 2006. Outstanding mosses and liverworts of Pennsylvania and nearby states. Sunbury Press, Mechanicsburg, PA. 89 p.

Myrbo, A., E.B. Swain, D.R. Engstrom, J.C. Wasik, J. Brenner, M.D. Shore, E.B. Peters and G. Blaha. 2017. Sulfide generated by sulfate reduction is a primary controller of the occurrence of wild rice (*Zizania palustris*) in shallow aquatic ecosystems. Journal of Geophysical Research: Biogeosciences 122(11):2736–2753.

NABCI (North American Bird Conservation Initiative). 2016. The state of North America's birds 2016. www.stateofthebirds.org/2016/ (11 July 2021).

Nagy, C. 2005. NY City Audubon 2005 Harbor herons monitoring program. New York City Audubon, New York, NY.

Naiman, R.J., H. Decamps and M. Pollock. 1993. The role of riparian corridors in maintaining regional biodiversity. Ecological Applications 3:209–212.

Naiman, R.J., J.M. Melillo and J.E. Hobbie. 1986. Ecosystem alteration of boreal forest streams by beaver (*Castor canadensis*). Ecology 67:1254–1269.

Nash, T.H. III and V. Wirth, eds. 1988. Lichens, bryophytes and air quality. Bibliotheca Lichenologica 30. J. Cramer, Berlin, Germany. 297 p.

National Weather Service. No date. Hurricane Sandy—October 29, 2012. www.weather.gov/okx/HurricaneSandy (24 November 2019).

Nebel, S. and J.M. Cooper. 2008. Least sandpiper (*Calidris minutilla*), Version 1.0, Birds of the World. Cornell Lab of Ornithology, Ithaca, NY. birdsoftheworld.org/bow/species/leasan/cur/introduction (1 September 2010).

Neill W.T. 1950. Reptiles and amphibians in urban areas of Georgia. Herpetologica 6(5):113–116.

Nelson, J.S., T.C. Grande and M.V.H. Wilson. 2016. Fishes of the world. 5th ed. John Wiley and Sons, Hoboken, NJ. 707 p.

Ner, S.E. and R.L. Burke. 2008. Direct and indirect effects of urbanization on diamond-backed terrapins of the Hudson River Bight: Distribution and predation in a human-modified estuary. P. 107–117 in J.C. Mitchell, R.E.J. Brown and B. Bartholomew, eds. Urban Herpetology. Society for the Study of Amphibians and Reptiles, Salt Lake City, UT.

Neuman, M.J., G. Ruess and K.W. Able. 2004. Species composition and food habits of dominant fish predators in salt marshes of an urbanized estuary, the Hackensack Meadowlands, New Jersey. Urban Habitats 2(1):62–82. http://www.urbanhabitats.org/v02n01/saltmarsh_full.html (23 January 2020).

Newman, C.E., J.A. Feinberg, L.J. Rissler, J. Burger and H.B. Shaffer. 2012. A new species of leopard frog (Anura: Ranidae) from the urban northeastern US. Molecular Phylogenetics and Evolution 63(2):445–455.

Newsome, S.D., H.M. Garbe, E.C. Wilson and S.D. Gehrt. 2015. Individual variation in anthropogenic resource use in an urban carnivore. Oecologia 178:115–128.

New Jersey Forest Service. 2019. Forest health in New Jersey. State of New Jersey Department of Environmental Protection, Trenton, NJ. www.state.nj.us/dep/parksandforests/forest/njfs_forest_health.html (7 September 2019).

Nikula, B., J.L. Loose and M.R. Burne. 2003. A field guide to the dragonflies and damselflies of Massachusetts. Massachusetts Division of Fisheries and Wildlife, Westborough, MA. 197 p.

Nilsson, G. 1980. River otter research workshop, 27–29 March 1980. Florida State Museum, Gainesville, FL. 26.

NJMC (New Jersey Meadowlands Commission). No date. Birds of the Meadowlands checklist. NJMC, Lyndhurst, NJ. meadowblog.typepad.com/files/njmc_bird_list_lores.pdf (17 October 2021).

NJDEP (New Jersey Department of Environmental Protection). 1998. 1997 Air quality report. NJDEP Bureau of Air Monitoring, Trenton, NJ. http://www.njaqinow.net / (23 January 2020).

NJDEP (New Jersey Department of Environmental Protection, Division of Fish & Wildlife). 2004a. Mammals of New Jersey. NJDEP, Trenton, NJ. www.state.nj.us/dep/fgw/chkmamls.htm (30 August 2010).

NJDEP (New Jersey Department of Environmental Protection, Division of Fish and Wildlife). 2004b. Birds of New Jersey. NJDEP, Trenton, NJ. http://www.state.nj.us/dep/fgw/chkbirds.htm (15 September 2008).

NJDEP (New Jersey Department of Environmental Protection). 2006. New Jersey marine mammal and sea turtle conservation workshop proceedings, April 17–19 2006. NJDEP, Endangered and Nongame Species Program, Trenton, NJ. http://www.state.nj.us/dep/fgw/ensp/pdf/marinemammal_seaturtle_workshop06.pdf (30 August 2010).

NJDEP (New Jersey Department of Environmental Protection). 2008. 2007 Air quality report. NJDEP Bureau of Air Monitoring, Trenton, NJ. http://www.njaqinow.net / (23 January 2020).

NJDEP (New Jersey Department of Environmental Protection, Division of Fish and Wildlife). 2012. New Jersey's endangered and threatened wildlife. NJDEP, Trenton, NJ. www.state.nj.us/dep/fgw/tandespp.htm (28 January 2020).

NJDEP (New Jersey Department of Environmental Protection). 2015. New Jersey Land Use/Land Cover (LU/LC) Data Set 2015. NJDEP, Division of Information Technology, Bureau of Geographic Information System. www.nj.gov/dep/gis/digidownload/zips/OpenData/Land_lu_2015_shapefile.zip (24 October 2021).

NJDEP (New Jersey Department of Environmental Protection). 2018a. New Jersey's Wildlife Action Plan. NJDEP Division of Fish and Wildlife. Trenton, NJ. 140 p. + appendices. www.njfishandwildlife.com/ensp/wap/pdf/wap_plan18.pdf (8 October 2021).

NJDEP (New Jersey Department of Environmental Protection). 2018b. 2017 New Jersey air quality report. NJDEP Bureau of Air Monitoring, Trenton, NJ. http://www.njaqinow.net/ (23 January 2020).

NJDEP (New Jersey Department of Environmental Protection, Division of Parks and Forestry). 2019. Special plants of New Jersey. NJDEP, Trenton, NJ. www.nj.gov/dep/parksandforests/natural/heritage/rarelist.html (15 October 2021).

NJDEP (New Jersey Department of Environmental Protection). 2021. Fish smart eat smart NJ. www.nj.gov/dep/dsr/njmainfish.htm (20 October 2021).

NJODES (New Jersey Odonata Survey). 2020. njodes.org/ (28 January 2020).

NJSEA (New Jersey Sports and Exposition Authority). 2019. Redevelopment. www.njsea.com/redevelopment/ (22 January 2020).

NJTA (New Jersey Turnpike Authority). 1986. New Jersey Turnpike 1985–90 widening, technical study. Volume II: Biological Resources. NJTA, Woodbridge, NJ.

Nordenson, C.S., G. Nordenson and J. Chapman. 2018. Structures of coastal resilience. Island Press, Washington, DC. 248 p.

Normandeau Associates. 2017. Environmental sample collection, Newark Bay study area. Final report to Tierra Solutions, Inc. 1197 p.

NYBG (New York Botanical Garden). No date. The C.V. Starr Virtual Herbarium of the New York Botanical Garden. http://sweetgum.nybg.org/science/vh/ (16 October 2021).

NYCEM (New York City Emergency Management). 2019. NYC's risk landscape: A guide to hazard mitigation. nychazardmitigation.com/wp-content/uploads/2019/06/risklandscape2.0_2019_r2_v1_digital_highres.pdf (4 August 2019).

Nyffeler, M., E.J. Olson and W.O. Symondson. 2016. Plant-eating by spiders. Journal of Arachnology 44:15–27.

NY-NJ-CT (NY-NJ-CT Botany Online). 2021. New Jersey natural areas: Piedmont (Newark Basin). nynjctbotany.org/njnewark/njnbtofc.html (28 September 2021).

Obbard, M.E., J.G. Jones, R.A. Newman, A. Booth, A.J. Satterthwaite, and G. Linscombe. 1987. Furbearer harvests in North America, 1600–1984. P. 1007–1034 in M. Novak, J.A. Baker, M.E. Obbard and B. Malloch, eds. Wild Furbearer Management and Conservation in North America. Ontario Ministry of Natural Resources.

Occi, J.L., A.M. Egizi, R.G. Robbins and D.M. Fonseca. 2019. Annotated list of the hard ticks (Acari: Ixodida: Ixodidae) of New Jersey. Journal of Medical Entomology 56(3):589–598.

Obropta, C., P. Kallin, B. Ravit, K. Buckley, R. Miskewitz, R. Lathrop, C. Phillipuk, C. Strickland, E. Giuliano and M.E. Cronk. 2007. Stormwater utility feasibility study. Report to New Jersey Meadowlands Commission, Lyndhurst, New Jersey. School of Environmental and Biological Sciences, Rutgers University, New Brunswick, NJ. 110 p.

Obropta, C., B. Ravit and S. Yeargeau. 2008. Kearny Marsh hydrology study final report. Report to New Jersey Meadowlands Commission, Lyndhurst, New Jersey. Rutgers Environmental Research Clinic, Rutgers University, New Brunswick, NJ. 107 p.

Olmstead, A.W. and G.L. LeBlanc. 2001. Low exposure concentration effects of methoprene on endocrine-regulated processes in the crustacean *Daphnia magna*. Toxicological Sciences 62(2):268–273.

O'Meara, T.E., W.R. Marion, O.B. Myers and W.M. Hetrick. 1982. Food habits of three bird species on phosphate-mine settling ponds and natural wetlands. Proceedings of the Annual Conference of the Southeastern Association of Fish and Wildlife Agencies 36:527–526.

O'Neill, J.M. 2010. Meadowlands bee variety has experts buzzing. The Record (9 July).

ONJSC (Office of the New Jersey State Climatologist). 2010. ONJSC at Rutgers University. Piscataway, NJ. http://climate.rutgers.edu/stateclim/?section =homeandtarget=home (21 August 2010).

Oprea, M., P. Mendes, T.B. Vieira and A.D. Ditchfield. 2009. Do wooded streets provide connectivity for bats in an urban landscape? Biodiversity and Conservation 18:2361–2371.

Oring, L.W., D.B. Lank and S.J. Maxson. 1983. Population studies of the polyandrous spotted sandpiper. Auk 100:272–285.

Orridge, J., J. Waldman and R.E. Schmidt. 2009. Genetic, morphological and ecological relationships among Hudson Valley populations of the clam shrimp, *Caenestheriella gynecia*. P. VI-1 to VI-36 in S.H. Fernald and D. Yozzo, eds. Final Reports of the Tibor T. Polgar Fellowship Program 2008. Hudson River Foundation, New York, NY. http://hudsonandharbor.org/wp-content/uploads/ library/Polgar_Orridge_TP_05_08_final.pdf (18 December 2021).

Ostergaard, E.C., K.O. Richter and S.D. West. 2008. Amphibian use of stormwater ponds in the Puget Lowlands of Washington, U.S.A. P. 259–273 in J.C. Mitchell, R.E.J. Brown and B. Bartholomew, eds. Urban Herpetology. Society for the Study of Amphibians and Reptiles, Salt Lake City, UT.

Ostfeld, R.S. and C.D. Canham. 1993. Effects of meadow vole population density on tree seedling survival in old fields. Ecology 74:1792–1801.

Ostry, M.E., M.E. Mielke and R.L. Anderson. 1996. How to identify butternut canker and manage butternut trees. U.S. Department of Agriculture, North Central Forest Experiment Station HT-70, St. Paul, MN. www.fs.usda.gov/naspf/sites/default/

files/naspf/pdf/9206howtoidentifybutternutcankermanagebutternuttrees_20180710_508.pdf (23 January 2020).

Ovaska, K. and C. Engelstoft. 2008. Conservation of the sharp-tailed snake (*Contia tenuis*) in urban areas in the Gulf Islands, British Columbia, Canada. P. 557–564 in J.C. Mitchell, R.E.J. Brown and B. Bartholomew, eds. Urban Herpetology. Society for the Study of Amphibians and Reptiles, Salt Lake City, UT.

Oxley, D.J., M.B. Fenton and G.R. Carmody. 1974. The effects of roads on populations of small mammals. Journal of Applied Ecology 11:51–59.

Pace, A.E. 1974. Systematic and biological studies of leopard frogs (*Rana pipiens* complex) of the United States. University of Michigan Museum of Zoology Miscellaneous Publication 148. 140 p.

Papp, B., A. Alegro, P. Erzberger, E. Szurdoki, V. Šegota and M. Sabovljević. 2016. Bryophytes of saline areas in the Pannonian region of Serbia and Croatia. Studia Botanica Hungarica 47(1):141–150.

Parkins, K.L. and J.A. Clark. 2015. Green roofs provide habitat for urban bats. Global Ecology and Conservation 4:349–357.

Parkins, K.L., M. Mathios, C. McCann and J.A. Clark. 2016. Bats in the Bronx: Acoustic monitoring of bats in New York City. Urban Naturalist 10:1–16.

Pattishall, A. and D. Cundall. 2009. Habitat use by synurbic watersnakes (*Nerodia sipedon*). Herpetologica 65(2):183–198.

Paulraj, M.G., P.S. Kumar, S. Ignacimuthu and D. Sukumaran. 2016. Natural insecticides from actinomycetes and other microbes for vector mosquito control. P. 85–99 in V. Veer and R. Gopalakrishnan, eds. Herbal Insecticides, Repellents and Biomedicines: Effectiveness and Commercialization. Springer, New Delhi, India.

Peacor, D.R. and P.J. Dunn. 1982. Petersite, a REE and phosphate analog of mixite. American Mineralogist 67:1039–1042.

Pedlar, J.H., L. Fahrig and H.G. Merriam. 1997. Raccoon habitat use at 2 spatial scales. Journal of Wildlife Management 61:102–112.

Pehek, E. 2007. Salamander diversity and distribution in New York City, 1820 to the present. Transactions of the Linnaean Society of New York 10:157–182. linnaeannewyork.com/wp-content/uploads/PDF/LSNY%20Transactions%20 10%202007%20NYC%20Parks%20and%20GGI.pdf#page=88 (19 June 2021).

Penn, S.L., S.T. Boone, B.C. Harvey, W. Heiger-Bernays, Y. Tripodis, S. Arunachalam and J.I. Levy. 2017. Modeling variability in air pollution-related health damages from individual airport emissions. Environmental Research 156:791–800.

Pennak, R.W. 1978. Fresh-water invertebrates of the United States. 2nd ed. John Wiley and Sons, New York, NY. 803 p.

Pennington, D.N., J. Hansel and R.B. Blair. 2008. The conservation of urban riparian areas for landbirds during spring migration: Land cover, scale and vegetation effects. Biological Conservation 141:1235–1248.

Perhans, K., L. Appelgren, F. Jonsson, U. Nordin, B. Söderström and L. Gustafsson. 2009. Retention patches as potential refugia for bryophytes and lichens in managed forest landscapes. Biological Conservation 142:1125–1133.

Perlmutter, G.B., G.B. Blank, T.R. Wentworth, M.D. Lowman, H.S. Neufeld and E.R. Plata. 2018. Highway pollution effects on microhabitat community structure of corticolous lichens. Bryologist 121:1–13.

Peterjohn, B.G. and J.R. Sauer. 1994. Population trends of woodland birds from the North American Breeding Bird Survey. Wildlife Society Bulletin 22(2):155–164.

Philpott, S.M., J. Cotton, P. Bichier, R.L. Friedrich, L.C. Moorhead, S. Uno and M. Valdez. 2014. Local and landscape drivers of arthropod abundance, richness, and trophic composition in urban habitats. Urban Ecosystems 17(2):513–532.

Pickens, B.A. and B. Meanley. 2018. King rail (*Rallus elegans*), Version 1.0, Birds of the World. Cornell Lab of Ornithology, Ithaca, NY. birdsoftheworld.org/bow/species/kinrai4/cur/introduction (24 January 2020).

Pickett, S.T.A., M.L. Cadenasso, J.M. Grove, P.M. Groffman, L.E. Band, C.G. Boone, W.R. Burch, C.S.B. Grimmond, J. Hom, J.C. Jenkins, N.L. Law, C.H. Nilon, R.V. Pouyat, K. Szlavecz, P.S. Warren and M.A. Wilson. 2008. Beyond urban legends: An emerging framework of urban ecology, as illustrated by the Baltimore Ecosystem Study. BioScience 58(2):139–150.

Pickett, S.T.A., M.L. Cadenasso and S.J. Meiners. 2009. Ever since Clements: From succession to vegetation dynamics and understanding to intervention. Applied Vegetation Science 12:9–21.

Poole, A.F., P.E. Lowther, J.P. Gibbs, F.A. Reid and S.M. Melvin. 2009. Least bittern (*Ixobrychus exilis*), Version 1.0, Birds of the World. Cornell Lab of Ornithology, Ithaca, NY. birdsoftheworld.org/bow/species/leabit/cur/introduction (22 January 2020).

Post, W. and J.S. Greenlaw. 2018. Seaside sparrow (*Ammospiza maritima*), Version 1.0, Birds of the World. Cornell Lab of Ornithology, Ithaca, NY. birdsoftheworld.org/bow/species/seaspa/cur/introduction (23 January 2020).

Powell, R.L., A.V. Crocker, S.L. Kumro, R.W. Rabe and E.O. Montalvo. 2015. *Siren intermedia* (lesser siren). Mass aestivation in artificial hibernacula. Herpetological Review 46(2):227–228.

Prange, S., S.D. Gehrt and E.P. Wiggers. 2003. Demographic factors contributing to high raccoon densities in urban landscapes. Journal of Wildlife Management 67:324–333.

Prange, S., S.D. Gehrt and E.P. Wiggers. 2004. Influences of anthropogenic resources on raccoon (*Procyon lotor*) movements and spatial distribution. Journal of Mammalogy 85:483–490.

Pratt, G.F., D.M. Wright and H. Pavulaan. 1994. The various taxa and hosts of the North American *Celastrina* (Lepidoptera: Lycaenidae). Proceedings of the Entomological Society of Washington 96(3):566–578.

PSE&G (Public Service Electric and Gas Company). 1998. Hudson generating station supplemental 316(b) report appendices. PSE&G, Newark, NJ. P. A1-A10 + figures and tables.

Puffer, J.H. 1984. Road log: Igneous rocks of the Newark Basin: Petrology, mineralogy, and ore deposits. P. 164–179 in J.H. Puffer, ed. Igneous Rocks of the Newark Basin: Petrology, Mineralogy, Ore Deposits and Guide to Field Trip. Geological

Association of New Jersey, 1st Annual Field Conference, Kean College, Union, NJ, October 19–20, 1984.

Puffer, J.H. and A.I. Benimoff 1997. Fractionation, hydrothermal alteration, and wall-rock contamination of an Early Jurassic diabase intrusion: Laurel Hill, New Jersey. Journal of Geology 105(1):99–110.

Puffer, J.H. and J. Husch. 1996. Early Jurassic diabase sheets and basalt flows, Newark Basin, New Jersey: An updated geological summary and field guide. P. 1–52 in A.I. Benimoff and A.A. Ohan, eds. Field Trip Guide for the 68th Annual Meeting of the New York State Geological Association. New York State Geological Association.

Pursell, R.A. 2007. Fissidentaceae. Flora of North America 27. http://www.efloras .org/florataxon.aspx?flora_id=1&taxon_id=10342 (14 March 2010).

Pyke, G.H. 2008. Plague minnow or mosquito fish? A review of the biology and impacts of introduced *Gambusia* species. Annual Review of Ecology, Evolution, and Systematics 39:171–191.

Quinn, J.R. 1997. Fields of sun and grass: An artist's journal of the New Jersey Meadowlands. Rutgers University Press, New Brunswick, NJ. 342 p.

Quinn, L. 2021. What's behind the unknown disease killing birds in New Jersey? Northjersey.com (9 July). www.northjersey.com/story/news/new-jersey/2021/07 /09/bird-disease-2021-mystery-new-jersey/7913582002/ (8 October 2021).

Raisley, E.J. 2001. Progress and status of river otter reintroduction projects in the United States. Wildlife Society Bulletin 29:856–862.

Rafter, D. 2013. Yellow-crowned night heron nesting in Lindenwood: A family of herons has made a home on an 84th street building. Queens Chronicle (25 July). www .qchron.com/editions/south/yellow-crowned-night-heron-nesting-in-lindenwood/ article_ed263708-2b49-53c1-8f26-2169bb13b5b2.html (11 July 2021).

Ranta, E., M.R. Vidal-Abarca, A.R. Calapez and M.J. Feio. 2021. Urban stream assessment system (UsAs): An integrative tool to assess biodiversity, ecosystem functions and services. Ecological Indicators 121:e106980.

Raphael, M.G. and R. Molina. 2007. Conservation of rare or little-known species. Island Press, Washington, DC. p. 375.

Rauchenberger, M. 1989. Systematics and biogeography of the genus *Gambusia* (Cyprinodontiformes, Poeciliidae). American Museum Novitates 2951.

Ravit, B. 2008. Treatment wetlands in the Hackensack Meadows? Hackensack Tidelines 11(2):10.

Ravit, B., K. Cooper, B. Buckley, M. Comi and E. McCandlish. 2014. Improving management support tools for reintroducing bivalve species (eastern oyster [*Crassostrea virginica* Gmelin]) in urban estuaries. Integrated Environmental Assessment and Management 10(4):555–565.

Reed, J.M., L.W. Oring and E.M. Gray. 2013. Spotted sandpiper (*Actitis macularius*), Version 1.0, Birds of the World. Cornell Lab of Ornithology, Ithaca, NY. birdsoftheworld.org/bow/species/sposan/cur/introduction (22 January 2020).

Reeds, C.A. 1933. The varved clays and other glacial features in the vicinity of New York City. P. 52–63 in C.P. Berey, ed. XVI International Geological Congress Guidebook 9.

Reich, L.M. 1981. *Microtus pennsylvanicus*. Mammalian Species 159:1–8.

Reichmuth, J.M., R. Roudez, T. Glover and J.S. Weis. 2009. Differences in prey capture behavior in populations of blue crab (*Callinectes sapidus* Rathbun) from contaminated and clean estuaries in New Jersey. Estuaries and Coasts 32:298–308.

Reid, F.A. 1985. Wetland invertebrates in relation to hydrology and water chemistry. P. 72–79 in M.D. Knighton, ed. Water Impoundments for Wildlife: A Habitat Management Workshop. Forest Service General Technical Report NC–100, U.S. Department of Agriculture, St. Paul, MN.

Reid, F.A. 1989. Differential habitat use by waterbirds in a managed wetland complex. PhD thesis, University of Missouri, Columbia, MO.

Reif, J. 2013. Long-term trends in bird populations: A review of patterns and potential drivers in North America and Europe. Acta Ornithologica 48:1–16.

Reinert, S.E. and F.C. Golet. 1979. Breeding ecology of the swamp sparrow in a southern Rhode Island peatland. Transactions of the Northeast Section of the Wildlife Society 1986:1–13.

Rey, J.R., W.E. Walton, R.J. Wolfe, C.R. Connelly, S.M. O'Connell, J. Berg, G.E. Sakolsky-Hoopes and A.D. Laderman. 2012. North American wetlands and mosquito control. International Journal of Environmental Research and Public Health 9(12):4537–4605.

Rhoads, A.F. and T.A. Block. 2000. The plants of Pennsylvania. University of Pennsylvania Press, Philadelphia, PA. 1061 p.

Richardson, D.H. and R.P Cameron. 2004. Cyanolichens: Their response to pollution and possible management strategies for their conservation in northeastern North America. Northeastern Naturalist 11(1):1–23.

Richardson, S.J. and L.R. Walker. 2010. Nutrient ecology of ferns. P. 111–139 in K. Mehltreter, L.R. Walker and J.M. Sharpe, eds. Fern Ecology. Cambridge University Press, Cambridge, U.K.

Riedel, P., M. Navrátil, I.H. Tuf and J. Tufová. 2009. Terrestrial isopods (Isopoda: Oniscidea) and millipedes (Diplopoda) of the city of Olomouc (Czech Republic). P. 125–132 in K. Tajovský, J. Schlaghamerský, V. Pižl, eds. Contributions to Soil Zoology in Central Europe III, České Budějovice, Czech Republic.

Riffell, J.A., E. Shlizerman, E. Sanders, L. Abrell, B. Medina, A.H Hinterwirth and J.N. Kutz. 2014. Flower discrimination by pollinators in a dynamic chemical environment. Science 344(6191):1515–1518.

Riley, S.P.D. and P.A. White. 2010. Gray foxes (*Urocyon cinereoargenteus*) in urban areas. P. 197–198 in S.D. Gehrt, S.P.D. Stanley and B.L. Cypher, eds. Urban Carnivores: Ecology, Conflict, and Conservation. Johns Hopkins University Press, Baltimore, MD.

Robbins, C.S., D.K. Dawson and B.A. Dowell. 1989. Habitat area requirements of breeding forest birds of the Middle Atlantic states. Wildlife Monographs 103. 34 p.

Robbins, S.D., Jr. 1991. Wisconsin birdlife: Population and distribution past and present. University of Wisconsin Press, Madison, WI. 702 p.

Roberts, M.R. and F.S. Gilliam. 2003. Response of the herbaceous layer to distur-
bance in eastern forests. P. 302–320 in F.S. Gilliam and M.R. Roberts, eds. The
Herbaceous Layer in Forests of Eastern North America. Oxford University Press,
Oxford, U.K.

Roberts-Semple, D. and Y. Gao. 2017. Pollution and climate-induced respiratory
health effects: Do national standards adequately provide a safe threshold for NO_x
and O_3 in urban New Jersey? Athens Journal of Sciences 4(4):301–322.

Robichaud, B. and M.F. Buell. 1989. Vegetation of New Jersey. Rutgers University
Press, New Brunswick, NJ.

Robinson, G.R. and S.N. Handel. 1993. Forest restoration on a closed landfill: Rapid
addition of new species by bird dispersal. Conservation Biology 7:271–278.

Robinson, G.R., S.N. Handel and J. Mattei. 2002. Experimental techniques for evalu-
ating the success of restoration projects. Korean Journal of Ecology 25(1):1–7.

Rogers, D.C. 2021. Conservation status of the large branchiopods (Branchiopoda:
Anostraca, Notostraca, Laevicaudata, Spinicaudata, Cyclestherida). P. 221–238
in T. Kawai and D.C. Rogers, eds. Recent Advances in Freshwater Crustacean
Biodiversity and Conservation. CRC Press, Boca Raton, FL.

Rohwer, F.C., W.P. Johnson and E.R. Loos. 2002. Blue-winged teal (*Spatula discors*),
Version 1.0, Birds of the World. Cornell Lab of Ornithology, Ithaca, NY. birdsofthe
world.org/bow/species/buwtea/cur/introduction (23 January 2020).

Rosatte, R.C., M.J. Power and C.D. MacInnes. 1991. Ecology of urban skunks, raccoons, and foxes in metropolitan Toronto. P. 31–38 in L.W. Adams and D.L. Leedy,
eds. Wildlife Conservation in Metropolitan Environments. National Institute for
Urban Wildlife, Columbia, MD.

Rosatte, R.C. and S. Larivière. 2003. Skunks (genera *Mephitis*, *Spilogale* and
Conepatus). P. 692–707 in G.A. Feldhamer, B.C. Thompson and J.A. Chapman,
eds. Wild Mammals of North America: Biology, Management, and Conservation.
Johns Hopkins University Press, Baltimore, MD.

Rosatte, R., K. Sobey, J.W. Dragoo and S.D. Gehrt. 2010. Striped skunks and allies
(*Mephitis* spp.). P. 94–106 in S.D. Gehrt, S.P.D. Stanley and B.L. Cypher, eds.
Urban Carnivores: Ecology, Conflict, and Conservation. Johns Hopkins University
Press, Baltimore, MD.

Rosenberg, K.V., A.M. Dokter, P.J. Blancher, J.R. Sauer, A.C. Smith, P.A. Smith, J.C.
Stanton, A. Panjabi, L. Helft, M. Parr and P.P. Marra. 2019. Decline of the North
American avifauna. Science 366:120–124.

Rosenberg, K.V., S.B Terrell and G.H. Rosenberg. 1987. Value of suburban habitats
to desert riparian birds. Wilson Bulletin 99:642–654.

Rosenwinkel, E.R. 1964. Vegetational history of a New Jersey tidal marsh, bog and
vicinity. Bulletin of the New Jersey Academy of Science 9(1):1–20.

Rosenzweig, C., W.D. Solecki, L. Parshall, M. Chopping, G. Pope and R. Goldberg.
2005. Characterizing the urban heat island in current and future climates in New
Jersey. Global Environmental Change Part B: Environmental Hazards 6(1):51–62.

Rosza, R., T. Halavik and R. Jacobson. 2001. Environmental management issues on
the lower Connecticut River. P. 68–76 in G.D. Dreyer and M. Caplis, eds. Living

Resources and Habitats of the Lower Connecticut River. Connecticut College Arboretum Bulletin 37.

Rottenborn, S.C. 1999. Predicting the impacts of urbanization on riparian bird communities. Biological Conservation 88:289–299.

Rue, D.J. and A. Traverse. 1997. Pollen analysis of the Hackensack, New Jersey Meadowlands tidal marsh. Northeastern Geology and Environmental Science 19(3):211–215.

Rush, S.A., K.F. Gaines, W.R. Eddleman and C.J. Conway. 2018. Clapper rail (*Rallus crepitans*), Version 1.0, Birds of the World. Cornell Lab of Ornithology, Ithaca, NY. birdsoftheworld.org/bow/species/clarai11/cur/introduction (22 January 2020).

Sabo, J.L., R. Sponseller, M. Dixon, K. Gade, T. Harms, J. Heffernan, A. Jani, G. Katz, C. Soykan, J. Watts and J. Welter. 2005. Riparian zones increase regional species richness by harboring different, not more, species. Ecology 86:56–62.

Safe, S. 1994. Polychlorinated biphenyls (PCBs): Environmental impact, biochemical and toxic responses, and implications for risk assessment. Critical Reviews in Toxicology 24:87–194.

Sakatos, M.J. 2006. The small mammal community in a restored area of the Fresh Kills Landfill. MS thesis, Rutgers University, New Brunswick, NJ. 52 p.

Šálek, M., L. Drahníková and E. Tkadlec. 2015. Changes in home range sizes and population densities of carnivore species along the natural to urban habitat gradient. Mammal Review 45:1–14.

Sandifer, P.A., J.V. Miglarese, D.R. Calder, J.J. Manzi, L.A. Barclay, eds. 1980. Ecological characterization of the Sea Island coastal region of South Carolina and Georgia. Volume 3. Biological features of the characterization areas. U.S. Fish and Wildlife Service office of Biological Services, Washington, DC. FWS/OBS-79/42.

Sankey, J.H.P. 1988. Provisional atlas of the harvest-spiders (Arachnida: Opiliones) of the British Isles. Biological Records Centre, Institute of Terrestrial Ecology, Abbots Ripton, Cambridgeshire, England. 42 p.

Savard, J.P.L. and J.B. Falls. 2001. Survey techniques and habitat relationships of breeding birds in residential areas of Toronto, Canada. P. 543–567 in J. Marzluff, R. Bowman and R. Donnelly, eds. Avian Ecology and Conservation in an Urbanizing World. Springer, Boston, MA.

Schlauch, F.C. 1978. Urban geographical ecology of the amphibians and reptiles of Long Island. P. 25–41 in C.M. Kirkpatrick, ed. Wildlife and People. Department of Forestry and Natural Resources and the Cooperative Extension Service, Purdue University, West Lafayette, IN.

Schlesinger, M.D., J.A. Feinberg, N.H. Nazdrowicz, J.D. Kleopfer, J.C. Beane, J.F. Bunnell, J. Burger, E. Corey, K. Gipe, J.W. Jaycox, E. Kiviat, J. Kubel, D.P. Quinn, C. Raithell, P.A. Scott, S.M. Wenner, E.L. White, B. Zarate and H.B. Shaffer. 2018. Follow-up ecological studies for cryptic species discoveries: Decrypting the leopard frogs of the eastern US. PloS One 13(11):e0205805.

Schmidt, R.E. 1993. Fish fauna of Manitou Marsh with comments on other aquatic organisms. Report to Museum of the Hudson Highlands. 63 p.

Schmidt, R.E. and A.B. Bragin. 2021. Extension of the range of the American fresh-water goby, *Ctenogobius shufeldti* (Pisces: Gobiidae) to northern coastal New Jersey. Bulletin of the New Jersey Academy of Science 66:1–2.

Schmidt, R.E. and E. Kiviat. 2007. State records and habitat of clam shrimp, *Caenestheriella gynecia* (Crustacea: Conchostraca), in New York and New Jersey. Canadian Field-Naturalist 121:128–132.

Schmidt, R.E. and P.A. Moccio. 2013. Description of the larvae of the feather blenny, *Hypsoblennius hentz* (Pisces: Blenniidae), from New York waters. Zootaxa 3646:581–586.

Schmidt, R.E. and J.J. Wright. 2018. Documentation of *Myrophis punctatus* (speckled worm eel) from marine waters of New York. Northeastern Naturalist 25:N1–3.

Schoettle, T. 1996. A guide to a Georgia barrier island. Watermarks Publishing, St. Simons Island, GA. 160 p.

Scholte, E.J., B.G. Knols, R.A. Samson and W. Takken. 2004. Entomopathogenic fungi for mosquito control: A review. Journal of Insect Science 4(1):e19.

Schuberth, C.J. 1968. The geology of New York City and environs. Natural History Press, Garden City, NY. 304 p.

Schuster, R.M. 1949. The ecology and distribution of Hepaticae in central and west-ern New York. American Midland Naturalist 42(3):513–712.

Schwab, D. and J. Shelton. 1981. The herpetofauna of the Brunswick area of Glynn County, Georgia. Bulletin of the Chicago Herpetological Society 16(3):67–69.

Schwartz, V. and D.M. Golden. 2002. Field guide to reptiles and amphibians of New Jersey. New Jersey Division of Fish and Wildlife, Endangered and Nongame Species Program, New Jersey Department of Environmental Protection, Trenton, NJ. 53 p.

Seewagen, C.L. and M. Newhouse. 2018. Mass changes and energetic condition of grassland and shrubland songbirds during autumn stopovers at a reclaimed landfill in the New Jersey Meadowlands. Wilson Journal of Ornithology 130(2):377–384.

Seigel, A., C. Hatfield and J.M. Hartman. 2005. Avian response to restoration of urban tidal marshes in the Hackensack Meadowlands, New Jersey. Urban Habitats 3:87–116. http://urbanhabitats.org/v03n01/avianresponse_pdf.pdf (9 September 2010).

Selbo, S.S. and A.A. Snow. 2004. The potential for hybridization between *Typha angustifolia* and *Typha latifolia* in a constructed wetland. Aquatic Botany 78:361–369.

Sewell, S.R. and C.P. Catterall. 1998. Bushland modification and styles of urban development: Their effects on birds in south-east Queensland. Wildlife Research 25(1):41–63.

Shapiro, A.M. 1970. The butterflies of the Tinicum region. P. 95–104 in J. McCormick, R.R. Grant and R. Patrick, eds. Two Studies of Tinicum Marsh. Conservation Foundation, Washington, DC.

Shapiro A.M. and A.R. Shapiro. 1973. The ecological associations of the butterflies of Staten Island. Journal of Research on the Lepidoptera 12:65–128.

Shaw, A.J. 1994. Adaptation to metals in widespread and endemic plants. Environmental Health Perspectives 102(12):105–108.

Sherwood, N.R. 2017. Risks associated with harvesting and human consumption of two turtle species in New Jersey. PhD thesis, Montclair State University, Montclair, NJ. 246 p.

Sheviak, C. and S. Young. 2010. Orchids of New York. New York State Conservationist 64(6):2–7.

Shih, J.G. and S.A. Finkelstein. 2008. Range dynamics and invasive tendencies in *Typha latifolia* and *Typha angustifolia* in eastern North America derived from herbarium and pollen records. Wetlands 28(1):1–16.

Shimshony, A. 2009. Tularemia: A threat to animals and humans. Infectious Disease News (August 2009). www.healio.com/infectious-disease/zoonotic-infections/news/print/infectious-disease-news/%7B548992f3-559d-433f-9f79-410b7cc0af58%7D/tularemia-a-threat-to-animals-and-humans (24 January 2020).

Shin, J.Y., F. Artigas, C. Hobble and Y.S. Lee. 2013. Assessment of anthropogenic influences on surface water quality in urban estuary, northern New Jersey: Multivariate approach. Environmental Monitoring and Assessment 185(3):2777–2794.

Shrestha, P., S. Su, S. James, P. Shaller, M. Doroudian, C. Firstenberg and C. Thompson. 2014. Conceptual site model for Newark Bay—Hydrodynamics and sediment transport. Journal of Marine Science and Engineering 2(1):123–139.

Silliman, B.R. and M.D. Bertness. 2004. Shoreline development drives invasion of *Phragmites australis* and the loss of plant diversity on New England salt marshes. Conservation Biology 18(5):1424–1434.

Sipple, W.S. 1972. The past and present flora and vegetation of the Hackensack Meadows. Bartonia (41):4–56 + folded map.

Skevington, J.H., M.M. Locke, A.D. Young, K. Moran, W.J. Crins and S.A. Marshall. 2019. Field guide to the flower flies of Northeastern North America. Princeton University Press, Princeton, NJ. 512 p.

Słaby, A. and M. Lisowska. 2012. Epiphytic lichen recolonization in the centre of Cracow (southern Poland) as a result of air quality improvement. Polish Journal of Ecology 60(2):225–240.

Slack, N.G. 1977. Species diversity and community structure in bryophytes: New York State studies. New York State Museum Bulletin 428. 70 p.

Slowik N. and A. Greller. 2009. Field trip reports: Palisades Interstate Park, Bergen Co., New Jersey, May 17, 2008. Journal of the Torrey Botanical Society 136(1):145–148.

Smiley, D. and C.J. George. 1974. Photographic documentation of lichen decline in the Shawangunk Mountains of New York. Bryologist 77:179–187.

Smith, K.J. and K.W. Able. 1994. Salt-marsh tide pools as winter refuges for the mummichog, *Fundulus heteroclitus*, in New Jersey. Estuaries 17:226–234.

Smith, D.G., T. Bosakowski and A. Devine. 1999. Nest site selection by urban and rural great horned owls in the Northeast. Journal of Field Ornithology 70:535–542.

Smith, S.B., S.A. DeSando and T. Pagano. 2013. The value of native and invasive fruit-bearing shrubs for migrating songbirds. Northeastern Naturalist 20:171–184.

Smith, K.G., S.R. Wittenberg, R.B. Macwhirter and K.L. Bildstein. 2011. Northern harrier (*Circus hudsonius*), Version 1.0, Birds of the World. Cornell Lab of

Ornithology, Ithaca, NY. birdsoftheworld.org/bow/species/norhar2/cur/introduction (22 January 2020).

Snell, S. 2010. Plant fact sheet seaside goldenrod (*Solidago sempervirens*). USDA-Natural Resources Conservation Service, Plant Materials Center. Cape May, NJ. plants.usda.gov/DocumentLibrary/factsheet/pdf/fs_sose.pdf (28 September 2021).

Soto-Cordero, L., A. Meltzer and J.C. Stachnik. 2018. Crustal structure, intraplate seismicity, and seismic hazard in the Mid-Atlantic United States. Seismological Research Letters 89:241–252.

Soulsbury, C.D., P.J. Baker, G. Iossa and S. Harris. 2010. Red foxes (*Vulpes vulpes*). P. 62–75 in S.D. Gehrt, S.P.D. Stanley and B.L. Cypher, eds. Urban Carnivores: Ecology, Conflict, and Conservation. Johns Hopkins University Press, Baltimore, MD.

Soulsbury, C.D., G. Iossa, P.J. Baker, N.C. Cole, S.M. Funk and S. Harris. 2007. The impact of sarcoptic mange *Sarcoptes scabei* on the British fox *Vulpes vulpes* population. Mammal Review 37:278–296.

Sousa, R., C. Antunes and L. Guilhermino. 2008. Ecology of the invasive Asian clam *Corbicula fluminea* (Müller, 1774) in aquatic ecosystems: An overview. Annales de Limnologie-International Journal of Limnology 44(2):85–94.

South, A. 1992. Terrestrial slugs: Biology, ecology and control. Chapman and Hall, London, U.K. 428 p.

Staniaszek-Kik, M., D. Chmura and J. Żarnowiec. 2019. What factors influence colonization of lichens, liverworts, mosses and vascular plants on snags? Biologia 74:375–384.

Statham, M.J., B.N. Sacks, K.B. Aubry, J.D. Perrine and S.M. Wisely. 2012. The origin of recently established red fox populations in the United States: Translocations or natural range expansions? Journal of Mammalogy 93(1):52–65.

Stein, B.A., L.S. Kutner and J.S. Adams. 2000. Precious heritage: The status of biodiversity in the United States. Oxford University Press, Oxford, U.K. 432 p.

Steinberg, N., D.J. Suszkowski, L. Clark and J. Way. 2004. Health of the harbor: The first comprehensive look at the state of the NY/NJ harbor estuary. A report to the NY/NJ Harbor Estuary Program. Hudson River Foundation, New York, NY. 82 p.

Stinnette, I., M. Taylor, L. Kerr, R. Pirani, S. Lipuma and J. Lodge. 2018. State of the estuary 2018. Hudson River Foundation. New York, NY. 79 p.

Stone, B.D., S.D. Stanford and R.W. Witte. 2002. Surficial geologic map of northern New Jersey. Map I-2540-C. U.S. Geological Survey Miscellaneous Investigations Series, Denver, CO. Large format map, scale 1:100,000.

Stone, W. 1908. The mammals of New Jersey. P. 33–110 in New Jersey State Museum Annual Report for 2007. New Jersey State Museum, Trenton, NJ.

Stotler, R.E. and B.J. Crandall-Stotler. 2006. Bryophytes. Southern Illinois University, Department of Plant Biology, Carbondale, IL. http://bryophytes.plant.siu.edu/index.html (3 April 2010).

Strayer, D.L. 2008. Freshwater mussel ecology. University of California Press, Berkeley, CA. 204 p.

Strayer, D.L. and K.J. Jirka. 1997. The pearly mussels of New York State. New York State Museum Memoir 26. 113 p. + 27 plates.

Stuckey, I.H. and L.L. Gould. 2000. Coastal plants from Cape Cod to Cape Canaveral. University of North Carolina Press, Chapel Hill, NC. 305 p.

Studlar, S.M. 1982. Host specificity of epiphytic bryophytes near Mountain Lake, Virginia. Bryologist 85(1):37–50.

Studlar, S.M. and J.E. Peck. 2009. Extensive green roofs and mosses: Reflections from a pilot study in Terra Alta, West Virginia. Evansia 26(2):52–63.

Su, T., Y. Jiang and M.S. Mulla. 2014. Toxicity and effects of mosquito larvicides methoprene and surface film (Agnique® MMF) on the development and fecundity of the tadpole shrimp *Triops newberryi* (Packard) (Notostraca: Triopsidae). Journal of Vector Ecology 39(2):340–346.

Sugihara, T., C. Yearsley, J.B. Durand and N.P. Psuty. 1979. Comparison of natural and altered estuarine systems: Analysis. Center for Coastal and Environmental Studies, Rutgers University, New Brunswick, NJ. CCES NJ/RU-DEP-11–9–79. 247 p.

Sukopp, H., H.P. Blume and W. Kunick. 1979. The soil, flora, and vegetation of Berlin's waste lands. P. 115–132 in I.C. Laurie, ed. Nature in Cities: The Natural Environment in the Design and Development of Urban Green Space. Wiley, Chichester, NY.

Sullivan, R. 1998a. I sing the Meadowlands. New York Times Magazine (15 February).

Sullivan, R. 1998b. The Meadowlands: Wilderness adventures at the edge of a city. Scribner, New York, NY. 220 p.

Surasinghe, T. 2013. Influences of riparian land-uses on habitat use and interspecific competition of stream-dwelling salamanders: Evidence from Blue Ridge & Piedmont. PhD thesis, Clemson University, Clemson, SC. 141 p.

Suszkowski, D.J. 1978. Sedimentology of Newark Bay, New Jersey: An urban estuarine bay. PhD thesis, University of Delaware, Newark, DE. 212 p.

Suthers, H.B., J.M. Bickal and P.G. Rodewald. 2000. Use of successional habitat and fruit resources by songbirds during autumn migration in central New Jersey. Wilson Journal of Ornithology 112(2):249–260.

Swanson, G.A., M.I. Meyer and J.R. Serie. 1974. Feeding ecology of breeding blue-winged teals. Journal of Wildlife Management 38:396–407.

Swarth, C. and E. Kiviat. 2009. Animal communities in North American tidal freshwater wetlands. P. 71–88 in A. Barendregt, D. Whigham and A. Baldwin, eds. Tidal Freshwater Wetlands. Backhuys Publishers, Leiden, The Netherlands.

Sweeney, J., L. Deegan and R. Garritt. 1998. Population size and site fidelity of *Fundulus heteroclitus* in a macrotidal saltmarsh creek. Biological Bulletin 195:238–239.

Swihart, R.K. and P.M. Picone. 1995. Use of woodchuck burrows by small mammals in agricultural habitats. American Midland Naturalist 133:360–363.

Sykes, L.R., J.G. Armbruster, W.-Y. Kim and L. Seeber. 2008. Observations and tectonic setting of historic and instrumentally located earthquakes in the greater

New York City-Philadelphia area. Bulletin of the Seismological Society of America 98:1696–1719.

Syvitski, J.P., A.J. Kettner, I. Overeem, E.W. Hutton, M.T. Hannon, G.R. Brakenridge, J. Day, C. Vörösmarty, Y. Saito, L. Giosan and R.J. Nicholls. 2009. Sinking deltas due to human activities. Nature Geoscience 2(10):681–686.

Takahashi, R.M., W.H. Wilder and T. Miura. 1984. Field evaluations of ISA-20 E for mosquito control and effects on aquatic nontarget arthropods in experimental plots. Mosquito News 44(3):363–367.

Tartaglia, E.S. 2013. Hawkmoth-flower interactions in the urban landscape: Sphingidae ecology, with a focus on the genus *Hemaris*. PhD thesis, Rutgers University, New Brunswick, NJ. 82 p.

Tashiro, J.S., R.E. Schmidt, E. Kiviat and D.R. Roeder. 1991. Baseline assessment of the Saugus River system, Massachusetts. Report to New England Natural Resources Research Center (Fund for New England) and Massachusetts Public Interest Research Group. Hudsonia Ltd, Annandale, NY. 2 volumes.

Taylor, N. 1915. Flora of the vicinity of New York: A contribution to plant geography. Memoirs of the New York Botanical Garden 5. 683 p.

Tedrow, J.C.F. 1986. Soils of New Jersey. Krieger Publishing Co., Malabar, FL. 512 p.

Telford, S.R. 1993. Forum: Perspectives on the environmental management of ticks and Lyme disease. P. 164–167 in H.S. Ginsberg, ed. Ecology and Environmental Management of Lyme Disease. Rutgers University Press, New Brunswick, NJ.

Tesky, J.L. 1992. *Calamagrostis canadensis*. Fire Effects Information System (FEIS), U.S. Department of Agriculture, Forest Service, Rocky Mountain Research Station, Fire Sciences Laboratory. www.feis-crs.org/feis/ (24 May 2010).

Theakstone, W.H. and A.D. Knighton. 1979. The moss *Aongstroemia longipes*, an environmentally sensitive colonizer of sediments at Austerdalsisen, Norway. Arctic and Alpine Research 11(3):353–356.

Thomson, J.W. 1967. The lichen genus *Cladonia* in North America. University of Toronto Press, Toronto, Ontario, Canada. 172 p.

Tigas, L.A., D.H. Van Vuren and R.M. Sauvajot. 2002. Behavioral responses of bobcats and coyotes to habitat fragmentation and corridors in an urban environment. Biological Conservation 108:299–306.

Tiner, R.W., Jr. 1985. Wetlands of New Jersey. U.S. Fish and Wildlife Service, Newton Corner, MA. 117 p.

Tiner, R. 1998. In search of swampland: A wetland sourcebook and field guide. Rutgers University Press, New Brunswick, NJ. 264 p.

Tiner, R.W., J.Q. Swords and B.J. McLain. 2002. Wetland status and trends for the Hackensack Meadowlands: An assessment report from the National Wetlands Inventory Program. U.S. Fish and Wildlife Service, Hadley, MA. 29 p. www.fws .gov/wetlands/Documents%5CWetland-Status-And-Trends-For-The-Hackensack -Meadowlands.pdf (14 September 2019).

Tiner R.W., J.Q. Swords and H.C. Bergquist. 2005. The Hackensack Meadowlands District: Wetland inventory and remotely-sensed assessment of "natural habitat" integrity. U.S. Fish and Wildlife Service, National Wetlands Inventory, Ecological

Services, Region 5, Hadley, MA. 31 p. + appendix. www.fws.gov/wetlands/Documents/Hackensack-Meadowlands-District-Wetland-Inventory-and-Remotely-sensed-Assessment-of-Natural-Habitat-Integrity.pdf (2 September 2019).

Toft, S. 2018. Ups and downs among Danish urban harvestmen. Arachnology 17(8):394–399.

Toland, B. 1986. Hunting success of some Missouri raptors. Wilson Bulletin 98(1):116–125.

Torrey, J.C. 1819. A catalogue of plants growing spontaneously within thirty miles of the City of New-York. Lyceum of Natural History of New-York. Websters and Skinners, Albany, NY. 100 p.

Tousignant, M.Ê., S. Pellerin and J. Brisson. 2010. The relative impact of human disturbances on the vegetation of a large wetland complex. Wetlands 30(2):333–344.

Toweill, D.E. and J.E. Tabor. 1982. The northern river otter. P. 688–703 in J.A. Chapman and G.A. Feldhamer, eds. Wild Mammals of North America: Biology, Management, and Economics. Johns Hopkins University, Baltimore, MD.

Townroe, S. and A. Callaghan. 2014. British container breeding mosquitoes: The impact of urbanisation and climate change on community composition and phenology. PloS One 9(4):e95325.

Trattner, R.B. and C.P. Mattson. 1976. Nitrogen budget determination in the Hackensack Meadowlands estuary. Journal of Environmental Science & Health Part A 11(8–9):549–565.

Trewhella, W.J. and S. Harris. 1990. The effect of railway lines on urban fox (*Vulpes vulpes*) numbers and dispersal movements. Journal of Zoology 221:321–326.

Trocki, C.L. and P.W.C. Paton. 2006. Assessing habitat selection by foraging egrets in salt marshes at multiple spatial scales. Wetlands 26:307–312.

Tsipoura, N., J. Burger, M. Newhouse, C. Jeitner, M. Gochfeld and D. Mizrahi. 2011. Lead, mercury, cadmium, chromium, and arsenic levels in eggs, feathers, and tissues of Canada geese of the New Jersey Meadowlands. Environmental Research 111:775–784.

Tsipoura, N., D. Mizrahi and K. Witkowski. 2006. Avian abundance and distribution in the New Jersey Meadowlands District: The importance of habitat, landscape, and disturbance. Interim report to the New Jersey Meadowlands Commission.

Tsipoura, N., K. Mylecraine and K. Ruskin. 2009. Ecology of colonial wading birds foraging in the Meadowlands District: 2008 Final Report. Submitted to New Jersey Meadowlands Commission. New Jersey Audubon Society, Bernardsville, NJ.

Turner, K., L. Lefler and B. Freedman. 2005. Plant communities of selected urbanized areas of Halifax, Nova Scotia, Canada. Landscape and Urban Planning 71(2–4):191–206.

Tuskes, P.M., J.P. Tuttle and M.M. Collins. 1996. The wild silk moths of North America: A natural history of the Saturniidae of the United States and Canada. Cornell University Press, Ithaca, NY. 250 p.

Tzilkowski, W.M., J.S. Wakeley and L.J. Morris. 1986. Relative use of municipal street trees by birds during summer in State College, Pennsylvania. Urban Ecology 9(3–4):387–398.

Uno, S., J. Cotton and S.M. Philpott. 2010. Diversity, abundance, and species composition of ants in urban green spaces. Urban Ecosystems 13(4):425–441.

Urffer, K. 2002. Wildlife profile: Diamondback terrapin. Hackensack Tidelines (Spring):8–9.

Urner, C.A. 1923. Notes on the short-eared owl. Auk 40(1):30–36.

USACE (US Army Corps of Engineers). 1997. Final Environmental Impact Statement on the Newark Bay Confined Disposal Facility. US Army Corps of Engineers, New York, NY. www.nj.gov/dep/passaicdocs/docs/NJDOTSupportingCosts/CDF -REPORT-EIS-FINAL-4-1997_USACE-0009153.pdf (20 January 2020).

USACE (US Army Corps of Engineers). 2000. Draft Environmental Impact Statement on the Meadowlands Mills Project proposed by Empire Ltd. US Army Corps of Engineers, New York District, New York, NY.

USACE (US Army Corps of Engineers). 2004. Meadowlands Environmental Site Investigation Compilation (MESIC). Hudson-Raritan estuary Hackensack Meadowlands, New Jersey. USACOE, New York, NY. meri.njmeadowlands.gov/ mesic/title-page-2/ (1 September 2019).

USDA (US Department of Agriculture, Natural Resources Conservation Service). 2021. The plants database. National Plant Data Team, Greensboro, NC. http:// plants.usda.gov/index.html (28 September 2021).

USFWS (US Fish and Wildlife Service). 2007. Hackensack Meadowlands Initiative: Preliminary conservation planning. US Fish and Wildlife Service, Pleasantville, NJ. 422 p. + appendix. http://www.fws.gov/northeast/njfieldoffice/PCP_2007/ Hack_Meadow_Initiative_PCP_MAR2007.pdf (9 September 2010).

USGS (US Geological Survey). 1967. Weehawken N.J. – N.Y. 7.5 minute topographic map sheet, scale 1:24,000, contour interval 10 feet. USGS, Washington, DC.

Utberg, G.L. and D.J. Sutherland. 1982. The temporal distribution of *Chironomus decorus* (Chironomidae) in northern New Jersey, 1979. New York Entomological Society 90(1):16–25.

Valiela, I., J.E. Wright, J.M. Teal and S.B. Volkmann. 1977. Growth, production and energy transformations in the salt-marsh killifish *Fundulus heteroclitus*. Marine Biology (Berlin) 40:135–144.

Van Baars, S. and I.M. Van Kempen. 2009. The causes and mechanisms of historical dike failures in the Netherlands. E-Water Official Publication of the European Water Association. 14 p.

Van Hassel, J.H. and J.L. Farris. 2007. A review of the use of unionid mussels as biological indicators of ecosystem health. P. 19–49 in J.L. Farris and J.H. Van Hassel, eds. Freshwater Bivalve Ecotoxicology. CRC Press, Boca Raton, FL.

van Heezik, Y., A. Smyth, A. Adams and J. Gordon. 2010. Do domestic cats impose an unsustainable harvest on urban bird populations? Biological Conservation 143:121–130.

van Houten, F.B. 1969. Late Triassic Newark Group, north central New Jersey and adjacent Pennsylvania and New York. P. 314–347 in S. Subitzky, ed. Geology of Selected Areas in New Jersey and Eastern Pennsylvania and Guidebook of Excursions. Rutgers University Press, New Brunswick, NJ.

VanDruff, L.W. and R.N. Rowse. 1986. Habitat association of mammals in Syracuse, New York. Urban Ecology 9:413–434.

Vellak, K., J. Liira, E. Karofeld, O. Galanina, M. Noskova and J. Paal. 2014. Drastic turnover of bryophyte vegetation on bog microforms initiated by air pollution in northeastern Estonia and bordering Russia. Wetlands 34(6):1097–1108.

Venn, S.J. and J.K. Niemelä. 2004. Ecology in a multidisciplinary study of urban green space: The URGE project. Boreal Environment Research 9:479–489.

Vennum, T., Jr. 1988. Wild rice and the Ojibway people. Minnesota Historical Society Press, St. Paul, MN. 357 p.

Vermeule, C.C. 1897. Drainage of the Hackensack and Newark tide-marshes. P. 287–318 in Annual Report of the State Geologist for 1896. MacCrellish and Quigley, Trenton, NJ. (Cited in Tedrow 1986; original not seen.)

Vezzani, D. 2007. Artificial container-breeding mosquitoes and cemeteries: A perfect match. Tropical Medicine & International Health 12(2):299–313.

Vilisics, F., D. Bogyó, T. Sattler and M. Moretti. 2012. Occurrence and assemblage composition of millipedes (Myriapoda, Diplopoda) and terrestrial isopods (Crustacea, Isopoda, Oniscidea) in urban areas of Switzerland. ZooKeys (176):199–214.

Vince, S., I. Valiela, N. Backus and J.M. Teal. 1976. Predation by the salt marsh killifish *Fundulus heteroclitus* (L.) in relation to prey size and habitat structure: Consequences for prey distribution and abundance. Journal of Experimental Marine Biology and Ecology 23:255–266.

Vitt, D.H., K. Wieder, L.A. Halsey and M. Turetsky. 2003. Response of *Sphagnum fuscum* to nitrogen deposition: A case study of ombrogenous peatlands in Alberta, Canada. Bryologist 106(2):235–245.

von Hake, C.A. 2009. New Jersey earthquake history. Abbreviated from Earthquake Information Bulletin 7(2), 1975. U.S. Geological Survey. pubs.er.usgs.gov/publication/70178962 (20 January 2020).

Vos-Wein, N. 2007. Ospreys fledge! Meadowlands history is made as chicks leave Jersey City nest. Hackensack Tidelines 10(3):1, 20.

Waddell, D.C. and M.L. Kraus. 1990. Effects of CuCl$_2$ on the germination response of two populations of the saltmarsh cordgrass, *Spartina alterniflora*. Bulletin of Environmental Contamination and Toxicology 44(5):764–769.

Wade, P.M. 1990. The colonization of disturbed freshwater habitats by Characeae. Folia Geobotanica 25(3):275–278.

Wakin, D.J. 2001. Brush fires in the Meadowlands disrupt road and train traffic. New York Times (30 April). http://www.nytimes.com/2001/04/30/nyregion/30FIRE.html (21 August 2010).

Waksman, S.A. 1942–43. The peats of New Jersey and their utilization. Bulletin 55, Parts A and B, Geological Series, New Jersey Department of Conservation and Development, Trenton, NJ.

Waldman, J.R., T.R. Lake and R.E. Schmidt. 2006. Biodiversity and zoogeography of the fishes of the Hudson River watershed and estuary. P. 129–150 in J.R. Waldman, K.E. Limburg and D. Strayer, eds. Hudson River Fishes and Their Environment. American Fisheries Society Symposium 51.

Walsh, J., V. Elia, R. Kane and T. Halliwell. 1999. Birds of New Jersey. New Jersey Audubon Society, Bernardsville, NJ. 704 p.

Walsh, G.C., M. Maestro, Y.M. Dalto, R. Shaw, M. Seier, G. Cortat and D. Djeddour. 2013. Persistence of floating pennywort patches (*Hydrocotyle ranunculoides*, Araliaceae) in a canal in its native temperate range: Effect of its natural enemies. Aquatic Botany 110:78–83.

Walsh, S. and R. Miskewitz. 2013. Impact of sea level rise on tide gate function. Journal of Environmental Science and Health Part A 48(4):453–463.

Wander, S.A. and W. Wander. 1995. Biological assessment of potential impacts to the peregrine falcon associated with the Special Area Management Plan for the Hackensack Meadowlands District. Report to Environmental Impacts Branch, U. S. Environmental Protection Agency, Region II, New York, NY.

Wargo, J.G. 1989. Avian species richness: A natural marsh vs. an enhanced marsh. MS thesis, Rutgers University, New Brunswick, NJ. 95 p.

Waters, D.P. and J.C. Lendemer. 2019. A revised checklist of the lichenized, lichenicolous and allied fungi of New Jersey. Bartonia (70):1–62.

Watts, A.W., T.P. Ballestero and K.H. Gardner. 2008. Soil and atmospheric inputs to PAH concentrations in salt marsh plants. Water Air and Soil Pollution 189:253–263.

Watts, B.D. 1989. Nest-site characteristics of yellow-crowned night-herons in Virginia. Condor 91:979–983.

Watts, B.D. 2011. Yellow-crowned night-heron (*Nyctanassa violacea*), Version 1.0, Birds of the World. Cornell Lab of Ornithology, Ithaca, NY. birdsoftheworld.org/b ow/species/ycnher/cur/introduction (22 January 2020).

Webber, H.M. and T.A. Haines. 2003. Mercury effects on predator avoidance behavior of a forage fish, golden shiner (*Notemigonus crysoleucas*). Environmental Toxicology and Chemistry 22:1556–1561.

Wegner, J.F. and G. Merriam. 1979. Movements of birds and small mammals between a wood and adjoining farmland habitat. Journal of Applied Ecology 16:349–357.

Weis, J.S. 2000. Habitat and food value of *Phragmites australis* and *Spartina alterniflora* for fiddler crabs, grass shrimp, and larval mummichogs. Water Resources Research Institute (Newsletter) (Fall):3, 6–7.

Weis, J.S. 2012. Walking sideways: The remarkable world of crabs. Cornell University Press, Ithaca, NY. 256 p.

Weis, J.S. and A. Candelmo. 2012. Pollutants and fish predator/prey behavior: A review of laboratory and field approaches. Current Zoology 58:9–20.

Weis, J.S., J. Skurnick and P. Weis. 2004. Studies of a contaminated brackish marsh in the Hackensack Meadowlands of northeastern New Jersey: Benthic communities and metal contamination. Marine Pollution Bulletin 49:1025–1035.

Weis, J.S. and P. Weis. 2000. Behavioral responses and interactions of estuarine animals with an invasive marsh plant: A laboratory analysis. Biological Invasions 2:305–314.

Weis, J.S., L. Windham, C. Santiago-Bass and P. Weis. 2002. Growth, survival, and metal content of marsh invertebrates fed diets of detritus from *Spartina alterniflora* Loisel. and *Phragmites australis* Cav. Trin. *ex* Steud. from metal-contaminated and clean sites. Wetlands Ecology and Management 10(1):71–84.

Weis, P. and J.T.F. Ashley. 2007. Contaminants in fish of the Hackensack Meadowlands, New Jersey: Size, sex, and seasonal relationships as related to health risks. Archives of Environmental Contamination and Toxicology 52:80–89.

Weissinger, M.D., T.C. Theimer, D.L. Bergman and T.J. Deliberto. 2009. Nightly and seasonal movements, seasonal home range, and focal location photo-monitoring of urban striped skunks (*Mephitis mephitis*): Implications for rabies transmission. Journal of Wildlife Diseases 45(2):388–397.

Weller, M.W. 1961. Breeding biology of the least bittern. Wilson Bulletin 73:11–35.

Werme, C.E. 1981. Resource partitioning in a salt marsh community. PhD thesis, Boston University, Boston, MA. 132 p.

Wetmore, C.M. 1988. Lichen floristics and air quality. P. 55–65 in T.H. Nash III and V. Wirth, eds. Lichens, Bryophytes and Air Quality. Bibliotheca Lichenologica 30. J. Cramer, Berlin, Germany.

Whitaker, J.O., Jr. 1972. *Zapus hudsonius*. Mammalian Species 11:1–7.

Whitaker, J.O., Jr. 2004. *Sorex cinereus*. Mammalian Species 743:1–9.

Whitaker, J.O., Jr. and W.J. Hamilton, Jr. 1998. Mammals of the eastern United States. Cornell University Press, Ithaca, NY. 583 p.

White, R. 2005. Letters to Riverkeeper. Hackensack Tidelines 8(1):13.

Whittier, P. 1996. Extending the viability of *Equisetum hyemale* spores. American Fern Journal 86(4):114–118.

Widmer, K. 1964. The geology and geography of New Jersey. Van Nostrand, Princeton, NJ. 193 p.

Wilcox, B.A. and D.D. Murphy. 1985. Conservation strategy: The effects of fragmentation on extinction. American Naturalist 125:879–887.

Wilk, S.J., D.G. McMillan, R.A. Pikanowski, F.M. MacHaffle, A.L. Pacheco and L.L. Stehlik. 1997. Fish, macroinvertebrates, and associated hydrographic observations collected in Newark Bay, New Jersey, during May 1993-April 1994. Northeast Fisheries Science Center, Reference Document 97–10. 91 p.

Will, R. and L.J. Houston. 1992. Fish distribution survey of Newark Bay, New Jersey, May 1987-April 1988. P. 428–445 in C.L. Smith, ed. Estuarine Research in the 1980s. SUNY Press, Albany, NY.

Will-Wolf, S., S. Jovan, P. Neitlich, J.E. Peck and R. Rosentreter. 2015. Lichen-based indices to quantify responses to climate and air pollution across northeastern U.S.A. Bryologist 118:59–82.

Williams, K.E., K.E. Hodges and C.A. Bishop. 2015. Hibernation and oviposition sites of Great Basin gophersnakes (*Pituophis catenifer deserticola*) near the northern range limit. Journal of Herpetology 49(2):207–216.

Williams, P.H., R.W. Thorp, L.L. Richardson and S.R. Colla. 2014. Bumble bees of North America. Princeton University Press, Princeton, NJ. 208 p.

Willis, B.H., J.R. Mahoney and J.C. Goodrich. 1973. Hackensack Meadowlands air pollution study—Air quality impact of land use planning. Prepared for U.S. Environmental Protection Agency, Research Triangle Park, NC. EPA-450/3–74–056-e. 139 p.

Willis, O.R. 1877. Catalogue of plants growing without cultivation in the state of New Jersey, with a specific description of all the species of a violet found therein. A.S. Barnes Co., New York, NY. (Cited in Yuhas et al. 2005; original not seen.)

Willner, G.R., G.A. Feldhamer, E.E. Zucker and J.A. Chapman. 1980. *Ondatra zibethicus*. Mammalian Species 141:1–8.

Wilson, D.E., F.R. Cole, J.D. Nichols, R. Rudran and M.S. Foster, eds. 1996. Measuring and monitoring biological diversity: Standard methods for mammals. Smithsonian Institution Press, Washington, DC.

Wilson, W.H., Jr. 2008. The behavior of the seaside dragonlet, *Erythrodiplax berenice* (Odonata: Libellulidae), in a Maine salt marsh. Northeastern Naturalist 15(3):465–468.

Windham, L., J.S. Weis and P. Weis. 2001a. Lead uptake, distribution, and effects in two dominant salt marsh macrophytes, *Spartina alterniflora* (cordgrass) and *Phragmites australis* (common reed). Marine Pollution Bulletin 42:811–816.

Windham, L., J.S. Weis and P. Weis. 2001b. Patterns and processes of mercury release from leaves of two dominant salt marsh macrophytes, *Phragmites australis* and *Spartina alterniflora*. Estuaries 24:787–795.

Winner, W.E. 1988. Responses of bryophytes to air pollution. P. 141–173 in T.H. Nash III and V. Wirth, eds. Lichens, Bryophytes and Air Quality. Bibliotheca Lichenologica 30. J. Cramer, Berlin, Germany.

Winner, W.E., C.J. Atkinson and T.H. Nash III. 1988. Comparisons of SO_2 absorption capacities of mosses, lichens, and vascular plants in diverse habitats. P. 91–107 in T.H. Nash III and V. Wirth, eds. Lichens, Bryophytes and Air Quality. Bibliotheca Lichenologica 30. J. Cramer, Berlin, Germany.

Winston, T. 2019. New York City Audubon's Harbor Herons Project: 2019 nesting survey report. New York City Audubon, New York, NY. 46 p. http://www .nycaudubon.org/images/2019_HH_Full_Survey_Report.pdf (11 January 2020).

Wirth, V. 1988. Phytosociological approaches to air pollution monitoring with lichens. P. 91–107 in T.H. Nash III and V. Wirth, eds. Lichens, Bryophytes and Air Quality. Bibliotheca Lichenologica 30. J. Cramer, Berlin, Germany.

Witztum, A. and R. Wayne. 2015. Variation in fiber cables in the lacunae of leaves in hybrid swarms of *Typha* ×*glauca*. Aquatic Botany 124:39–44.

Wolfe, P.E. 1977. The geology and landscapes of New Jersey. Crane, Russak & Co., New York, NY. 351 p.

Wolfenden, J., P.A. Wookey, P.W. Lucas and T.A. Mansfield. 1992. Action of pollutants individually and in combination. P. 72–92 in J.R. Barker and D.T. Tingey, eds. Air Pollution Effects on Biodiversity. Springer US, New York, NY.

Wolseley, P., and P. James. 2002. Assessing the role of biological monitoring using lichens to map excessive ammonia (NH_3) deposition in the UK. P. 68–86 in Effects of NO_x and NH_3 on Lichen Communities and Urban Ecosystems. A Pilot Study. Imperial College & The Natural History Museum, London, U.K.

Woodhead, P.M.J. 1987. The structure of the fish community and distribution of major species in the lower Hudson estuary and New York Harbor. Final Report to the Hudson River Foundation, New York, NY. 178 p + appendix. http://

hudsonandharbor.org/wp-content/uploads/library/Woodhead_007_87R_014_final
_report.pdf (19 December 2021).

World Bank. 2019. Urban population (% of total population). The World Bank Group,
data.worldbank.org/indicator/SP.URB.TOTL.IN.ZS (8 December 2019).

Wright J. 2018. Birdwatcher: Plenty of ducks and more to see at Mehrhof Pond.
North Jersey (1 November) www.northjersey.com/story/life/columnists/2018/11
/01/birdwatchinbbbbg-new-jersey-mehrhof-pond/1773183002/ (3 July 2021).

Wyman, R.L. 1998. Experimental assessment of salamanders as predators of detrital food webs: Effects on invertebrates, decomposition and the carbon cycle.
Biodiversity and Conservation 7(5):641–50.

Yates, T.L. and D.J. Schmidly. 1978. *Scalopus aquaticus*. Mammalian Species
105:1–4.

Yozzo, D.L. and F. Ottman. 2003. New distribution records for the spotfin killifish,
Fundulus luciae (Baird), in the lower Hudson River estuary and adjacent waters.
Northeastern Naturalist 10:399–408.

Yu, C.H., L. Huang, J.Y. Shin, F. Artigas and Z. Fan. 2014. Characterization of concentration, particle size distribution, and contributing factors to ambient hexavalent
chromium in an area with multiple emission sources. Atmospheric Environment
94:701–708.

Yuhas, C., J.M. Hartman and J.S. Weis. 2005. Benthic communities in *Spartina
alterniflora*– and *Phragmites australis*–dominated salt marshes in the Hackensack
Meadowlands, New Jersey. Urban Habitats 3(1):158–191. http://www.urbanhabitats
.org/v03n01/benthic_full.html (23 January 2020).

Yurlina, M.E. 1998. Bee mutualists and plant reproduction in urban woodland restora-
tions. PhD thesis, Rutgers University, New Brunswick, NJ. 121 p.

Zappalorti, R.T and J.C. Mitchell. 2008. Snake use of urban habitats in the New Jersey
Pine Barrens. P. 355–359 in J.C. Mitchell, R.E.J. Brown and B. Bartholomew, eds.
Urban Herpetology. Society for the Study of Amphibians and Reptiles, Salt Lake
City, UT.

Zhang, J., J. Zhang, R. Liu, J. Gan, J. Liu and W. Liu. 2016. Endocrine-disrupting
effects of pesticides through interference with human glucocorticoid receptor.
Environmental Science & Technology 50(1):435–443.

Zhao, P. and K.E. Woeste. 2011. DNA markers identify hybrids between butternut (*Juglans cinerea* L.) and Japanese walnut (*Juglans ailantifolia* Carr.). Tree
Genetics & Genomes 7:511–533.

Zhou, T., H.B. John-Alder, J.S. Weis and P. Weis. 2000. Endocrine disruption:
Thyroid dysfunction in mummichogs (*Fundulus heteroclitus*) from a polluted habitat. Marine Environmental Research 50:393–397.

Zimmerman, G.L. and K.A. Mylecraine. 2003. Reconstruction of an old growth
Atlantic white cedar stand in the Hackensack Meadowlands of New Jersey:
Preliminary results. P. 125–135 in R.B. Atkinson et al., eds. Atlantic White Cedar
Restoration Ecology and Management. Proceedings of a Symposium held May 31
– June 2, 2000 at Christopher Newport University, Newport News, VA.

Zolotarev, M.P. and A.V. Nesterkov. 2015. Arachnids (Aranei, Opiliones) in meadows: Response to pollution with emissions from the Middle Ural copper smelter. Russian Journal of Ecology 46(1):81–88.

Index

Species lists in the tables and appendices are not indexed.

About the Authors

Erik Kiviat, wetland ecologist and conservation scientist, is Executive Director of Hudsonia, a nonprofit institute for scientific research and education based in the Bard College Field Station in Annandale, New York. He has studied the Hudson River since 1970 and the Meadowlands beginning in 1999. He holds the PhD in ecology (Union Institute), and has published *The Northern Shawangunks: An Ecological Survey* (1988), *Biodiversity Assessment Manual for the Hudson River Estuary Corridor* (with Gretchen Stevens, 2001), and *Biodiversity Assessment Handbook for New York City* (with Elizabeth Johnson, 2013).

Kristi MacDonald, conservation scientist, is Director of Science at Raritan Headwaters, a nonprofit conservation organization based in Bedminster, New Jersey. Her research interests focus on understanding how species respond to human-caused stressors, especially in urban landscapes, and she works to guide local communities and private landowners on science-based conservation and planning practices to sustain healthy ecosystems, biodiversity, and clean water. She holds a PhD in ecology from Rutgers University, where she studied birds in small urban forests and human-built environments in the Arthur Kill watershed of New York and New Jersey.